LIBRARIES AND COPYRIGHT

LIBRARIES AND COPYRIGHT:

A GUIDE TO COPYRIGHT

LAW IN THE 1990s

by
Laura N. Gasaway
and Sarah K. Wiant

Special Libraries Association

Copyright © 1994 by Special Libraries Association
1700 Eighteenth Street, N.W.
Washington, D.C. 20009-2508

Library of Congress Cataloging-in-Publication Data

Gasaway, Laura N.
      Libraries and copyright: a guide to copyright law in the 1990s/
by Laura N. Gasaway and Sarah K. Wiant.
          p. cm.
          Includes bibliographical references and index.
          ISBN 0-87111-407-0
          1. Photocopying processes--Fair Use (Copyright)--United States.
2. Photocopying services in libraries--United States.   I. Wiant,
Sarah K. II. Title.
KF3030.1.G37   1994
346.7304'82--dc20   94-8694
[347.306482]   CIP

CONTENTS

Preface   vii

Chapter 1   Introduction to Copyright
    Introduction   1
    History of Anglo-American Copyright Law   4

Chapter 2   The 1976 Copyright Act: An Overview
    Introduction   17
    Scope, Subject Matter and Exclusive Rights   17
    Term of Copyright, Formalities and Registration   21
    Infringement of Copyright   24
    Fair Use   26
    Defenses to Infringement and Remedies   31

Chapter 3   Library Photocopying and Other Reproduction
    Introduction   43
    Section 108(a)   44
    Copying for Library Use - Sections 108(b)-(c)   46
    Copying for Users - Sections 108(d)-(e)   48
    Section 108(f): General Exemptions   51
    Section 108(g): Systematic Reproduction   53
    Section 108(h) and (i) (repealed)   56
    Copying for Library Reserves   57
    Relationship of Section 107 to Section 108   57
    Conclusion   59

Chapter 4   Licensing Agencies and Collectives
    Introduction   67
    Copyright Clearance Center   68
    International Music and Film Licensing   73
    Miscellaneous Licenses   75
    Conclusion   76

Chapter 5   Audiovisual and Nonprint Works
    Introduction   81
    Section 108: Library Exemptions   82
    Section 107: Fair Use   83
    Library Duplication of Audiovisual and Nonprint Materials   87
    Performance of Audiovisual and Musical Works in Libraries   103
    Conclusion   105

Chapter 6   Computers, Software, Databases and Copyright
    Introduction   113
    Library Practices   115
    Computer Software Use by Libraries   116
    Loaning Computer Software   123
    Databases   127
    Electronic Publishing and Other Issues   132
    Conclusion   134

Chapter 7   Special Problems: Unpublished Works, Educational Copying and Library
             Reserves
                         Fair Use of Unpublished Works   141
                         Educational Photocopying   144
                         Library Reserves   148
                         Conclusion   151

Chapter 8   International Copyright
                         Introduction   157
                         The Berne Convention   162
                         The Universal Copyright Convention   166
                         European Economic Community Copyright Law   168
                         Conclusion   173

Chapter 9   Canadian and British Copyright
                         Canadian Copyright Law   181
                         British Copyright Law   187
                         Conclusion   193

Chapter 10   A Public Lending Right for the United States
                         Introduction   199
                         Legislative Action   200
                         Public Lending Right Issues   202
                         Theoretical Issues   207
                         International Influence   209
                         Conclusion   210

Chapter 11   Conclusion   217

Appendix   A.   Answers to Questions Posed in Chapter 1   219
             B.   Answers to Questions Posed in Chapter 6   225
             C.   Agreement on Guidelines for Classroom Copying in Not-for-Profit 231
                  Educational Institutions with Respect to Books and Periodicals
             D.   Guidelines for Educational Use of Music   235
             E.   Interlibrary Loan Guidelines   237
             F.   Guidelines for Off-Air Recording of Broadcast   239
                  Programming for Educational Purposes
             G.   Library Reserve Guidelines   241
             H.   Sample Permission Letters   243
             I.   List of Case Citations   245

Bibliography Books and Chapters   247
             Articles   251

Index   257

PREFACE

We dedicate this work to our colleagues in special libraries and information centers who are struggling to comply with the copyright laws, and who are trying to educate other members of their institutions about copyright law compliance. It is from the numerous questions about copyright applications raised to us at meetings and workshops, by phone, fax and internet that the idea for a book came about. Both of us have served as chair of the Special Libraries Association Copyright and Implementation Committee and have had opportunities to respond to questions about libraries and copyright over the years. In working with our colleagues we realized the need for a book that would provide sufficient background for a real understanding of the copyright issues that are settled. Because new technologies provide alternatives to photocopying and offer more efficient means to respond to patrons' requests, new issues of copyright compliance are raised. Our goal is to address the fundamentals of the copyright law and its application to traditional requests for photocopying and other reproduction, to answer some of the frequently asked questions and to provide guidance in those areas of library uses of copyrighted works that have not been resolved. Another goal is to encourage those institutions which have not done so to develop copyright policies.

Developing and implementing a copyright policy, however, is not enough. Librarians must stay informed. The explosion of the publishing industry, increasing costs of publications, permissions programs which are neither timely in response nor realistic in cost, whistle-blower awards, the pending *Texaco* litigation mentioned throughout the book and the narrowing interpretations of fair use, or even the abolition of fair use in some circumstances, are areas of considerable concern among information professionals. Software use, electronic publishing and downloading, together with the developing information highway and the National Information Infrastructure, raise more issues than we perhaps have answered. As information professionals we find that these issues will require our full attention.

We are indebted to our research assistants John Titus, Meg Wood, Christopher Enloe and Kacy O'Brien for their many contributions in the preparation of this book and for their outstanding work on the endnotes. The secretarial staffs at the law libraries of University of North Carolina and Washington and Lee University spent countless hours preparing drafts and the final manuscript, and we are grateful for the efforts.

Laura N. Gasaway

Sarah K. Wiant

CHAPTER 1

INTRODUCTION TO COPYRIGHT

I. INTRODUCTION

This book covers the function and uses of U.S. Copyright law;[1] it also includes a chapter on international copyright and a chapter on Canadian and British copyright law.[2] It is geared to the areas of law of immediate interest primarily to special librarians, but also to other librarians, students of library science and anyone engaged in the lending of and dissemination of copyrighted works. Because the complexity of the present law creates numerous pitfalls for the unwary, an understanding of copyright law in these fields is vital.

This chapter affords an overview of the history and purpose of the field of copyright law. In addition, it discusses the present U.S. statutory law, codified as Title 17 of the U.S. Code, which, since its most recent amendments, conforms in general to the predominant international copyright law embodied in the Berne Convention.

A. *Copyright Law and the Role of Librarian*

Why should a librarian be concerned with copyright law? Because copyright is identified with written works, that connection is immediately apparent especially in the era of photocopiers, fax machines, optical scanners, and the like. This is merely the tip of the iceberg when it comes to the importance of the law of copyright, however. Consider these scenarios:

■ A research librarian is asked by an advertising executive to locate and photocopy a magazine advertisement and cartoon that first appeared in 1939. The executive is both writing an article on the history of advertising psychology and preparing a speech for a group of local public relations professionals. Are the ad and cartoon still protected by copyright or are they now in the public domain? May the executive reproduce the entire ad and/or cartoon in his book without the permission of original copyright holder? May he reproduce the ad on a transparency to use to illustrate his speech?

■ A music librarian is asked by a vocal artist to copy a rare recording of arias by Maria Callas so the performer can study at home. What if the recording were a new release on Classic Compact Discs which was readily available in any record store? Does it matter if the student wanted to copy only a few bars to study a difficult phrase?

■ A motion picture archivist is asked to screen *The Chimes at Midnight* for a university Shakespeare society. Are royalties owed? What if the society wants to charge admission?

■   An aerospace librarian has completed a complicated database search for researchers in the company. Is downloading on a disk permitted? May the librarian retain the search results for later use and update the results on a regular basis?

■   A school librarian wants to make a videotape to interest younger students in reading and wishes to include slides of illustrations by Dr. Seuss and Maurice Sendak. May she make these slides without infringing the artists' rights? What if she reproduces them to create a pamphlet for a first-grade teacher's use as a primary reader?

■   A pharmaceutical company purchases an original work from a famous sculptor to place in its reception hall. After several years, the sculpture is moved to the library. In order to fit the sculpture in the available space, the librarian has it removed from its base. Does this infringe the sculptor's copyright?

■   The librarian in a technical library is asked to photocopy material from a technical report published under the auspices of NASA. May she do so? Does it make a difference that the report was republished by a commercial publisher?

■   An archivist fears that the public lending of sound recordings of the poets Dylan Thomas and T. S. Eliot reading their own works will endanger the recordings because of the fragility of the long-playing phonorecords. Instead, he retains the albums as masters and copies the recordings onto cassette tapes for lending. Has he infringed copyright? What if the recordings are still available on the original label, but at a very high price because of their historical value?

These examples illustrate some fairly typical situations that librarians face on a daily basis which are covered by the copyright law as codified in the 1976 Act. For answers to these questions see Appendix A. Most of the commercially produced information resources that a librarian encounters in his work is copyrightable; therefore, the loan, reproduction, performance or transformation of works from the original may have ramifications in copyright law. Thus, in addition to the more common questions concerning photocopying of protected works, special librarians encounter other copyright issues daily.

B. What Is Copyright?

Perhaps the best way to start a discussion of copyright law is with a basic question: what is a copyright? A copyright grants to its owner the right to control an intellectual or artistic creation, to prohibit other persons from using that work in specific ways without permission and to profit from the sale and performance of the work. Copyright sometimes is called a monopoly,[3] which it undeniably may be, at least in its purest form. This is not the meaning of the word "monopoly" in the copyright sense, however. Rather, it means exclusive rights to something granted by the crown or the government. Despite what today is an unfavorable categorization, the copyright monopoly provides vital rights to authors, artists, scholars and all others whose income derives from the products of creative and intellectual efforts. There is considerable fear that if monopoly-like rights are not given to these individuals, there will be no economic enticement for people to create scholarly works or art or literature. This concern with ensuring profit for the copyright owner, however, must be balanced against the needs of the public, scholars and critics for access to use these works. The purpose of copyright law, therefore, is to balance the rights of the author with the rights of the public. The statute represents a compromise between the owner's rights to control the work and to receive income from the sale or performance from the work and the public's right to use the work in certain specified ways.[4]

What constitutes a copyrightable work? Under the current statute, copyright protection extends to "original works of authorship fixed in any tangible medium of expression, now known or later developed, from which they can be perceived, reproduced, or otherwise communicated, either directly or with the aid of a machine or device."[5] The term "original works of authorship" is left undefined in the Act. Case law, however, has identified "originality" as the extent to which a work has *not* been copied from a previous work.[6] Originality in this context does not require uniqueness or novelty or any attempt at artistic merit.[7] Interestingly, labor to create the work is not expressly required either, although labor may be seen as a corollary to the creation of the work.[8]

No work that has been plagiarized in whole or part is eligible for protection.[9] This statement is not so simple as it seems. Generally, the amount of originality necessary to claim copyright protection is relatively slight: minor variations from a famous model, such as changing the size or using a different medium, will sometimes protect the second author from a charge of plagiarism.[10]

"Authorship" is the second component of copyright.[11] It is defined as "independent creation" and traditionally is used as though synonymous with originality.[12] New technology, however, has made authorship itself a subject for debate. Does the requirement for authorship demand *human* creation or can a computer generate a copyrightable work that satisfies the statutory criteria?[13] Although the debate likely would prove fascinating, it seems probable that the courts will avoid the issue by crediting the program's owners or creators with the authorship of computer-generated works unless the operator of the computer is required to exercise independent judgment to create the work. Then, the copyright in the end product would belong to the operator of the computer. The final requirement for copyright protection is that the work be "fixed in any tangible medium." Fixation is defined as

> [a] work is "fixed" in a tangible medium of expression when its embodiment in a copy or phonorecord, by or under the authority of the author, is sufficiently permanent or stable to permit it to be perceived, reproduced or otherwise communicated for a period of more than transitory duration. A work consisting of sounds, images, or both, that are being transmitted, is "fixed" for the purposes of the title if a fixation of the work is being made simultaneously with its transmission.[14]

This definition covers not only copies of the written word and recordings of sound, but visual images, either static, as in photographs or illustrations, or animated, or as in motion pictures or videotapes. It even extends to live performances that are taped as they are broadcast.[15]

One last concept is vital to the understanding of the substance of copyright. No protection is available for an "idea, procedure, process, system, method of operation, concept, principle, or discovery, no matter how original or unique."[16] Copyright protection is available only for an *expression* of the idea. At first blush, this provision may seem harsh, but if copyright were available for an idea, no writer could publish a novel concerning a Southern woman's struggle to keep her family's plantation during and after the Civil War without infringing the copyright on *Gone With the Wind*. Generally, if an idea can be expressed in more than one way, the courts will protect an expression even though the expression and the idea may be very close.[17] The corollary exists that if an idea can be expressed only in *one* way, it is an unprotectable idea in spite of the author's effort to express it. Courts have had particular difficulty with computer software and the idea/expression dichotomy.[18]

What purpose is served by granting copyright protection? For the copyright holder the purpose is predominantly commercial. There are other reasons an author seeks to protect works he created besides an interest in commercial gain, however. An author may want to ensure that there are no alterations in the work, or she may seek to protect her reputation or to control whether the work is translated into a particular language. The drafters of the U.S. Constitution apparently thought that artists, scholars and other producers of coprightable works would best be encouraged to continue their labors if ensured commercial benefit.[19] Not only are copyright holders granted rights to control their works and receive royalties, but they also may assign rights to another person or organization. For this reason, a copyright holder can sell the right to dramatize his novel in whole or part to a motion picture production company. The right to dramatize, which technically is the preparation of a derivative work, is separate and distinct from the right to reproduce and distribute, which the author presumably has sold to a publishing house. It is the option of the proprietor to separate and assign her rights to a variety of persons. In this manner, copyright law permits an owner to maximize the potential gain from the creation of a copyrighted work.

II. HISTORY OF ANGLO-AMERICAN COPYRIGHT LAW

An author's work is protected by copyright from the time the original work is fixed in a tangible medium of expression, e.g., typewritten, videotaped or preserved on a computer disk.[20] This protection survives 50 years after the death of the author, or about 75 years.[21] These provisions are the basis of the present United States copyright law, which incorporates most of the international copyright agreement codified by the Berne Convention. Such innovations in the U.S. are very recent, however, dating only since 1988. In order to understand fully the present state of the law, it is advantageous to know something about the laws which preceded it. This section provides the historical background of U.S. copyright law and its predecessors which led to the 1976 Copyright Act.

A. Brief History of the English Copyright Law

Like most U.S. laws, the concept of regulating works through a system of copyright was borrowed from the English common and statutory law. The idea that a proprietary right in the intellectual and artistic labor which results in a concrete product appears to have been long standing in the British Isles.[22] The earliest British story involving proto-copyright law concerns an ancient Irish king named Diarmait who found himself arbitrating a dispute between two Irish clerics, later to become saints, Columba of Iona and Fennian, Abbot of Moville. Columba had made a copy of the Abbot's precious Psalter without permission and, not surprisingly, the Abbot was unhappy about the situation, particularly since Columba, in decidedly unsaintly fashion, refused to return the original. Diarmait brought the issue to a close by presenting both the original and the copy to Fennian, with the words, "To every cow her calf, to every book its copy."[23]

Despite the apparent age of the concept, the need for an organized law of copyright did not arise until the fifteenth or sixteenth century, after the invention of the printing press. William Caxton introduced the printing press to England in 1476,[24] but printing remained an intimate operation performed for the clergy and the aristocracy, because the bulk of the public was illiterate, unleisured or both. In addition, printing was still so new that the procedure was prohibitively expensive. The economic incentive to engage in literary piracy, therefore, was miniscule.[25]

One may wonder what prompted artists to work if no market for literature existed. The economic incentive that existed had nothing to do with mass publishing. The Tudor period, particularly during the reigns of Henry VIII and Elizabeth I, was the heyday of patronage.[26] Competition existed among aristocrats to inspire the greatest of authors and among authors to curry favor with the most powerful nobles.[27] Although an author could expect no direct financial gain from a poem or book, an aristocratic patron would provide bed, board and expenses in exchange for a masterpiece dedicated to the patron by name and title, which he could dangle before the noses of his friends and adversaries.[28]

Those writers who attempted to live by their pens without the aid of patronage faced grim times. Most of these writers were failed university students. Although they wrote poetry and plays, they remain best remembered for their pamphlets, nasty little tracts, self-published, in which they vented their envy and despair. The most famous of these is notable not for any virtue of its own, but for the venom it unleashes on a new writer in the London literary arena: Shakespeare. An attempt to live by publishing writings frequently resulted in death by starvation (a similar fate is faced by many contemporary writers).

With its emphasis on learning, the Tudor period saw the Renaissance and the rise of the English middle class and the beginning of the modern trend of mass published literature. Plagiarism and literary piracy soon followed.[29] Copyright, however, was not created to stem piracy, but rather to censor seditious literature.[30] To combat this presumed threat, the Crown, through the Court of Star Chamber, issued a Royal Charter in 1557, that created a new guild known as the Stationers' Company, which could approve or deny the printing or works of authorship.[31] It was granted a monopoly on printing, so that only guild members had the legal right to publish works which were listed by title and author in a registry maintained by the company. The registry contained a complete list of all literary works published in England since the issuance of the charter.[32] Since it was usual for an author to sell a manuscript outright to a printer, these two powers gave printers a sort of copyright in the works they owned. The members could prohibit a nonmember from publishing any work through the monopoly, and through the registry, a member could deny another member or any author the registration of a duplicate title.[33] Although the Stationers' Company may have had little immediate interest in censorship, the monopolistic benefit was extremely valuable, because the right to print under the charter was perpetual.[34]

The Stationers' Company administered the registry and the printing industry; the company survived the Civil War and the restoration of the monarchy.[35] The printing industry was rocked by disasters like the Great Plague of 1665 and the Fire of London in 1666, which decimated both the printers and their stocks.[36]

By the turn of the eighteenth century, literary piracy was rampant. The members of the old Stationers' Company petitioned Parliament for relief from their woes, but the resulting act was far from what the guild had anticipated. Instead, the Parliament of the new Queen Anne created the first modern copyright statute, known as the Act, or Statute of Anne.[37] Its terms were revolutionary. The Act provided that a literary work was the property of its author or the author's assigns for a term of 21 years if the work was already in publication. If the work was new, the term was 14 years with an option to renew for another 14 years if the author was alive at the end of the first term.[38] Penalties for infringement were provided for all works registered in the Stationers' Hall. The Act gave continued authority to the Stationers' Company since it effectively made registration of titles a requirement of copyright.[39] Sometime later, an addition to the Statute made "innocent infringement" of a copyright a legal impossibility by requiring every copyrighted work to carry a notice of the fact that it was protected by copyright.[40]

Printers were distressed by the passage of the Statute, since it severely curtailed their traditional rights, the most important of which was perpetual ownership. They fought the term of years in the courts and claimed that the Statute of Anne did not divest them of the perpetual rights they held under common law copyright.[41] For some time, that argument swayed the lower courts. Then, in the landmark case of *Donaldson v. Becket*[42] the House of Lords held that the term of years set down in the Statute of Anne was effective as to all published works.[43]

The generous rights granted authors were not extended to artists, because the Statute of Anne dealt exclusively with written works.[44] After the adoption of the Statute of Anne, the protection of visual works was delegated to the murky realm of common law property rights. Prints, engravings and other visual arts became easy marks for pirates who copied the subject matter and form of a famous work and tried to pass it off as the original to the unsophisticated public.[45] The pirates were remarkably successful at eviscerating a legitimate artist's market, so much so that the artists began to fight for legislative protection of their rights in their creations. The leader of this movement was the noted artist and engraver William Hogarth, whose 1732 print series, *The Harlot's Progress*, was one of the noteworthy victims of the plagiarism craze.[46] Before issuing a sequel series called *The Rake's Progress*, Hogarth petitioned Parliament for protective legislation.[47] Hogarth delayed the issuance of his series, until Parliament finally issued the Engraver's Act in 1735.[48] This Act created rights in artists equivalent to those enjoyed by authors under the Statute of Anne.[49] Eventually, the Statute of Anne became the basis of a broad series of copyright acts, which protected all manner of artistic creation.[50]

C. History of the U.S. Copyright Laws

Copyright law in the United States derived both from the Statute of Anne and from the common law. During the colonial period, copyright law was administered by the British Empire and was, therefore, federal in form. That pattern changed after the colonies broke with England.[51]

During the time the rebellion leaders were sequestered in Philadelphia debating the rhetoric of the Declaration of Independence, legislative bodies of the individual states were laying the groundwork for their local governments and laws.[52] All the colonies with the exception of Delaware passed laws protecting the creative works of authors.[53] None of these laws, however, were reconcilable with the law of any other state, which had the unfortunate effect of forcing an author who published in Massachusetts to comply with copyright requirements not just in Massachusetts, but in eleven other colonies. The author could only hope for the best in Delaware.[54]

The situation was not helped by the first tentative stabs at creating a centralized government. While the war was still in progress, the first draft of the Articles of Confederation was presented to the Constitutional Congress; the Articles were adopted in March 1781.[55] The Articles created a weak federal government, which had little control over the states' autonomy.[56] It even lacked the power to tax.[57]

By the end of the 1780s, it became apparent that this disjointed federation of states organized in name only was not a viable arrangement.[58] The tenor of the new country shifted away from favoring a decentralized government to demanding a federal model. A new convention was convened in Philadelphia in May 1787 for the purpose of redrafting the Articles of Confederation.[59]

The outcome of the Federal Convention, as it became known, was the Constitution of the United States.[60] Not only did the Constitution establish a stronger and more centralized model of government, it reserved to the federal legislature authority to create and administer certain areas of law, notably federal taxation, interstate commerce and intellectual property. In reference to the latter, the Constitution states:

> The Congress shall have power...to promote the progress of science and useful arts...by securing for limited times to authors and inventors the exclusive rights to their respective writings and discoveries.[61]

This provision empowers the Congress to control patent and copyright by enacting laws. In keeping with the Statute of Anne, the granting of rights is viewed as constitutional only if limited to a number of years; Congress has no authority to grant perpetual rights to any individual in these fields. The term "discoveries" has been held to cover inventions to be patented. "Writings" refers to literary works and other written endeavors; the term has been construed further to include, among other categories, visual and sculptural arts, videotaped and audiotaped performance and filmed works and notated choreography.[62]

Under the authority of the Constitution, copyright laws became almost purely federal. The Congress enacted a series of copyright acts, first in 1790,[63] with general revisions following in 1831,[64] 1870,[65] 1909,[66] and most recently, 1976.[67]

The 1909 Act expressly protected published works by a term of years in which authors enjoyed exclusive federal rights. This is not to say, however, that unpublished works were unprotected; in fact, in a certain respect, an unpublished work was sacrosanct. While the federal courts consistently held that federal law preempted state law in suits brought concerning the infringement of published works,[68] a presumption arose that unpublished works belonged to the author in perpetuity and were defensible under a states' common laws of property. This idea, which seems to be a descendant of the law the Stationers' Company attempt to defend against the Statute of Anne, became known as common law copyright.[69]

Common law copyright is a judicial doctrine, developed to protect works of authorship that did not fall within the federal statutory boundaries. It allowed protection to be granted to copyrightable works, either fixed or unfixed in a tangible medium of expression, and to works that did not fall with the scope of the copyright statute.[70] So, common law copyright applied to unpublished works, since these were exempted from the language of the statute; to works not "fixed" in a tangible medium, for example, an extemporaneous speech; and to works beyond the scope of the copyright statute.[71] Examples of the latter are growing increasingly difficult to imagine, since the scope of the statute is grown so broad: architectural works might have fit that category, had the current Act not been amended in 1990.[72] Section 301 of the 1976 Act preempted common law copyright almost completely.[73]

D. The 1909 Act

A major revision of the U.S. Copyright Act was completed in 1909.[74] Although its scope was broader than that of the earlier Acts,[75] it shared the requirement that the formalities of registration be met,[76] as well as the requirement of deposit[77] before a work could receive copyright protection. Some of the major improvements wrought by the revision of the copyright law were: (a) an extension of the numbers of years in a renewal term, from 14 to 28, with the result that copyright protection could be had for a total of 56 years;[78] (b) making the certificate of registration prima facie evidence of the facts recorded in relation to any work;[79] and (c) exempting foreign works in the original language from the need to be reprinted in the United States.[80]

The 1909 Act remains important because some extant works are still subject to its tenets. The newer Act changes the duration of copyright protection; therefore, there was concern that contracts entered into prior to the new Act's effective date would be rendered invalid. Section 304(a) of the 1976 Act, however, accommodates copyrights subsisting since the 1909 Act into the provisions of the 1976 Act, by affording renewal rights to the older works in their first term of copyright to increase their terms to equal the new norms.[81]

The current copyright act is discussed in detail in the chapters that follow. Much of the reasoning for certain changes that were embodied in this revision was a desire to harmonize the U.S. law with that of other developed nations which are major producers of copyrighted works. This recognition of the increasingly global nature of copyright forced the United States to reexamine its position on the primary European copyright treaty, the Berne Convention.

E. The Berne Convention of 1886

The Berne Convention, which convened in 1886, was the outcome of three prior meetings of the Swiss Federal Council, beginning in 1884.[82] The Convention was a diplomatic joinder of European nations seeking to establish a mutually satisfactory uniform copyright law to replace the need for separate registration in every country. The general result of the Berne Convention was a simplifying of standing copyright law.

There were five articulated purposes of the Convention: (a) to develop copyright law favoring authors in all civilized countries; (b) to eliminate reciprocity among nations as the basis of multinational copyright protection; (c) to end discriminatory practices by nations in favoring a local author above a foreign author; (d) to abolish the formalities of registration and deposit as prerequisites to copyright protection; and, the ultimate purpose, (e) to promote uniform international legislation for the protection of literary and artistic property.[83] The original Convention produced a document which had two important effects. The first established a union of the signatory nations.[84] The second created the Rule of National Treatment, where all union members are obligated to grant the same rights and privileges to a foreign Berne author as to any local author.[85]

The Berne Convention has been revised five times since its inception and has had two amendments of significance. The 1908 Berlin Act prohibited formalities as a prerequisite of copyright protection, set the duration of a copyright at the life of the author plus 50 years and expanded the scope of the Act to embrace products of the newer technologies, like photography and cinematography, as well as giving composers the exclusive right to adapt their music.[86] The 1928 Rome Act first recognized the moral rights of authors and artists — this right permits the artist to object to modifications to or destruction of a work in a way that might prejudice or decrease the artist's reputation.[87] The 1948 Brussels Act recognized an optional resale royalty, which extended an artist's ability to demand royalties past the first sale.[88] The final revisions, those of Stockholm in 1967 and of Paris in 1971, are most concerned with protocol among the member nations.[89]

The United States became a Berne signatory in the 1980s and implemented the treaty into its statutory law through the Berne Implementation Act of 1988.[90] It is important to note that Berne is not treated as a normal treaty. Most treaties are self-executing, meaning that the terms of the treaty, once entered by the signatory, have full effect and will be superior to any conflicting local law.[91] Although Congress stated it to be so, this was not to be the case with Berne, mostly because the concepts of moral rights and resale royalty were foreign to prior U.S. law. For that reason, section 104(c) was added to the 1976 Act.[92] In spite of this section, U.S. law is slowly being drawn into the Berne liberality to authors. Congress recently implemented the Visual Artists' Rights Act of 1990[93] which permits visual artists to exercise rights similar to moral rights under Berne.[94]

For a full discussion of the Berne Convention see Chapter 8 on international copyright.

ENDNOTES - CHAPTER 1

1   17 U.S.C. § 101 (1988). (Codification of section 102 of Pub. L. 94-553, Title 1, § 101, Oct. 19, 1976, 90 Stat. 2541.) This revision preempted all previous copyright law. The revised title became effective January 1, 1978. It has been amended several times since its adoption, most notably by the Berne Implementation Act of 1988.

2   Canada, Copyright Act, 1 R.C.S., c. C-42 (1991); United Kingdom, Copyright, Designs, and Patents Act, 1988, ch. 48 (Eng.).

3   Barbara A. Ringer, *The Demonology of Copyright* 13 (1974) [hereinafter Ringer].

4   For a historical treatment of copyright law that focuses on users' rights, *see* L. Ray Patterson & Stanley W. Lindberg, *The Nature of Copyright; A Law of Users' Rights* (1991).

5   17 U.S.C. § 102(a) (1988). Unless otherwise noted, citations to the Copyright Act are to the statute as enacted in 1976.

6   Paul Goldstein, 1 *Copyright: Principles, Law and Practice* § 2.2.1 (1989) [hereinafter Goldstein].

7   *Id*. Goldstein cites Judge Learned Hand to illustrate what parameters define originality. Judge Hand said, "... if by some magic a man who had never known it were to compose anew Keats' *Ode on a Grecian Urn*, he would be an "author," and if he copyrighted it, others might not copy the poem, though they might, of course, copy Keats'." (*citing*, Sheldon v. Metro-Goldwyn-Mayer Pictures Corp., 81 F.2d 49 (2d. Cir.), *cert. denied*, 298 U.S. 669 (1936). Despite the quality of this explanation, it probably falls short of being a good defense against a charge of infringement. *See* Granite Music Corp. v. United Artists Corp., 532 F.2d 7618 (9th Cir. 1976).

8   *Id*. § 2.2.1.3, n.43. (*citing*, Hearn v. Meyer, 664 F.Supp. 832, 839, (S.D.N.Y. 1987) which held that an artist's exact reproductions of public domain illustrations lacked sufficient originality for copyright in spite of the time and effort expended).

9   *Id*. § 2.2.1. Original additions to older works, like a new line to an old song, may be copyrighted on their own but not as a part of the previously know work.

10   *Id*. § 2.2.1.3. (*citing*, Alva Studios v. Winninger, 177 F. Supp. 265 (S.D.N.Y. 1959) which held that an artist's exactly copied miniature of a Rodin sculpture held was sufficiently original for copyright. *See also*, Original Appalachian Artworks, Inc. v. Toy Loft, Inc., 797 F.2d 1222 (3d Cir. 1986), *cert. denied*, 684 F.2d 821 (1987). Where the court found no plagiarism by a toy maker whose line of dolls, marketed as "The Little People" was similar to the famous "Cabbage Patch" line.

11   *Id*. § 2.2.2.

12   *Id*. § 2.2.2.

13   *Id*. § 2.2.2.

14   *Id*. § 2.4.

15   17 U.S.C. § 101 (1988).

16   *Id*. § 102(b).

17   1 Goldstein, *supra* note 6, § 2.3.2.

18   See Chapter 6 for a discussion of computer programs.

19   *See* R. F. Whale, *Copyright: Evolution, Theory and Practice* 12 (1972) [hereinafter Whale]. This opinion is not universal. Other commentators believe that restriction of copyright laws actually limit authors. *See*, Ringer, *supra* note 3, at 15-16.

20   17 U.S.C. § 102(a) (1988).

21   *Id*. § 302(a).

22   William F. Patry, *Latman's The Copyright Law* 2 (6th ed. 1986) [hereinafter Latman]. *See* Chapter 2 for a detailed discussion of the duration of copyright.

23   *Id*. The story is generally held to be apocryphal, although Latman, in note 3, states that a book identified as Columba's Psalter was displayed in the British Museum in the mid-1800s.

Not surprisingly, Alban Butler neglected to mention the story in his biography on St. Columba. Alban Butler, *The Lives of the Fathers, Martyrs and Other Principal Saints*, July 9 (1883) (Butler's book is organized as a calender of Saints' Days). He does say, however, that Columba was born in the 600s in the North of Ireland, a member of the ruling Niall clan. He is known as the "Apostle of the Picts" because of the conversions he accomplished of the Pictish tribes in what is now Scotland. Butler does mention Columba was expelled from Ireland by Diarmait (from whence he went to Scotland) because the king was displeased at the saint's "zeal ... in reproving public vices."

Eleanor Shipley Duckett gives a more detailed version of the Psalter story. She mentions that both Columba and Diarmait were O'Niall's, although Diarmait, then the High King of Ireland, was of the southern branch of the family. The two branches of the clans were engaged in a rather bitter feud and the incident of the copy gave the king an opportunity to strike a blow at a blood enemy. Eleanor Duckett, *The Wandering Saints of the Early Middle Ages* 80-81 (1964). She also raises the stakes a bit by implying that the book in question was not a mere prayerbook, but was, in fact, the first copy of the Vulgate Gospel of St. Jerome to be brought to Ireland. Whatever the truth or untruth of the tale, Columba goes into the annals, not simply as the Apostle of the Picts, but as the first known literary pirate in the British Isles.

24   Latman, *supra* note 22, at 3. *See also* Victor Bonham-Carter, *Authors by Profession* 11 (1978) [hereinafter Bonham-Carter].

25   Whale, *supra* note 19, at 3.

26   *Id.* at 2. Whale traces the artristocratic practice of patronage back to ancient Greece and Rome.

27   For a discussion of the artistic and literary life in the courts of Elizabeth I and her followers, *see* G.P.V. Akrigg, *Shakespeare and the Earl of Southhampton* 201-15, 228-39 (1968).

28   *Id.* at 201-15. William Shakespeare succeeded in earning Southampton's favor with the very popular Romanesque poems, *Venus and Adonis* and *The Rape of Lucrece*.

29   Whale, *supra* note 19, at 3.

30   Bonham-Carter, *supra* note 24, at 11. The Renaissance and the Reformation brought with them an influx of humanist philosophy and new religious thought, much of which, it was feared, would create dissent in the populance if allowed dissemination.

31   *Id.*

32   *Id.* at 12.

33   *Id.*

34   Latman, *supra* note 22, at 3.

35   Bonham-Carter, *supra* note 24, at 15.

36   *Id.*

37   8 Anne, ch. 19 (1710) (Eng.). The Act, which became effective in April, 1710, is said to have been created from drafts written by Jonathan Swift. Bonham-Carter, *supra* note 23, at 16. For a partial text of the Statute of Anne, *see* Whale, *supra* note 19, at 8.

38   Latman, *supra* note 22, at 4.

39   *Id.*

40   *Id.*

41   *Id.*

42   4 Burr. 2303, 98 Eng. Rep. 201 (K.B. 1774).

43   Latman, *supra* note 22, at 4.

44   David Kunzle, "Hogarth Piracies and the Origin of Visual Copyright," in John Sheldon Lawrence and Bernard Timberg, *Fair Use and Free Inquiry* 21 (2d ed. 1989).

45   *Id.* at 25-27.

46   *Id.* at 23-24.

47   *Id.* at 23.

48   *Id.* at 29.

49   *Id.* at 23; Engravers' Act of 1735, 8 Geo. 2, ch. 13 (Eng.).

50   *See* Whale, *supra* note 19, at 14. Following the Engraver's Act (which Whale dates at 1734) Parliament passed a Sculpture Copyright Act in 1814 and a Fine Arts Copyright Act in 1862 (which included paintings, drawings and photographs). Music and dramatic composition were included in the Statute of Anne, under the definition of "books" as of 1842. A copyright in dramatic performances had been recognized since 1833 and by 1842 extended to the performance of music. All of these rights were codified *in toto* under the Copyright Act of 1911, which is now subject to the Berne Convention of International Copyright of 1886, of which the U.K. is a signatory.

51   For background into the history of the American Colonies, *see* Samuel Eliot Morison, *The Oxford History of the American People: Vol. 1, Prehistory to 1789* 84-155 (1972) [hereinafter Morison].

52   *Id.* at 356-58.

53   Latman, *supra* note 22, at 5.

54   *Id.*

55   Morison, *supra* note 51, at 363. John Dickinson, Pennsylvania delegate and the chair of a congressional committee to draft a confederation presented his draft before Congress in July, 1776. Because of the necessities of the war and because Dickinson's draft provided for representation by state population in a unified Congress, a suggestion that outraged the smaller states, ratification was delayed. It was not until November, 1777 that Congress formulated the final draft and submitted the proposal to the states. All but Maryland ratified the Articles of Confederation by February, 1779.

56   *Id.* The powers of the federal congress were limited to the declaration of war, coinage, the setting of weights and measures, the creation of a postal service and the administration of Indian affairs.

57   *Id.*

58   *Id.* at 391-94.

59   *Id.* at 394.

60   The U.S. Constitution was ratified by the thirteen states; Rhode Island was the last to ratify in 1790.

61   U.S. Const., art. I, § 8, cl. 8.

62   *See* H.R. 94-1476, 94th Cong., 2d Sess. (1976) *reprinted in* 17 *Omnibus Copyright Revision Legislative History* (1977).

63   Latman, *supra* note 22, at 6 (*citing* Act of May 31, 1790, 1st Cong. 2d Sess., 1 Stat. 124). The Act of 1790 provided that any author who formally registered his "map, chart or book," by recording the title in the clerk of district courts' office, publishing notice in a newspaper and depositing a copy with the Secretary of State, was assured protection for a renewable term of 14 years. Included in this protection were the exclusive rights to "print, reprint, publish or vend" the work. It provided a one year statute of limitations on the bringing of an infringement for published works and set damages. *Id.* An amendment regarding expanding the scope of protected subject matter and setting further formalities of registration was passed in 1802. *Id.* at 7 (*citing* Act of April 29, 1802, 7th Cong., 1st Sess., 2 Stat. 171).

64   Latman, *supra* note 22, at 6 (*citing* Act of February 3, 1831, 21st Cong., 2d Sess., 4 Stat. 436). The revision expanded the protections. Not only was the scope of protection granted to musical compositions, but the term of original copyright was doubled to 28 years (with the 14 year renewal option allowed to the author and the author's spouse and children) and the statute of limitations on bringing of an action was increased to two years.
The revision was amended in 1834, 1846 and 1865. These Acts generally had the effect of broadening the scope or demanding more formalities. *Id.* at 7-8 (The 1846 Act, for example, required deposit with the Smithsonian Institute and with the Library of Congress. *See* Act of August 10, 1846, 29th Cong., 1st Sess., 9 Stat. 106). Latman, *supra* note 22, at 6. 17 U.S.C. §§ 1-219 (1970), revised to December 1, 1977.

65   Latman, *supra* note 22, at 6 (*citing* act of July 8, 1970, 41st Cong., 2d Sess., 16 Stat. 198). Protection was granted to works of visual art. Administration of the Act was vested in the Library of Congress,

through the creation of the new Copyright Office. Amendments made in 1874 further specified procedures of the Copyright Office. Protection was first offered to foreign authors wishing to copyright their works in the U.S. by an amendment in 1891.

66   *Id.* at 6. 17 U.S.C. §§ 101-2319 (1976).

67   *Id.*

68   *See* Wheaton v. Peters, 33 U.S. 591 (1834).

69   *See*, Latman, *supra* note 22, at 6. The common law copyright refers to rights of ownership granted in perpetuity, presumably under state law, since the Constitution expressly limits Congress's power to enact perpetual rights in intellectual property. Section 6 the Act of 1790 specifies damages for the infringement of unpublished works but gives no term of years for the protection of these works.

70   2 Goldstein, *supra* note 6, § 15.5.

71   *Id.* § 15.5.1.1.

72   17 U.S.C. § 102(a)(8) (Supp. III 1990).

73   17 U.S.C. § 301(a) (1988). Common law copyright may still exist, to a very limited extent, in copyrightable works that are not in fixed medium or which are not within the purview of section 102(a). *See*, 2 Goldstein *supra* note 6, §§ 15.5.1.1-.2.

74   The revision was adopted by Congress in 1909 (*see* H.R. 28192, 60th Cong., 2d Sess. (1909); S. 1108, 60th Cong., 2d Sess. (1909)) and made part of the Revised Statutes. It codified as 17 U.S.C. §§ 1-219, July 30, 1947 (61 Stat. 652). Thereafter, it was amended in 1948, 1949, 1951, 1952, 1954, 1956, 1957, 1962, 1971 and 1974, before being revised by the Copyright Act of 1976. For the complete text of the Act, *see* 17 U.S.C. §§ 1-219 (1970), revised to December 1, 1977.

75   The 1909 Act originally protected the same categories developed in the preceding acts, for example, books (although the category was broadened to include "all the works of an author," *see* 1909 Act section 4), periodicals, dramatic and musical works, maps, works of art, photographs and prints; however, it added categories in keeping with the new technology of the twentieth century: motion picture photo-plays, motion pictures and sound recordings. *Id.* § 13.

76   17 U.S.C. §§ 1-219 (1970) (repealed 1978).

77   *Id.* § 13.

78   *Id.* § 24. The statute provided for the total 56 year term to any kind of copyright owner, an individual, a corporation, etc.

79   *Id.* § 209.

80   *Id.* § 9(c). This provision articulates the adoption of Universal Copyright Convention into U.S. law. The treaty, which was originally signed in Geneva in 1952, granted some reciprocity in the recognition of copyrights issued by signatory nations. Unlike the Berne Convention, it did not attempt to recraft the law on an international scale, but left each country with its own law. This provision took effect in the U.S. on September 16, 1955.

81   17 U.S.C. § 304 (1988). For further discussion on the duration of copyright under the 1976 Act, see Chapter 2.

82   The Berne Convention Implementation Act of 1988, (Pub. L. No. 100-568, 100th Cong., 2d Sess.), 102 Stat. 2853 (1988) (codified at various sections in 17 U.S.C.).

83   Melville B. Nimmer & David D. Nimmer, 4 *Nimmer on Copyright*, app. at 32-12 (1991).

84   *Id.*

85   *Id.*

86   *Id.*

87   *Id.*

88   *Id.* app. at 32-12 & 13. To comprehend this right, imagine purchasing a Phillip Perlstein painting and then selling it, only to discover one owes Perlstein a percentage of the sale price!

89   *Id.* app. at 32-13.

90   *Id.* app. at 32-1. In doing so, the U.S. established copyright treaty relations with the 24 other nations presently signatories of Berne.

91   *Id.* app. at 32-28.
92   17 U.S.C. § 104(c) (1988). This subsection reads:
     No right or interest in a work eligible for protection under this title may be claimed by virtue of, or in reliance upon, the provisions of the Berne Convention, or the adherence of the United States thereto. Any rights in a work eligible for protection under this title that derive from this title, other Federal or State statutes, or the common law, shall not be expanded or reduced by virtue of, or in reliance upon, the provisions of the Berne Convention, or the adherence of the United States hereto.
93   Codified as 17 U.S.C. § 106A (Supp. III 1990).
94   *Id.* § 106A(a)(1) and (a)(3).

CHAPTER 2

THE 1976 COPYRIGHT ACT: AN OVERVIEW

I. INTRODUCTION

The 1976 revision of U.S. copyright law[1] was necessary for two primary reasons. First, technological changes dictated a fresh look at copyright and how such developments affected what works might be copyrighted and whether certain conduct would constitute infringement. The second reason an entirely new statute was needed was to begin to bring the United States into accord with international copyright law policies and practices. This chapter addresses general provisions of the Act, the subject matter covered, rights of the copyright holder, exemptions on exclusive rights and the doctrine of fair use.

By its own terms, the 1976 Act preempted all prior copyright law in the United States.[2] Works created before 1976 and still under copyright were brought into the purview of the new Act to ensure consistent protection. Thus, not only registered works were covered by the Act, but also works that had never been published nor protection sought. Since the new Act generally rejected the notion of common law copyright,[3] all preexisting works, unpublished as well as those registered under the 1909 Act, were given new terms based on the 1976 limitations. After January 1, 1978, virtually all copyright protection became federal. The total duration of copyright was increased from a maximum of 56 years to life of the author plus 50 years.[4]

II. SCOPE, SUBJECT MATTER AND EXCLUSIVE RIGHTS

Like most statutory law, the copyright law is quite interconnected, so that information vital to an understanding of one section may be found in a completely different section. This section focuses on the scope of copyright law, what works are eligible for copyright protection and rights of the copyright holder.

A. Subject Matter of Copyright

The statutory definition of copyrightable works is "original works of authorship fixed in any tangible medium of expression, now known or later developed, from which they can be perceived, reproduced, or otherwise communicated..."[5] The Act lists several categories of works for which copyright protection is available are listed in the Act. These categories are more generic than those found in earlier Acts:

1. Literary works,
2. Musical works, including accompanying words,
3. Dramatic works, including accompanying music,
4. Pantomimes and choreographic works,
5. Pictorial, graphic and sculptural works,
6. Motion pictures and other audiovisual works,
7. Sound recordings, and
8. Architectural works.[6]

Literary works constitute the important category for librarians since this category includes most books, journal articles and other printed works. The term "literary work" refers to the underlying work and not the format in which the work is embodied.[7] Thus, regardless of whether the format is printed book, a microform, a CD-ROM or a "talking book" or phonorecord, it is still a literary work for copyright purposes. Copyright also extends to compilations and derivative works.[8] A compilation is "formed by the collection and assembling of pre-existing materials or of data that are selected, coordinated or arranged in such a way that the resulting work as a whole constitutes an original work of authorship."[9] The term includes collective works.[10] Most periodicals are collective works while many reference books, such as directories, are compilations. These nonfiction works are fact based but nonetheless are copyrightable if there is sufficient originality in the gathering, selection, arrangement and indexing of the work.[11]

Electronic databases also are compilations that may be subject to copyright protection. The fact that a work is stored in electronic format does not change the *nature* of the work relative to copyright. If the work in printed form is a compilation, the same work in electronic format remains a compilation and is subject to protection. Originality is found in the selection, arrangement and access to the information just as in printed works. Works originally in electronic format are protected without regard to whether they have ever been in print.

Computer programs are also protected as literary works. While software generally does not conform to the general understanding of the meaning of the term "literary work," Congress declared that programs fall into this category.[12] See Chapter 6 for a discussion of library uses of computer software and copyright considerations relevant to electronic databases.

Audiovisual works are of increasing importance in all types of libraries. Librarians tend to view all works as information, but the copyright law treats these categories of works differently. Likewise, musical works are important, not only to special music libraries, but also to public, academic and school libraries. Pictorial, graphic and sculptured works comprise a huge portion of the collections of museum libraries. Audiovisual and non-print works are discussed in Chapter 5.

B. Scope of Copyright

Some works are excluded from copyright even though the work fits within one of the enumerated categories. For example, fraudulent works are excluded.[13] In the past, obscene works likewise were excluded. Although older cases deny protection based on a finding of obscene matter,[14] the modern view is that just because a work may be viewed as obscene is not a relevant reason to refuse copyright protection since courts do not exercise aesthetic or literary judgment. This is because there is no objective standard for determining obscenity and because neither the Constitution nor the Copyright Act specifically denies protection to obscene works.[15]

Another exclusion is any U.S. government document. A copyright cannot vest in any work created by the federal government.[16] Under this provision, informational pamphlets, legislative proposals, statutes, legal documents, judicial opinions, and the like cannot be copyrighted either by the federal government or by an employee of the federal government who authors a work within the course of her employment.[17] The prohibition against copyright on government works used to be absolute, but beginning about 1978 or 1979, the National Science Foundation began to offer some grants in which the researcher was permitted to claim copyright in articles or other works that report results under the grant. Since that time, some works published by the National Technical Information Service (NTIS) contain a notice a copyright. For works produced under the terms of a federal grant to an outside agency, the terms of the contract control whether an author may claim copyright in the work produced under the grant. With increasing frequency, the principal researcher is permitted to hold copyright in her own name. This section does not prevent the government from owning copyrights that are assigned to it, however.[18]

Works published by state governments are not generally excluded from federal copyright protection. Because the Act is a federal statute, it is silent as to whether states and municipalities may claim copyright protection for their works. Some states claim copyright in all of their publications and some explicitly do not. For example, the state of California claims copyright in its Administrative Code while Virginia does not appear to make a claim of copyright in any of its state documents. A few states seem to claim copyright in some documents they publish but not in all publications. For other states, one cannot easily determine whether they claim copyright in state-sponsored publications or not. The U.S. Copyright Office states that it does not believe that edicts of government such as legislative enactments, judicial opinions and administrative rulings are eligible for federal copyright protection, for public policy reasons, and this is true whether the works are federal, state or local government works.[19] Clearly, some state governments disagree and claim copyright in their state documents including their official legal documents. When the copyright status of a state government-produced work is not clear, a librarian should contact the agency responsible for producing the work to inquire about its status.

The Copyright Act allows U.S. copyrights to be extended to foreign works, if such works meet the requirements of sections 102 and 103.[20] This is complicated by the fact that the law now embraces two competing international treaties: the Universal Copyright Convention (UCC)[21] and the Berne Convention (Berne).[22] The Copyright Act authorizes that (a) unpublished works should be treated in the same manner regardless of the country of origin[23] and (b) published works are permitted full U.S. copyright protection if they fall within one of several categories.[24] A published foreign work may be copyrighted under the Copyright Act if one of its authors is a domiciliary or national of the United States or of a country which is a signatory to one of the copyright conventions to which the Unite States is a member.[25] Works originally published in UCC, Berne or United Nations countries also are granted protection.[26] Finally, the Copyright Act can be made applicable to a work if required by presidential proclamation.[27]

C. *Exclusive Rights*

The owner of a copyright is granted five exclusive rights to ensure the opportunity to exploit the work for profit. These rights are:

1. Reproduction,
2. Distribution,
3. Adaption,
4. Performance, and
5. Display[28]

The rights to reproduce the work in copies and to distribute the work refers to the act of copying the work into material objects[29] and to distribute copies publicly.[30] The most common manifestation of the reproduction and distribution rights occurs when an author transfers to a publisher the right to reproduce a novel in book copies and to distribute the copies through sale to bookstores and libraries. The adaptation right is the right to prepare derivative works, i.e., works that are derived from an existing copyrighted work. Common examples of derivative works are new editions, translations and condensations. The adaptation right also involves the right to create new arrangements of copyrighted musical compositions, the right to prepare the motion picture script from a novel and the right to transform the format of an audiovisual work such as converting a phonorecord to audiotape, a 16mm film to videotape or a 3/4-inch videotape to 1/2-inch format.[31]

The right to perform the work publicly means to recite, render, play or dance the work.[32] The definition covers performance whether done directly or by means of a machine or device. For motion pictures or other audiovisual works, performance means to show its images sequentially or to make its sounds audible.[33] Display is defined as the showing of a copy of a work either directly or by means of a television image, slide, etc., or if the work is a motion picture or other audiovisual work, to show the images nonsequentially.[34] The performance and display right is limited to *public* performance or display which is defined as a performance or display in a place either open to the public or at any place where a substantial number of persons outside the normal circle of family and friends might be gathered. Transmission or other communication to the public of a performance or display also is included in the definition.[35] These rights together encompass all economically significant uses of copyrighted works.[36]

On its face, section 106 grants copyright owners total and complete control of the public dissemination of their works. These rights, however, are balanced against the public's right to use and experience works that fall within the copyright statute. Therefore, the owner's rights are limited to allow public access for certain socially beneficial purposes and even for personal, noncommercial use.[37]

The rights enumerated in section 106 are divisible; therefore, the owner may retain the rights en masse, assign or license them individually to several persons, or transfer the entire copyright.[38] The divisibility maximizes the economic benefits derived from the ownership of a copyright. A neophyte novelist, without a solid expectation of success in his book, may opt to transfer the copyright to a publishing house for a lump sum. An established writer may prefer to sell the hardback rights to one publisher and retain the paperback rights in order to renegotiate a more lucrative second contract if the book is a best-seller. The writer also may retain derivative rights, such as authorizing a performance or the scripting of a screenplay, so that she has the opportunity to negotiate for the performing or taping of the work in some form.

When any or all of the rights are assigned to an individual of a business concern specializing in the dissemination of copyrighted works, either by sale or gift, the assignee receives the same rights of the author.[39]

D. *Ownership and Transfer*

Ownership of any of the works described in sections 102 and 103 usually belongs to the author,[40] the employer of the creator or the individual who has commissioned the work if it is a "work for hire."[41] Ownership also may rest in a person or entity to which a previous owner or the author has voluntarily transferred his rights.[42] Because the copyright is separate from the physical object in which the work is embodied, the sale of the work itself does not imply a sale of the underlying copyright unless that transfer is expressly provided.[43] Authors of a joint work are considered co-owners. A copyright owner is permitted to transfer the copyright either by conveyance, by operation of law or by will.[44] If the owner dies without a will, the copyright interest passes by intestate succession.[45]

A transfer of rights in a copyright is valid only if it is made in writing and signed by the owner or the owner's agent.[46] Although a certificate that officially acknowledges the transfer by the assignee is not required, such a document does serve as *prima facie* evidence[47] of the execution of the transfer if (a) in the United States, the certificate is issued by a person authorized to administer oaths or (b) in a foreign country, the certificate is issued by a diplomatic or consular officer of the United States or by a person authorized to administer oaths.[48] Likewise, recordation of transfers is not mandatory.[49] After the transfer is completed, it cannot be terminated for 35 years following the date of execution.[50] After that period, however, a former owner or her heirs or assigns[51] may institute a termination of the grant within five years after the 35-year term ends.[52] On the terminal date, all rights previously bestowed on the assignee revert to the previous owner or the present holder of right.[53]

III. *TERM OF COPYRIGHT, FORMALITIES AND REGISTRATION*

A. *Duration*

With the enactment of the 1976 Act, the term of copyright increased from 56 years to a roughly 75-year model. For works created on or after the effective date of the statute, January 1, 1978, the general rule is that a copyright remains in effect until 50 years after the death of the author.[54] If the identity of the author is not known or revealed, if the work is made for hire, or if it is a collective work or compilation, the copyright term is 75 years from the publication date.[55] If the work is jointly authored, the 50-year term begins on the death of the last surviving author.[56] If for whatever reason, no notice of an author's death is announced 75 years after the publication date or 100 years from the creation of the work, there is a presumption that the author has been dead for 50 years and the work passes into the public domain.[57] This presumption may be rebutted by proof to the contrary, but if an alleged infringer relied on this presumption in good faith prior to proof to the contrary, such reliance serves as a complete defense against any action for infringement.[58] Countries in the European Union (formerly European Economic Community) currently are considering expanding the term from life plus 50 years to life plus 70. The issue also has been raised in the United States,[59] and it is likely that within the next few years, the term will be increased to life of the author plus 70 years.

For unpublished works in existence prior to January 1, 1978, the term is somewhat different. Under earlier Acts, unpublished works enjoyed common law copyright protection and an author had perpetual proprietary interest in any unpublished work. The current Act preempts the common law copyright concerning unpublished works, however,[60] and subjects unpublished preexisting works to terms equivalent to those contained in section 302, with the provision that no copyright on unpublished material will expire before December 31, 2002.[61] Further, if one of these works is published on or before December 31, 2002, its term will be extended for an additional 25 years from the publication date, but it will not expire before December 31, 2027.[62] Thus, for all works that were unpublished as of January 1, 1978, the publication status is immaterial in determining the term of protection. After 2027, the same will be true of all works protected by common law copyright before the 1976 Act became effective.

Because the change in the term of copyright was so dramatic under the new Act, the statute aims to correlate works protected under the 1909 Act with the new Act. Until 1992, works in the first term under the 1909 Act were allowed to run the full 28-year term, as under the prior law, but the renewal term is extended to 47 years.[63] When a work is in its renewal term or was registered for renewal before January 1, 1978, the term was extended to embrace a termination date 75 years from the date the copyright was originally secured.[64] In 1992 the statute was amended to create an automatic 47-year renewal of copyright on works in their first term under the 1909 act. Any copyright secured between January 1, 1964 and December 31, 1977 is eligible for the automatic renewal.[65] Thus, there is no longer any necessity to renew copyrights secured under the 1909 Act.

After the copyright expires, the work enters the public domain. Public domain includes works after copyright protection is exhausted.[66] Neither the previous copyright owner nor any of his assigns retain any rights to royalty payments or to the restriction of use of the work. Thus, a work in the public domain may be freely used by any individual in any manner; it may be reproduced, adapted or performed publicly by anyone. Many theater groups reap the benefits of public domain material when they opt to perform older works; a would-be Blanche DuBois will owe royalties to the estate of Tennessee Williams; a would-be Eliza Doolittle has no such obligation (at least so long as she is content to speak G. B. Shaw's lines and does not long to sing Alan J. Lerner's songs, since *My Fair Lady* presumably remains under copyright as of this writing). Further, works of Shakespeare may be freely performed publicly or photocopied, even in multiple copies.

Despite the lack of restrictions on works in the public domain, separate copyright protection is available for compilations or arrangements of public domain works, if the finished product displays a sufficient amount of originality.[67] For example, a person who compiles *The Poe Book of Days* and uses a different poem or story by the writer to highlight each week will undoubtedly produce a copyrightable work. The copyright in such work clearly does not extend to Edgar Allen Poe's poems but rather to new matter and the work as a whole, i.e., the calendar, any artwork and the poems as a whole. By the same rationale, a musician who adapts the orchestral score of *Aiida* for Native American instruments will be able to copyright the arrangements. Copyright protection is limited to the new material contributed, however. Thus, copyright *does not attach* to public domain works themselves, but only to the adaptations, the compilations, the arrangements, the new material that was original. So, even if the copyright on *The Poe Book of Days* restricts someone from using *The Raven* on a calender during Halloween week, it will not prohibit her from including the work in collection of macabre nineteenth-century American poetry.

B. Notice

In the 1909 Act and the prior laws including the 1976 Act as originally adopted, copyright registration was mandatory and notice of copyright was required to be placed on publicly distributed copies of the protected work to give notice to potential infringers. In fact, before the 1976 Act, failure to include notice resulted in loss of protection.

There are four reasons that U.S. law traditionally required notice: (a) to place in the public domain, material in which no one has an interest in protecting; (b) to inform the public whether a particular work is protected; (c) to identify the name of the copyright holder; and (d) to indicate the publication date.[68] Copyright notice consists of three elements: the symbol ©, the word "copyright" or the abbreviation "copr.;" the year of first publication of the work and the name of the author or copyright holder.[69] Notice is a term of art in copyright law, and whenever the term is used it means these three elements and not a general warning statement. Under the 1909 Act, failure to include notice on all publicly distributed copies could result in the loss of copyright protection.[70] The 1976 Act liberalized considerably the notice requirement and provided a mechanism by which copyright owners could correct the accidental omission of notice on published copies.[71] When the United States joined the Berne Convention, notice ceased to be a requirement for copyright protection.[72]

The lack of a notice requirement is troubling to most librarians. In the past, if a book contained no notice of copyright, one was relatively safe in assuming that the work was within the public domain. Today this is no longer true. In fact, the opposite is the case; thus, a librarian must assume that a work which contains no notice of copyright still is protected.

Despite the fact that notice is no longer required, U.S. authors are likely to continue to include notices on their works for some important reasons. First, it places good-faith users on notice that copyright protection is claimed in the work, and persons in good faith will seek permission before attempting to use the work in any manner that might infringe the rights of the copyright holder. Second, the statute encourages copyright owners to include the notice of copyright by denying remedies against innocent infringers who relied on a copy of a copyrighted work which did not contain a notice.[73]

C. Registration and Deposit

Under the 1976 Act, registration ceased to be a requirement to perfect copyright protection. Today, copyright attaches at the time a work is fixed in tangible media and not on registration.[74] Registration consists of completing a form, filing a form with the U.S. Copyright Office and paying the requisite fee.[75] Registration is said to be permissive because it may occur at any time during term of copyright[76] and it is not necessary for protection.

Although registration is no longer necessary to perfect the copyright, it continues to play a role in U.S. law. The Register of Copyrights issues a registration certificate to the owner whenever the proper form is filed and the fee is paid.[77] Registration made within five years after publication of the work is prima facie evidence of the primacy of the copyright holder.[78] Registration also simplifies the major burden of proving ownership of the copyright. If the work were not registered, a complainant might have a difficult task proving that his work predated the allegedly infringing work. But most importantly, registration gives the copyright holder the right to sue infringers in federal court. No action for infringement may be filed until the work is formally registered with the Copyright Office. Currently, Congress is considering elimination of this requirement, however.[79]

Another benefit of registration is that remedies are limited if the work is not registered. Unless the work is registered within three months after first publication, no award of statutory damages or attorneys fees is permitted.[80] Congress is also considering amending the Act to permit suit in federal court without registration and to allow copyright owners to claim statutory damages regardless of whether the work is registered. It appears likely that this proposal will become law. [81]

In order to complete the registration, deposit of two copies or phonorecords of the best edition of a published work is required. Copies of the work must be submitted to the Library of Congress within three months of publication,[82] although this deposit is not a condition of copyright protection. If the owner of the copyright or proprietor of the right of publication fails to make the deposit, however, she may be the recipient of a demand for deposit from the Register of Copyrights.[83] If an additional three months pass following the demand letter without deposit being accomplished, the owner becomes liable to the Copyright Office or the Library of Congress, or both, for fines.[84] Since deposit is required for valid registration and registration is advisable, if not officially required in order to secure full copyright protection, most scholars recommend deposit to avoid any future difficulties in proving ownership of a copyrighted work. The same proposal to eliminate registration as a prerequisite to filing suit or to obtaining statutory damages may also eliminate the deposit requirement, however. [85] Library associations have been very concerned about the effect elimination of the deposit requirement will have on the Library of Congress since it has greatly assisted the Library in building its collection of U.S. imprints. At present, a report has been issued by the Advisory Committee on Copyright Registration and Deposit (ACCORD) created by the Librarian of Congress that would make registration voluntary but would include inducements to owners to encourage registration and deposit.[86]

IV. INFRINGEMENT OF COPYRIGHT

The general rule governing copyright is that any person, other than the copyright owner, her heirs, assigns or transferees, who uses of the copyrighted work in any manner reserved under section 106(a) without the owner's permission infringes the owner's copyright.[87] In 1990 the Act was amended to make it clear that states and employees of states acting in their official capacities may be found liable of copyright infringement just as any nongovernmental entity may be.[88] The legal or beneficial owner of a right may institute an action for an infringement, so long as that person had ownership at the time the infringement is claimed to have occurred.[89] The owner may be required to serve written notice with a copy of the complaint, not only on the alleged infringer, but also on any other person with an interest in the copyright or whose interests may be affected by a decision in the case (i.e., any entity who has a contract with the alleged infringer).[90] A party who brings suit for infringement must prove to the court that he owned the copyright in a work which was copied in some impermissible manner.[91]

A. *Tests for Infringement*

Copying may be proved by establishing two elements: (a) that the alleged infringer had *access* to the copyrighted work, and (b) that the works are *substantially similar*.[92] "Access" refers to the alleged infringer's or defendant's familiarity with the copyrighted work.[93] This factor can be established by inference; for example, if a copyrighted song had received a substantial amount of air play on the radio, courts infer that a defendant had access.[94] Direct access also may be proved, particularly if the work alleged to have been infringed was created by a person involved with the earlier work.[95]

Substantial similarity between the two works may in itself establish an inference of access.[96] Similarity can assume three forms. The first two are virtually conclusive: (a) if the second work is substantially identical to the first,[97] or (b) where there is "striking similarity," for example, shared sections "so idiosyncratic in...treatment as to preclude coincidence."[98] The third type of similarity may be referred to as abstract similarity. Here, the second work invokes the first, but it does not borrow any concrete elements from the first. The copyright owner must prove that the infringer copied the *expression* and not merely the idea underlying the expression.[99] Judge Learned Hand offered the so-called "abstractions" test to decide if an infringement occurred, when it is less than clear whether the expression has been infringed. He stated,

> [W]hen the plagiarist does not take out a block in situ, but an abstract of the of the whole, decision is more troublesome. Upon any work, and especially upon a play, a great number of patterns of increasing generality will fit equally well, as more and more of the incident is left out. The last may perhaps be no more than the most general statement of what the play is about, and at times might consist only of its title; but there is a point in this series of abstractions where they are no longer protected, since otherwise the playwright could prevent the use of his "ideas," to which, apart from their expression, his property is never extended.[100]

Although problematic to prove, this form of similarity is undeniably common. Often the easiest way to establish or disprove an allegation of abstract similarity is with an "audience" or "ordinary observer" test; under this test, if the ordinary person, upon reading the second work, believes that it was copied from the first, it probably was.[101]

B. *General Limitations on and Exceptions to Exclusive Rights*

Even if a copyright owner is able to prove the elements of infringement, there are a number of limitations and exceptions to the exclusive rights granted under the Copyright Act.[102] These exceptions provide that the individuals, under certain circumstances, may use, copy, perform, etc., the copyrighted material without the necessity of paying royalties or securing the permission of the owner. Probably the most common of these is one in which there is seldom little discussion and which is not even mentioned in the Act — personal use. Personal use is a private use of a copyrighted work such as reading a literary work for one's own enjoyment or learning. It might also involve sharing the work with a friend or colleague. The hallmark of such personal use is that there is no commercial motive.[103] Personal use has no page length or musical bar restriction; however, it is limited to a single copy for which there is no intention to distribute.[104]

The statutory limitations cover a wide variety of uses, but they generally serve one of several purposes: scholarly inquiry which includes instruction, research, criticism and newsworthiness; and performances and displays by educational, charitable, religious or government groups. In addition to the approved purposes, the use normally cannot be for direct commercial gain. Several of the limitations on exclusive rights require the user to obtain a license from the copyright holder or a licensing agency, such as the Copyright Clearance Center or the American Society of Composers,

Authors and Publishers (ASCAP) which provides access to numerous copyrighted works for the payment of royalties through a "blanket license." [105] The advantage is that the user does not need to pay a separate royalty for the public performance of each work.

Not all of the limitations are pertinent to libraries, and specific library uses that are relevant are discussed in the appropriate chapter.

V. FAIR USE

A. *Introduction*

The limitation on the copyright owner's rights which provides the widest public exploitation of copyrighted works is known as the fair use exception.[106] Anyone who makes a "fair use" of a work, even though the use involves one or more of the exclusive rights listed in section 106, does not infringe the copyright.[107] The role of the fair use doctrine has been described as a mechanism to ensure that copyright does not become an obstacle that impedes learning.[108] The underlying purpose of fair use is to provide copyrighted materials freely to individuals engaged in criticism, news reporting, scholarship and research; in other words, if the copyrighted work is of importance to the public and the proposed use will not unduly devalue the copyright, it is in the public interest to suspend an otherwise exclusive right for that use. By this exception, the television film reviewer is permitted to include clips of a new movie on a newscast and the critical biographer of a living writer may print passages from novels in the biography, both without paying royalties.

Fair use, however, is open to potential misuse by the very breadth of the exception. While the drafters of the statute understood the advantage of allowing critical and scholarly research unimpeded need to compensate the owner in the form of royalty payments, neither did they want to rob the copyright holder of due profit. Consider for a moment a biographer. She may be working on a highly detailed inquiry into the work of the author Harold Brodkey. Such a book probably would be of limited appeal outside the literary world, but it might have a great deal of importance inside that community. It would seem logical to grant her the right to quote extensively from Brodkey's work under the fair use exception, since her use of Brodkey's words are instructional and not likely to decrease the audience for his work. In fact, the biography likely will increase demand for his work. If the biography is about the autobiographical qualities of Jackie Collins's work, however, with long steamy passages included to illustrate the points, the book may be of lesser scholarly interest than the first. The second work, however, could well make it to the best-seller list. Is this a fair use of Ms. Collins's words, despite of the fact it is a biography? Maybe not, because the purpose of such an effort appears to be profit. If sexy novels sell, why should sexy nonfiction not be similarly blessed? On the other hand, such a biography should not adversely affect Collins's profits. If anything, it may actually increase her readership. Then, if the book were written by a noted researcher into sexual behavior as part of a study of the psychology of libido, perhaps there would be justification to consider it a fair use exception. There are, however, many works that fall between scholarship and sensationalism, such as a book regarding Salman Rushdie and *The Satanic Verses.*

The concept of fair use as it exists today was at least a century in the making.[109] What appears to be the earliest fair use case is the English case of *Gyles v. Wilcox*,[110] in which an infringement suit was denied jury trial and referred instead to a master for determination of whether the "abridgement" of a copyrighted work was fair.[111] Subsequent cases in the English courts honed the permissible exceptions to the exclusive rights of copyright.[112] In 1841, the modern articulation of fair use was pronounced in the United States in *Folsom v. Marsh*.[113] Justice Joseph Story, perhaps without realizing the full ramifications of his relatively modest statement, announced:

> In short, we must often, in deciding questions of this sort, look to the nature and objects of the selections made, the quantity and value of the materials used, and the degree in which the use may prejudice the sale or diminish the profits, or supersede the objects, of the original work.[114]

This statement became the basis of the statutory fair use exemption. Until the 1976 Act, fair use remained a judicially created doctrine, but today it is included in the copyright statute along with its four-pronged test. Fair use is a defense to copyright infringement; it is almost always raised by a defendant in a copyright suit. The four-pronged test is:

1. the purpose and character of the use, including whether such use is of a commercial nature or is for nonprofit educational purposes;
2. the nature of the copyrighted work;
3. the amount and substantiability of the portion used in relation to the copyrighted work as a whole; and
4. the effect of the use upon the potential market for or value of the copyrighted work.[115]

In order to determine fair use, the factors must be balanced to form a complete picture of the effect of the use on the copyright. Despite this balancing, the greatest weight today is on the fourth factor,[116] since that factor reflects the relative monetary gain to the user or loss to the copyright holder if the use is allowed. It ultimately includes the other three factors within its breadth, since misuse in any of those will adversely affect the potential market.[117]

The legislative history of the 1976 Act contains considerable discussion of fair use. Most of the examples used relate to uses of copyrighted works within nonprofit educational institutions; a few of the examples relate directly to libraries. Concerning fair use generally, however, examples in the legislative history include quotation of excerpts in a review or criticism of a work, inclusion for purposes of illustration, short passages as quotations in a scholarly or technical work, use in parody, reproduction of a work in a judicial or legislative proceeding and reproduction of a work by a teacher to illustrate a lesson.[118] Because fair use is such an important defense, it will be discussed in many of the chapters as it pertains to particular works and specific uses. A general understanding of the four fair use factors also is important, however.

B. The Purpose and Character of the Use

While most discussions of the purpose and character of the use focus on whether the use commercial or not,[119] this is not the only important consideration. If one views copyright as existing first to benefit the public and secondarily the author, then the use should be examined in terms of its relevance to the public welfare. From this viewpoint, three activities are involved: learning, free speech and the advancement of knowledge. If the use enhances any of these activities, some argue that it should be a fair use regardless of whether the use is commercial or competitive.[120]

Cases in which a secondary user's activity to be fair use on the basis of the first factor tend to focus on two issues: (a) transformative or productive uses and (b) noncommercial use. Historically, courts have preferred secondary uses of works that did not merely copy the protected work and thus becomes a substitute for the original. Works that took something from an original work for a new purpose or result that differed from the original were favored. Perhaps the best example of a transformative use is when a reviewer quotes from a copyrighted literary work in a book review.[121] Although every commercial use may be said to be presumptively unfair, this presumption is fairly easy to overcome if the use is a transformative one.[122] When courts find a productive nonsuperseding use, they tend to give scant consideration to any profit motivation.[123] So, when a textbook author quotes from another but gives proper credit to the first author, courts find fair use even though the motivation of the second author is profit. The court found a nonproductive use but held that such a finding did not end the fair use inquiry, and that other considerations still could lead to a finding of fair use in the 1984 case, *Sony Corporation of America v. Universal Studios, Inc.*[124]

Nonetheless, if a use is commercial, it weighs against a finding of fair use while the combination of "nonprofit" and "educational purposes" balances toward a finding of fair use.[125] Because of this strong interrelationship between a finding of fair use and nonprofit educational uses, Congress encouraged a compromise on multiple copying for classroom use. *The Guidelines on Multiple Copying for Classroom Use* were negotiated by representatives of publishers' and authors' groups and representatives of education and library associations to govern photocopying to support both the preparation of teachers and the learning experiences of students.[126] The guidelines do not have the force and effect of law, but they certainly have considerable validity. The guidelines were published in the House Report that accompanied the Act[127] and frequently have been cited by courts.[128]

The Guidelines on Multiple Copying for Classroom Use appear as Appendix C; they focus initially on the making of a single copy of a copyrighted work for a teacher for either scholarly research or for teaching. Under the guidelines, such single copying of a chapter from a book, a periodical article, an illustration or graph, poem, etc., is permissible. The remainder of the guidelines focus on multiple copying, i.e., one photocopy of a copyrighted work not to exceed one copy for each student in the class. There are four tests that such copying must meet: brevity, spontaneity, cumulative effects and each copy must include a notice of copyright. Some works by their nature are meant to be consumed, such as workbooks, standardized tests, answer sheets, etc. Copying of these is not permitted under the guidelines. Additionally, copying to create anthologies is prohibited.[129]

Brevity is defined very specifically with word and percentage limitations.[130] Spontaneity means first that the decision to copy and use the work was made by the individual teacher, and second, that the decision to use the work was made so late in the class term that there is no opportunity to obtain permission of the copyright owner.[131] The cumulative effects test has three parts: (a) the copying is done only for one course in the school, (b) no more than one poem, article or story from an author or more than three from the same periodical volume or collective work during the class term,[132] and (c) copying cannot be repeated with respect to the same item from term to term, since such repetition could not meet the spontaneity test.[133]

Such copying should be limited to no more than nine instances during the class term. Further, there must be no charge to students beyond the actual cost of the photocopying.[134]

There are also guidelines on the copying of music for academic purposes;[135] they are discussed in Chapter 5 and reproduced in Appendix D.

C. The Nature of the Copyrighted Work

The first criterion cannot be considered in a vacuum, because to do so could injure either the holder of the copyright or the user.[136] Suppose, for example, that the copyrighted work's only source of profit is academic use, such as a college textbook. To allow a fair use for the otherwise meritorious purpose of education could unduly injure the copyright owner. Therefore, it is advantageous to consider the second statutory factor, the nature of the work.[137] If the work is educational in nature, a fair use exception for an educational purpose might not be justified logically. Alternatively, to penalize a serious scholar merely because his work is on a popular figure such as Ernest Hemingway, rather than on a less notorious individual like Harold Brodkey, appears unjust and may discourage critical study of popular artists.

In considering the nature of the copyrighted work, courts usually focus on whether the work is factual or creative. Traditionally, courts have afforded greater protection to creative works and are more likely to find fair use if the work involved is a factual one. Because promotion of learning is the constitutional purpose of copyright,[138] a didactic work is given preference over creative ones in fair use terms. This creates an interesting paradox: the less value a work has to society (a creative work intended for entertainment purposes), the greater the protection it receives.[139] Whether a work is published or unpublished also is a consideration. Unpublished works are given somewhat greater protection since uses of these works affect the right of first publication which secures a very important market interest.[140] Another consideration is whether the work is out-of-print, but at least one court recently held that royalty payments for reproduction of out-of-print works were even more important since such royalties were the only income out-of-print works produced for the author.[141]

The nature of the work has another interpretation as well, which relates to the originality of the copyrighted work. The greater the creative effort the copyright owner exerted in creating the copyrighted work, the less likely that fair use is found to be an appropriate defense.[142] Under this concept, it is easier to argue successfully that use of a section of book containing statistical information collected from a number of independent studies is a fair use than it is for the verbatim use of a chapter of a novel or an entire poem.

D. The Amount and Substantiality Used

The proportional amount and substantiality factor relates to both the quantity and the quality of the portion used. In applying this test, courts frequently count words and pages of literary works and bars of music. The amount copied often is expressed in percentages of the copyrighted work that is copied.[143] In the *American Geophysical Union v. Texaco* case the court found that each article was a separately copyrightable item, so that even if a scientist copied only one short article from a volume of a journal, it was too much to be a fair use.[144] If the amount appropriated is minimal, it is a fair use. This is related to the purpose of the use, however.[145] The amount and substantiality of the amount copied also is relevant in determining the degree of harm to the copyright holder.[146] At its most basic, a user usually cannot claim a fair use for republishing another's work in its totality (for example, a letter or a poem), although even such a extensive usage may be justified by other factors.

Another pertinent consideration for this factor is the qualitative importance of the piece taken for the user's work. In *Meeropol v. Nizer*,[147] the defendant published a book containing 28 of the plaintiff's copyrighted letters. Although these letters constituted less than one percent of the entire work; they were key to the book's thesis and formed a significant part of the advertising campaign for the book. Thus, the court held that fair use was not an appropriate defense to the plaintiff's charge of copyright infringement.[148] The Court held in *Harper & Row Publishers, Inc. v. Nation Enterprises*[149] that *Nation* had copied the qualitative portion of President Gerald Ford's memoirs, that is, the material on his pardon of President Richard Nixon.[150] Although readers of the memoirs might be interested in everything the book contained, the direct quotations of President Ford's description and explanation of why he pardoned Nixon were the heart of his memoirs, the portion likely to be of most concern to readers.[151] Although the circuit court held that the amount and substantiality used was infinitesimal, the qualitative value of the Nixon pardon material meant that the amount copied exceeded the bounds of fair use.[152] Even paraphrasing copyrighted materials may not forestall an infringement charge, particularly where there are other similarities between the copyrighted work and the alleged infringing work.[153]

E. The Market Effect

The market effect or value of the work test is satisfied if the use does not have an adverse impact on the market for the work. The following factors are considered in this determination: (a) accessibility of the work, (b) date of the work, (c) economic life of the work, (d) availability of copies on the market, (e) price of the work, and (f) evidence of abandonment.[154] The effect of the use on the original author's market or the value of the work has been called "the single most important element of fair use."[155] Courts appear willing to find injury to the market when the defendant's work directly competes with that of the owner.[156]

If a use is for profit, clearly there is the opportunity that the use will diminish or supplant the original author's market. The extent to which the diminution of the market may injure the author incorporates all the other elements of the fair use defense. If the purpose of the use is commercial, that profit motive may manifest in the choice of the work infringed or in the amount and substantiality of the portion used, but ultimately the profit motivated use will have the effect of making the work less valuable for the copyright holder.

Harper & Row[157] illustrates these principles. The plaintiff was a book publisher licensed by former President Ford to publish his forthcoming autobiography and to engage in publicizing the book. As a part of the latter effort, the plaintiff contracted with *Time* magazine to print excerpts (principally, a section discussing Ford's pardon of former President Nixon) in exchange for royalties for first publication rights. *Time* agreed to pay $25,000 for the right to publish this excerpt which would appear a week or so before the book was available in bookstores. *The Nation*, another news magazine and the defendant, however, acquired illicit galley prints of the book and published the same excerpts before *Time* could do so. In other words, it scooped the *Time* publication. *Time* dissolved the contract because the value of its excerpt publication was lost.[158] Although the defendant raised various defenses that ranged from the First Amendment right to disseminate newsworthy information to the relatively small amount of the autobiography it had reproduced, the U.S. Supreme Court found that the use was not fair use but infringement.[159] Although the purpose of the defendant's use may have been journalistic, it was also for profit from the sale of its magazine. In addition, the nature of the work was highly original, which appealed to persons interested in modern history; the licensing of first publication rights to a magazine similar to the defendant's evidences the nature of the interest in the book.[160] It was generally acknowledged that the section concerning the Nixon pardon, although short in length, was the single most important part of the book from the standpoint of public interest, so the use infringed the very heart of the copyrighted work. The combination of these factors robbed the licensing of the autobiography of its value and directly affected the plaintiff's market, since *Time Magazine* canceled its contract for prepublication rights.[161]

F. Unpublished Works

In 1992 section 107 was amended[162] to add the following sentence: "The fact that a work is unpublished shall not itself bar a finding of fair use if such finding is made upon consideration of all of the above factors." The amendment ended several years of debate concerning the use of excerpts from unpublished letters and diaries by historians, critics and biographers. A series of cases principally from the Second Circuit U.S. Court of Appeals had seriously restricted the ability of scholars to quote extensively from unpublished works.[163] After 1992 it is clear that both published and unpublished works are subject to the fair use exemption.

G. Libraries and Fair Use

Section 107 fair use rights are available to libraries, but there is little guidance on applicability. Despite some publisher assertions to the contrary, section 108(f) states that nothing in the section "... in any way affects the right of fair use as provided by section 107, or any contractual obligations assumed at any time by the library or archives when it obtained a copy or phonorecord of a work in its collections."[164] Additionally, some of the examples of fair use included in the legislative history are library uses.[165]

VI. DEFENSES TO INFRINGEMENT AND REMEDIES

A. Defenses

Fair use or the section 108 limitation on exclusive rights are the defenses most likely to be claimed by a librarian accused of copyright infringement. A common defense claimed by other authors accused of infringement, however, is independent creation. If an author creates a very similar or even identical work but does not copy from a previous work, then there is no infringement of copyright.[166] Proving independent creation when defending against a claim of copyright infringement is difficult unless there is tangible evidence as to independent creation and if there is no access to the earlier work (which must be a little-known work). If neither fair use nor other exceptions or limitations apply, an alleged infringer still may have legal or equitable defenses available to exculpate herself. Probably the easiest to prove is that the statute of limitations on bringing the action has run; therefore, the copyright owner would have no cause of action remaining. The statute of limitations for copyright infringement is three years, and the time begins to run as soon as the infringing act occurs.[167] If the infringement is ongoing, however, the statute of limitations period begins with the initiation of the last act of infringement.[168] Because copyright infringement may involve a plaintiff and a defendant, both of whom are innocent of any intentional wrongdoing, the court has a good deal of discretion in balancing the statutory law with the equities of the situation to achieve a just conclusion.

Two defenses rest on legal technicalities. For the defendant who has been previously involved in a suit that decided the same claims or issues as the instant case, two potential defenses are *res judicata* and collateral estoppel.[169] Related equitable defenses are laches and estoppel.[170] Laches requires a defendant to prove that the copyright owner delayed in bringing the suit and that this delay seriously prejudiced the defendant.[171] To succeed under an estoppel defense, the defendant must prove that the plaintiff lulled the innocent defendant into the infringement, and the defendant, ignorant of the existence of the copyright, relied on the plaintiff's inducement to his detriment.[172]

A more tenable defense, albeit not complete, is that of innocent infringement particularly since the Berne implementation removed the necessity of placing notice on copyrighted works. To succeed, the defendant must prove that she acted in good faith, but was misled by the lack of notice.[173] While this defense will not result in a verdict in the defendant's favor, it may influence the court to lower statutory damages to the absolute minimum. The Act contains a special provision concerning library employees and innocent infringement. The statute directs the court to remit the damages to the lowest level, $200 per act of infringement where the infringement was innocent and the infringer is an employer or agent of a nonprofit educational institution who was acting within the scope of his employment.[174]

Other defenses are based on impropriety on the part of the copyright holder. For example, the defense of misuse may be raised when a copyright holder has conducted herself in a manner antithetical to competition; therefore, the defense is similar to charging the plaintiff with an antitrust violation.[175] Considering the monopolistic aspect of copyright, this defense is effective only if the defendant can prove that the plaintiff is using his copyright to injure competition in another market.[176] This occurs most often in computer situations, where certain hardware requires the use of specific copyrightable chips or software, so the exercise of the copyrights has a monopoly effect in another business area. To be successful with a misuse defense, however, the defendant must prove that the equities are substantially, if not totally, in her favor; in the example of the computers, the question lurks, who created hardware that required this particular program or chip? Are there no alternatives that could make the hardware viable? Thus, misuse is a problematic defense and one that seldom succeeds.

Other types of misconduct on the part of the copyright holder, known en masse as "unclean hands," are a defense; such conduct may take one of three forms. Misleading conduct, which may consist of such actions as a misrepresentation to the Copyright Office regarding the originality of a work, does not have to rise to the level of fraud to serve as the basis of a defense.[177] A bad faith defense requires that the copyright holder and the alleged infringer have had a previous relationship involving the work, which is destroyed by a self-serving act of the owner such as publishing a book on his own that was to have been a collaborative work.[178] The final defense is abuse of the judicial process, which might constitute a falsified order or release of misleading information about pending infringement suits that discourage the defendant's customers.[179]

B. Remedies

The owner of a copyright who brings a successful suit for infringement has a number of potential remedies available. The most common remedy is an injunction,[180] one of several coercive remedies available through the Act.[181] This remedy allows the court in its discretion to issue a restraining order to stop the alleged infringer from producing or vending any more of the infringing goods. An injunction may be either temporary or permanent[182] and it is enforceable anywhere in the U.S.[183] While this is a very effective remedy in controlling mass piracy, it may be oppressive in cases where the alleged infringement is a small part of an otherwise original work.[184]

Impoundment and destruction of works are also coercive remedies and are closely related to injunctions. While an action is pending, the court may order the impoundment of all copies made or used in alleged violation of the plaintiff's copyright.[185] The court may order these impounded items destroyed if, in the final judgment, the items are found to infringe the copyright.[186]

Monetary awards also are available: damages, attorneys fees and court costs. The successful copyright owner has the choice of two forms of damages: actual damages and profit or statutory damages.[187] Actual damages and profits allow the copyright owner to recover any profits lost because of the infringing use, plus any profit earned by the defendant above the actual damages.[188] The problem with actual damages is the proof of the amount of damages suffered or the defendant's profits earned; the copyright owner has the burden of proof,[189] which is in itself problematic. In a library situation, it is unlikely that a copyright holder would seek actual damages and profits. Normally, these damages are sought against highly successful record and tape pirates when millions of dollars are in controversy. In most libraries, any infringement likely has netted no profit and caused little direct economic harm to the owner; thus, statutory damages are more likely. If the copyright owner chooses statutory damages, the parameters of monetary awards are set by the court. In its discretion, the court determines on the amount of the award for each act of infringement within those parameters.[190] Courts tend to approximate the amount of statutory damages to the rough amount of the actual damages to the plaintiff, if such evidence has been offered.[191]

Statutory damages range from $500 to $20,000 per act of infringement. If the infringement is innocent infringement, the court may lower the damage award to $200 per act of infringement.[192] An example of innocent infringement is when a librarian contacts an author and requests the right to make multiple copies of an article. The author graciously grants permission but forgets that she transferred the copyright to the publisher. Thus, when the librarian made the copies, he innocently infringed the copyright assuming he had the proper permission. If the court finds that the infringement is willful, or where there is reckless disregard for the existence of a copyright, it may raise the damage award to $100,000 per act of infringement.[193] A librarian who contacts a copyright owner to request permission to make multiple copies of a work may be guilty of willful infringement if the owner denies permission but the librarian makes copies anyway.

Attorneys fees also may be awarded to the prevailing party in a suit at the discretion of the court.[194] No attorneys fees may be obtained unless the work was formally registered with the Copyright Office, however.[195] Court costs also may be awarded to the prevailing party in any civil suit brought under the Act.[196]

The Act further provides that criminal charges may be brought in connection with copyright infringement. Criminal penalties are available only if the infringement was willful and done in expectation of commercial gain.[197] There has been some speculation that a librarian in a profit-seeking company who infringes copyright in performing his job always does it for commercial gain, at least for the corporation's commercial advantage. Fraud, either by forging a copyright notice,[198] removing a valid notice mark[199] or intentionally misrepresenting information in an application for copyright registration,[200] may be grounds for criminal prosecution. An action for criminal infringement must be commenced within three years after the original cause of action arose, or in other words, three years after the first infringing event.[201]

ENDNOTES - CHAPTER 2

1 17 U.S.C. §§ 101-810 (1988). The title was revised in its entirety by Pub. L. 94-553, Title I, Oct. 19, 1976, 90 Stat. 2598; the Act became effective January 1, 1978.

2 Id. § 301(a). The 1976 Act preempted all previous copyright statutes under the provisions of section 301, although no other federal laws were subjugated to it. This means that applicable laws of misappropriation, unfair competition and other areas of the law which may extend to copyrightable materials may be applied although no specific provision for their application is made in the Act. Sections 301(b)-(e) set some limitations on the preemptive power of the statute.

3 This statement is true regarding the concept of common law copyright that refers to perpetual protection of works which had never been published. There is, however, a second aspect to the common law copyright, a right of first publication, in which the author has right to control the work's initial public dissemination. This aspect continues to exist as section 202 of the Act.

4 17 U.S.C. § 302(a) (1988).

5 Id. § 102(a).

6 Id. This section was amended and part (8) was added by Pub. L. 101-650, Title VII 703, Dec. 1, 1990, 104 Stat. 5133 (codified at 17 U.S.C. § 102 (Supp. IV 1991)).

7 Id. § 202.

8 Id. § 103(a).

9 Id. § 101.

10 Id.

11 Feist Publications, Inc. v. Rural Telephone Service Co. Inc., 499 U.S. 340, 111 S.Ct. 1282, 1289 (1991).

12 H.R. Rep. No. 1476, 94th Cong., 2nd Sess. (1976) reprinted in 17 Omnibus Copyright Revision Legislative History 54 (1977) [hereinafter House Report].

13 Paul Goldstein, 1 Copyright: Principles, Law and Practice § 2.5 (1989) [hereinafter Goldstein]. Fraudulent works are those that have been passed off or misrepresented as legitimate copyrighted works of someone not the author, or distributed, copied, licensed, etc., by a person falsely purporting to have the right to use the work in those manners.

14 Id. § 2.5.1 (citing, Edward S. Rogers, Copyright and Morals, 18 Mich. L. Rev. 390 (1920)).

15 Id. § 2.5.1.

16 17 U.S.C. § 105 (1988). See, 1 Goldstein, supra note 13, § 2.5.2.1 (legal documents) and § 2.5.2.2 (other federal works).

17 Id. § 2.5.2.2. The work of a federal employee outside her official employment is not subject to this prohibition and may be copyrighted.

18 17 U.S.C. § 105 (1988).

19 See U.S. Copyright Office, Compendium II; Compendium of Copyright Office Practices § 206.01 (1984).

20 17 U.S.C. § 104 (1988).

21 For the text of the UCC as revised in 1971, see 17 U.S.C. § 104 (Appendix, Treaties and Conventions, 1992), at 37-53. See also Chapter 8.

22 For information on Berne, see Chapter 8.

23 17 U.S.C. § 104(a) (1988).

24 Id. § 104(b)(1)-(5).

25 Id. § 104(b)(1). This section applies also to a stateless person and to a work that is the product of a foreign government which is a treaty signatory.

26 Id. § 104(b)(2)-(4).

27 Id. § 104(b)(5). This provision allows the President to extend to a nation not within any of the other categories rights under the Copyright Act by proclamation. This provision presumably will come into

play in developing relations with the new nations arising from the former Soviet Union and in Eastern Europe.

28   Id. § 106(1)-(5). Subject to sections 107 through 118, the owner of copyright under this title has the exclusive right to do and to authorize any of the following:
(1) to reproduce the copyrighted work in copies or phonorecords;
(2) to prepare derivative works based upon the copyrighted work;
(3) to distribute copies or phonorecords of the copyrighted works to the public by sale or other transfer of ownership, or by rental, lease or lending;
(4) in the case of literary, musical, dramatic, and choreographic works, pantomimes, and motion pictures and other audiovisual works, to perform the copyrighted work publicly; and
(5) in the case of literary, musical, dramatic, and choreographic works, pantomimes, and pictorial, graphic, or sculptural works, including the individual images of a motion picture, or other audiovisual works, to display the copyrighted work publicly.

29   1 Goldstein, supra note 13, § 5.1.

30   17 U.S.C § 106(3) (1988).

31   Id. § 101.

32   Id.

33   Id.

34   Id.

35   Id.

35   1 Goldstein, supra note 13, § 5.0.

37   L. Ray Patterson & Stanley W. Lindberg, The Nature of Copyright: A Law of User's Rights 193-95 (1991) [hereinafter Patterson].

38   1 Goldstein, supra note 13, § 4.4.1.

39   Id.

40   17 U.S.C. § 201(a) (1988). The copyright vests immediately in the author unless it is a work made for hire which requires a separate written agreement outside of the employment context. See also 1 Goldstein, supra note 13, § 4.1.

41   Id. § 201(b). A "work made for hire" is:
(1) a work prepared by an employee within the scope of his or her employment; or
(2) a work specially ordered of commissioned for use as a contribution to a collective work, as a part of a motion picture or other audiovisual work, as a translation, as a supplementary work, as a compilation, as an instructional text, as a test, as answer material for a test, or as an atlas, if the parties expressly agree in a written instrument signed by them that the work shall be considered by them a work made for hire. Id. § 101.

42   Id. § 201(d).

43   Id. § 202. In this section that the last vestiges of the common law copyright "right of first publication" survive. Unless the artist expressly transfers the copyright, he is presumed to have retained the right to publish the work first, even though the work itself was sold. This situation is generally limited to works of fine art, like paintings and sculpture, where the artist has the right show the work first or render it as a print or photograph, even though it may be sold to a collector before it is finished form. For a discussion of this situation, see, e.g., Latern Press Inc. v. American Publishers Co., 419 F. Supp. 1267 (E.D.N.Y. 1976); Pushman v. New York Graphic Society, Inc., 39 N.E.2d 249 (1942).

44   Id. § 201(d).

45   Id. § 204(d).

46   Id. § 204(a).

47   Prima facie evidence is evidence sufficient in itself to establish a fact, if not rebutted or contradicted. It is also sufficient to sustain a judgment in favor of the issue it supports, again if not contradicted. Black's Law Dictionary 1190 (6th ed. 1990).

48   17 U.S.C. § 204(b)(1)-(2) (1988). Recordation of transfers as a prerequisite to infringement suits and developing a system of priorities between conflicts of transfers are found in section 205.

49   Id. § 205(a).

50   Id. § 203(a)(3).

51   See id. § 203(a)(2) for the intestate descent of copyright interest.

52   Id. § 203(a)(3). The actual wording of the section foresees more complication. The statute of limitations on a termination action tolls either after five years from the end of the 35-year grant, or if the grant covers the right of publication, the period begins either 35 years from the date of publication under the grant or 40 years from the date the grant was executed, whichever is earlier.

53   Id. § 203(b). There are some limitations on this general rule that are listed in this section.

54   Id. § 302(a).

55   Id. § 302(c). Provision for the revealing of an anonymous or pseudonymous author, who comes forward after the date of publication is included.

56   Id. § 302(b).

57   Id. § 302(e).

58   Id.

59   See Hearings Held on Possible Extension of Copyright Term, U.S. Copyright Office, 46 Pat. Trademark & Copyright J. (BNA) 466-67 (1993).

60   Id. § 301(a).

61   Id. § 303.

62   Id.

63   Id. § 304(a). If no renewal of a 1909 work was made, however, the copyright terminated after 28 years.

64   Id.

65   Id. § 304, as amended by Act of June 16, 1992. Pub. L. No. 102-307, (codified at 17 U.S.C. (Supp. IV 1992)).

66   Materials in the public domain are of three types: (a) works on which the copyright has run, (b) works published by the U.S. government and (c) works on which the author never claimed copyright.

67   17 U.S.C. § 103(a)-(b) (1988).

68   House Report, supra note 12, at 143.

69   17 U.S.C. § 401(b) (1988).

70   17 U.S.C. § 21 (1970) (repealed 1978).

71   17 U.S.C. § 405 (1988).

72   Id. § 401(a). Until the Berne Implementation Act was adopted in 1988, the text of this subsection read: "... a notice of copyright shall be placed on all publicly distributed copies ..." The current version of the text is "... a notice of copyright may be placed on all publicly distributed copies ..." (Italics added for emphasis.)

73   Id. § 405(b).

74   Id. § 102(a).

75   Id. §§ 408-09. The current registration fee is $20; it may be raised from time to time by the Register of Copyrights.

76   Id. § 408.

77   Id. § 410(a).

78   Id. § 410(c).

79   Id. § 411(a). See also H.R. 897, 103d Cong., 1st Sess. (1993). The bill passed the House without objection on November 20, 1993.

80   Id. § 412.

81   H.R. 897, 103d Cong., 1st Sess (1993). The bill passed the House without objection on November 20, 1993.

82   Id. § 407(a)-(b).

83   Id. § 407(d).

84   Id.

85   On November 4, 1993, a substitute bill to H.R. 897 was introduced to ensure that repeal of the mandatory registration requirement would not reduce deposits. See House Passes Copyright Bill Ending Registration As a Condition for Suing, 47 Pat. Trademark & Copyright J. (BNA) 78-79 (1993).

86   Library of Congress Committee Issues Report on Registration and Deposit, 46 Pat. Trademark & Copyright J. (BNA) 439-41 (1993). The Librarian of Congress is soon expected to issue a report based on the advice from this committee for submission to Congress.

87   Id. § 501(a) (1991). This subsection reads:
      Anyone who violates any of the exclusive rights of the copyright owner as provided by sections 106 through 118 or of the author as provided in section 106(a), or who imports copies or phonorecords into the United States in violation of section 602, is an infringer of the copyright or right of the author as the case may be. For purposes of this chapter (other than section 506), any reference to copyright shall be deemed to include the rights confereed by section 106A(a).

88   Id. (as amended by Pub. L. 101-553, 104 Stat. 2749 (1990); 17 U.S.C. § 501(a) (Supp. III 1990)).

89   17 U.S.C § 501(b)(1988).

90   Id.

91   See e.g., Jones v. CBS, Inc., 733 F. Supp.748 (S.D.N.Y. 1990); Spectravest, Inc. v. Mervyn's, Inc., 673 F. Supp. 1486 (N.D. Cal. 1987). "Copying" in this context refers to the exploitation of the work in any manner foreseen by the exclusive rights listed under section 106(a). See 2 Goldstein, supra note 13, § 7.2.

92   2 Goldstein, supra note 13, § 7.2.1.

93   Id. § 7.2.1.1.

94   Id. § 7.2.1.1, n.13 (citing, ABKCO Music, Inc. v. Harrisongs Music, Inc., 722 F.2d 988, (2d Cir. 1983). George Harrison's My Sweet Lord was held to infringe the copyright on the musical composition of He's So Fine; dicta of the court established that the popularity and familiarity of the earlier song would have been sufficient to establish an inference of access, even if direct access had not been proven.

95   Id. § 7.2.1.1. (citing, Kamar Int'l., Inc.v. Russ Berrie & Co., 657 F.2d 1059, 1062 (9th Cir. 1981)).

96   Id. § 7.2.1.1.

97   Id. § 7.2.1.1. Proof of this may also lie in a studied dissimilarity, where the copied work obviously deviates from the original to circumvent a charge of plagiarism. Id. § 7.2.1.2.

98   Id. § 7.2.1.1 (citing Heim v. Universal Pictures Co., 154 F.2d 480, 488, (2d Cir. 1946)). These may include shared errors or instances of startling uniqueness.

99   Id. § 7.3.1.

100   Id. § 7.3.1 (citing, Nichols v. Universal Pictures, Corp., 45 F.2d 119, 121 (2d Cir. 1930), cert. denied, 282 U.S. 902 (1931)).

101   For a more articulate expression of this test and its history, see id. § 7.3.2.

102   17 U.S.C. §§ 107-18 (1988).

103   Patterson, supra note 37, at 193.

104   Id. at 194.

105   See Chapter 4 on licensing.

106   17 U.S.C. § 107 (1988).

107   See Sony Corp. of Am. v. Universal City Studios, Inc., 464 U.S. 417 rehearing denied, 465 U.S. 1112 (1984).

108   Patterson, supra note 37, at 196.

109   William F. Patry, The Fair Use Privilege in Copyright Law 3 (1985) [hereinafter Patry].

110   Id. at 7 (citing, 2 Atk. 141 (1740)).

111 Id. at 7. At trial, Lord Chancellor Hardwicke stated, "If I should extend the rule so far as to restrain all abridgements, it would be of mischievous consequence, for the books of the learned, les Journels des Scavans, and several others the might be mentioned would be brought within the meaning of this act [Statute of Anne] of parliament."

112 For an overview of case law pertaining to the historical development of the fair use exception, see id. at 6-19. The term "fair use" was first applied to these exceptions in Britain in Lewis v. Fullerton, 2 Beav. 6 (1869).

113 Patry, supra note 109, at 19 (citing, 9 F. Cas. 342 (C.C.D. Mass. 1841)). The controversy surrounded the use of George Washington's private and public letters in a work of fictionalized autobiography, without permission of the proprietor of the letters, who had used them earlier as the basis of a biography of Washington.

114 Id. at 20 (citing, 9 F. Cas. at 348.).

115 17 U.S.C. § 107(1)-(4) (1988).

116 See Harper & Row Publishers, Inc. v. Nation Enters., 471 U.S. 539, 566 (1985).

117 2 Goldstein supra note 13, § 10.1.

118 House Report, supra note 12, at 65.

119 17 U.S.C. § 107(1) (1988).

120 Patterson, supra note 37, at 201.

121 American Geophysical Union v. Texaco, Inc., 802 F. Supp. 1 (S.D.N.Y. 1992).

122 Sony, 464 U.S. 417, 448.

123 Texaco, 802 F. Supp. at 12.

124 464 U.S. at, 448-54.

125 2 Goldstein, supra note 13, § 10.2.2.1.

126 House Report, supra note 12, at 68.

127 Id. at 68-70.

128 See Basic Books, Inc. v. Kinko's Graphics Corp., 758 F. Supp. 1522 (S.D.N.Y. 1991).

129 House Report, supra note 12, at 69.

130 Id. at 68.

(iii) Illustration: One chart, graph, diagram, drawing, cartoon or picture per book or per periodical issue.

(iv) "Special works": Certain works in poetry, prose or in "poetic prose" which often combine language with illustrations and which are intended sometimes for children and at other times for a more general audience fall short of 2,500 works in their entirety. Paragraph (ii) above notwithstanding such "special works" may not be reproduced in their entirety; however, an excerpt comprising more than two of the published pages of such special work and containing not more than two of the published pages of such special work and containing not more than 10% of the words found in the text thereof, may be reproduced.

(i) Poetry: (a) A complete poem if less than 250 words and if printed on not more than two pages or, (b) from a longer poem, an excerpt of not more than 250 words.

(ii) Prose: (a) Either a complete article, story or essay of less than 2,500 words, or (b) an excerpt from any prose work of not more than 1,000 words or 10% of the work, whichever is less, but in any event a minimum of 500 words.

[Each of the numerical limits stated in (i) and (ii) above may be expanded to permit the completion of an unfinished line of a poem or of an unfinished prose paragraph.]

131 Id. at 69.

132 Id.

133 Id.

134 Id. at 70.

135 Id.

136   2 Goldstein, supra note 13, § 10.1.1.

137   17 U.S.C. § 107(2) (1988).

138   U.S. Const. art. 1, § 8, cl. 8 "To promote the progress of science and useful arts, by securing for limited times to authors and inventors, the exclusive right to their respective writings and discoveries."

139   Patterson, supra note 37, at 210, but see Landes & Posner, An Economic Analysis of Copyright Law, 18 J. Legal Studies 325 (1989).

140   2 Goldstein, supra note 13, § 10.2.2.

141   In Basic Books, Inc., v. Kinko's Graphics Corp., 758 F. Supp. 1522 (S.D.N.Y. 1991), the court held that the copying of articles and book chapters for course packets infringed the rights of the copyright holder. The impact on the market for the value of the work was ..." more powerfully felt by authors and copyright owners of out-of-print books ..."

142   Goldstein, supra note 13, § 7.3.1. This idea harks back to the concept that expression is protected but ideas are not. The more originality a work contains, such as fiction or original philosophy, the more likely it is that anything taken from it would be expression. On the other hand, when the work is historical, critical or interpretive and therefore based on hard facts that exist outside the copyrighted work, it grows less likely that a casual use of the work will infringe on expression.

143   See Kinko's, 758 F. Supp. 1522.

144   Texaco, 802 F. Supp. at 17.

145   Patterson, supra note 37, at 204.

146   2 Goldstein, supra note 13, § 10.2.2.3.

147   560 F.2d 1061 (2nd Cir. 1977), cert. denied, Nizer v. Meeropol, 434 U.S. 1013 (1977).

148   Id. at 1070-71.

149   471 U.S. 539 (1985).

150   Id. at 564-65.

151   Id.

152   Id. at 567.

153   2 Goldstein, supra note 13, § 7.3.1, n.15.

154   Patterson, supra note 37, at 204-05.

155   Harper & Row, 471 U.S. at 566.

156   2 Goldstein, supra note 13, § 10.2.2.4.

157   471 U.S. 539 (1985).

158   Id. at 543.

159   Id. at 549.

160   Id. at 553.

161   Id. at 562. The Court also discussed at length the issue of first publication rights and the value of such. Those issues, however important to the resolution of the case, are ancillary to the fair use issues involved therein.

162   Act of Oct. 24, 1992, Pub. L. 102-492, 106 Stat. 3145, (codified at 17 U.S.C. § 107 (Supp. IV 1992)).

163   See generally, Salinger v. Random House, Inc., 811 F.2d 90 (2d Cir. 1987); New-Era Publ. Int'l, ApS v. Henry Holt and Co., Inc., 873 F.2d 576 (2d Cir. 1990); and Wright v. Warner Books, Inc., 953 F.2d 731 (2d Cir. 1991). See also Chapter 7 for further discussion of this issue.

164   17 U.S.C. § 108(f)(4) (1988).

165   For example, the Senate Report discusses the preparation of works in special forms needed by blind persons which usually is done by libraries. S. Rep. No. 473, 94th Cong., 1st Sess. (1975) reprinted in 13 Omnibus Copyright Revision Legislative History 66 (1977).

166   See Goldstein, supra note 13, § 7.2.2.

167   17 U.S.C. § 507 (1988).

168   2 Goldstein, supra note 13, § 9.1.1 (citing, e.g., United States v. Shabazz, 724 F.2d 1536, 1540 (11th Cir. 1984).

169 Id. § 9.2. Although res judicata and collateral estoppel are often discussed together, the two are not at all synonymous. It is easier to conceive of the differences if res judicata is thought of as "claim preclusion" and collateral estoppel as "issue preclusion." To illustrate, imagine a copyright suit in which the defendant is a writer charged with having copied the plaintiff's plotline into his play. The court finds that, although the works are similar, no more than an idea is shared and, therefore, no suit for infringement will stand. A final judgment is issued to that effect. Because the decision is binding on the parties, the plaintiff cannot bring a similar suit against the original defendant's co- author because the matter is barred by res judicata. By the same force, even if the plaintiff won, he still would be barred against bringing a later suit against the co-writer because the claim was decided and res judicata applies. Id. § 9.2.1.

Collateral estoppel applies only to specific issues decided in an interlocutory order, which lacks the effect of a final judgment. Although the issues cannot be relitigated, that will not necessarily bar a broader cause of action. So, although a civil plaintiff may not be able to relitigate a charge of fraud against a person found guilty of that charge in a criminal action, that will not bar the plaintiff from suing the infringer for other aspects of the infringement. (Notice, however, that because the standard of proof is greater in criminal actions than in civil cases, issues decided in a civil suit will not stop the issue from being relitigated in a criminal suit.) Id. § 9.2.1 n.9.

170 Id. § 9.5.

171 Id. § 9.5.1. Unlike the statute of limitations, laches is not based on a specific period of time. Anything from a couple of days to a couple of years can constitute an inexcusable delay. Laches is not often a successful defense, however; in addition to the difficulty to the defendant of proving that the delay was prejudicial, many states bar the intentional infringer from using the defense altogether.

172 Id. § 9.5.2.

173 Id. § 9.4.2.

174 17 U.S.C. § 504(c)(2) (1988).

175 2 Goldstein, supra note 13, § 9.6.

176 Id. § 9.6.1.

177 Id. § 9.6.2.

178 Id.

179 Id.

180 17 U.S.C. § 502(a) (1988).

181 2 Goldstein, supra note 13, § 11.0.

182 17 U.S.C. § 502(a) (1988).

183 Id. § 502(b).

184 2 Goldstein, supra note 13, § 11.0.

185 17 U.S.C § 503(a) (1988). "Copies" also refers to phonorecords, molds, masters, tapes, film negatives, etc.

186 Id. § 503(b).

187 Id. § 504(a).

188 Id. § 504(b).

189 Id. § 504(b). If the plaintiff chooses defendant's profits as the measure of damage, all she need prove is the defendant's gross profits. The burden of proof then shifts to the defendant who has the opportunity to prove that the profits he reaped were toattributable to the infringing use, and therefore not required to be paid as damages to the plaintiff.

190 Id. § 504(c).

191 2 Goldstein, supra note 13, § 12.2.11.

192 17 U.S.C. § 504(c)(2) (1988).

193 Id.

194 Id. § 505.

195   Id. § 412.

196   Id. § 505.

197   Id. § 506(a). The punishment for any person found guilty of infringement will be in keeping with the provisions of 18 U.S.C. § 2319. In addition, all copies and copying devices of the guilty party will be seized and destroyed. Id. § 506(b).

198   Id. § 506(c).

199   Id. § 506(d).

200   Id. § 506(e).

201   Id. § 507(a). The statute of limitations is the same for civil copyright infringement suits. Id. § 507(b).

CHAPTER 3

LIBRARY PHOTOCOPYING AND OTHER REPRODUCTION

I. INTRODUCTION

Several sections of the Copyright Act are important to libraries, but section 108 is clearly the most important for several reasons. First, it covers the photocopying of printed copyrighted works, and this is the primary connection libraries have with the reproduction right that belongs to copyright owners. Second, section 108 serves as the basis for determining whether other forms of copying of printed works, such as microfilming, electronic copying, etc., conform with the Copyright Act. Third, if the conditions under each subsection are met for that type of copying, the library may reproduce the work without seeking permission or paying fees to the copyright holder. Finally, the library section grants considerable rights to libraries that were either unclear prior to 1976 or may not have existed at all.

The 1909 Act[1] made no mention of libraries, but as revision of the copyright law began to be discussed, it was clear that publishers considered library photocopying a major problem, one that deprived them of considerable revenue. Thus, from the 1970s, Congress indicated that libraries would be especially included in the new law; section 108 was the result. It represented compromise among the needs of libraries, their users and the owners of copyrighted materials.

Section 108 allows a library or archives, or any employee of a library or archive acting within the scope of that employment, to make no more than one copy of a copyrighted work, if certain conditions are met.[2] In many cases section 108 permits only what is also allowed by fair use, but this section is independent of fair use.

During the revision process Senate Report 94-473 was published, and it essentially prohibited library copying.[3] Subsequently, House Report 94-1476, recognized as the most influential of the accompanying congressional reports, addressed the issue of library photocopying and focused on systematic as opposed to isolated copying.[4] As a result of the House Report, all section 108 copying is governed by a clear, simple and concrete rule: a library may make no more than one copy at a time.[5] Moreover, this limitation is just that: single copying is all that is allowed under section 108, but is not necessarily always allowed. "Copies" are defined as material objects other than phonorecords in which a work is fixed by any method now known or later developed.[6] So, although section 108 is silent as to electronic copying, it is clear that its provisions encompass works that exist only in electronic format, in both printed copies and electronically or which are works that traditionally have appeared in print. If it is permissible for a library to make a photocopy of a work under section 108, it is permissible to make an electronic copy.[7]

As libraries move beyond photocopying for a user, then it is permissible to scan a copy and transfer it electronically to the user. Retention of the scanned copy in a database by the library is an infringement if retaining a photocopy of the photocopied item delivered to a user is infringement. To determine what types of single copies may be made under the section, libraries must satisfy several qualifications.

II. SECTION 108(a)

Only certain libraries and archives qualify under the preliminary requirements of section 108. In order to take advantage of the exemption, a library or archives must meet each one of these preconditions. First, the reproduction or distribution must be made without the purpose of any direct or indirect commercial advantage. Second, the library or archive must be either (a) open to the public or (b) available to unaffiliated researchers working in that particular field. Third, the copy must contain a notice of copyright.[8]

Although each of these provisions seems straightforward, each one has been the subject of much debate and at least some controversy.

A. *Direct or Indirect Commercial Advantage*

This requirement is fairly easy to meet for a nonprofit library, such as a public library or university library. So long as the library does not duplicate and sell copies of copyrighted works for profit, nonprofit libraries should satisfy this provision. Just because a library charges users a fee to cover the cost of making the copy does not mean that the library has violated this requirement, as the library gains no commercial advantage in such a transaction. Even nonprofit libraries could become involved in activities for which they would lose this exemption, however. For example, if a university library establishes a document delivery center that serves business and industry and if it operates this center as a profit-making enterprise, that portion of the library's activity clearly does not meet the "no direct or indirect commercial advantage" requirement. Further, if a library contracted with a commercial copying business to provide all of the photocopying services for the library, the exemption is lost even though such reproduction and distribution would be exempted if performed by the library itself.[9]

The controversy over this requirement involves the eligibility of libraries and archives that are located in or are part of a profit-making enterprise, such as a company or law firm library. The weight of legislative history explicitly includes for-profit entities, and examples of copying done by such libraries which would not run afoul of the direct or indirect commercial advantage requirement are included in the House Report.[10] Only the earlier Senate Report suggests otherwise.[11] Both the House Report and the later Conference Report stress that "the advantage referred to must attach to the immediate commercial reproduction itself, rather than to the ultimate profit-making motivation behind the enterprise in which the library is located";[12] that is, the reproduction and distribution of a copy must be noncommercial, not the enterprise which makes the copy. The Conference Report goes even further and states that "The isolated, spontaneous making of single photocopies by a library or archives in a for-profit organization without any commercial motivation ... would come within the scope of section 108."[13]

In 1992 the U.S. District Court for the Southern District of New York issued a long-awaited decision in *American Geophysical Union v. Texaco*.[14] The court ruled that commercial companies which make copies of copyrighted articles published in scientific and technical journals violate fair use under the Copyright Act of 1976. The question before the court was whether a research scientist infringes copyright if he makes copies of articles from copyrighted journals to which the company subscribes. Texaco claimed that such activity was permitted under section 107 fair use. The court answered the question with a resounding yes.[15] The opinion contains a detailed description of the

Copyright Clearance Center (CCC) and indicates there are several options by which a company may provide its scientists with copies of articles for use in their research.[16] Although decided under section 107, the judge discussed library copying under section 108. The court said that section 108(a)(1) limits library photocopying to copies made without any purpose of direct or indirect commercial advantage. Texaco made copies solely for commercial advantage as it is involved in profit-motivated scientific research and not in an exercise in philanthropy.[17] Further, section 108 permits the making of no more than one copy, which the court says means one per user, i.e., one copy of a particular item to anyone in Texaco would exhaust this limit.[18]

Library and information associations disagree with the court's interpretation and application of section 108.[19] As the legislative history states, section 108 applies to libraries in commercial companies as well as to libraries in not-for-profit organizations.[20] Moreover, the term "user" has a long-accepted common meaning. Libraries generally view the customer as the individual who requests the copy rather than the organization which employs her. Finally, library associations believe that libraries may make authorized copies under section 107 as well as under section 108.[21] The *Texaco* case currently is on appeal to the Second Circuit U.S. Court of Appeals.[22]

B. *Open to the Public or Available to Researchers*

Certainly many nonprofit libraries are open to the public at large and meet this requirement. The legislative history is silent, however, as to whether open to the public means that access to the collection must be free of charge. The general consensus on this matter holds that a nominal fee, of the type university libraries sometimes charge local residents for their use, would not disqualify a library.[23]

Once again, the main debate on this clause centers on for-profit libraries. The Register of Copyrights has stated that many for-profits could have problems meeting this requirement, since the condition of openness (or availability) dictates that libraries in profit-seeking entities would have to provide access to competitors in their field.[24] Library associations maintain that this requirement is met if other libraries have access to a for-profit collection, by means of either visiting rights or interlibrary loan. This means that the library's holdings of published materials is thus available to researchers through interlibrary lending.

Neither Congress nor the courts have spoken definitively on this issue.

C. *Notice of Copyright*

For any copy a library reproduces and distributes, notice is required; however, there is some disagreement about what this means. All parties agree that inclusion of the formal notice will satisfy this requirement. Such notice includes (a) ©, the word "copyright" or the abbreviation "copr.," (b) the year of first publication and (c) the name of the copyright holder.[25] The debate surrounds how much less, if anything, will suffice. Publishers maintain that only formal notice is adequate.[26] Thus formal notice should be included. The American Library Association advises its members to use "Notice: This material is subject to the copyright law of the United States."

If however, the notice of copyright is not found on the original document after reasonable investigation, the following statement seems a logical alternative: "This material is subject to the U.S. Copyright Law; further reproduction in violation of that law is prohibited." No library has been sued for using less than the formal notice, and clearly some notice is required. Since the United States joined the Berne Convention, which precludes signatory nations from requiring copyright holders to place formal notice on their works, the requirement that libraries include a notice seems to conflict with Berne. It seems peculiar that copyright holders are relieved of the responsibility to give notice but libraries are still required to do so on any copies reproduced. Thus, over time, the more general notice may be found to be increasingly adequate.

D. *Other General Requirements*

Section 108(g)(1) is similar to section 108(a) in that it contains general requirements that must be met before any copying may be done under the section. The subsection reiterates that copying under section 108 is limited to "the isolated and unrelated reproduction or distribution of a single copy."[27] As stated earlier, no more than one copy at a time may be made, and instances of single copying must not be related.

Section 108(g)(1) imposes an additional requirement, however; even a single, isolated copy cannot be made if the library or its employee "is aware or has substantial reason to believe that it is engaging in the related or concerted reproduction or distribution of multiple copies."[28] Thus, if a library or its employee believes that, despite its otherwise innocuous appearance, a copy request is in fact a related or concerted reproduction of multiple copies,[29] then section 108 offers no protection, and any copying done by the library is infringement. An example of such copying might be found in an academic library when a large number of students in one class come to the library to request a copy of the same article within a one-hour period and indicate that their instructor told them to request a copy of the particular article. At that point, the library employee is aware that the copying is no longer isolated or spontaneous.

The House Report emphasizes another limitation. Related or concerted reproduction of multiple copies of the *same* materials is not permitted.[30] While this seems to preclude much interlibrary loan practice and almost all copying for library reserve collections, those activities are permitted under the Act and are dealt with elsewhere in the statute (interlibrary loan at section 108(g)(2) and reserve copying under section 107).

III. COPYING FOR LIBRARY USE

Photocopying material for a library's or archive's own use is limited to subsections (b) and (c); the former permits copying of unpublished works for archival preservation and the latter concerns reproduction of lost, damaged, stolen or deteriorating published matter. Even when acting under these two subsections, reproduction is restricted to certain situations.

A. *Section 108(b): Archival Reproduction of Unpublished Works*

Copying of an unpublished work may be done only for purposes of preservation or security, or for deposit for research use in another library or archive. If for deposit in another library, the library making the copy must currently have a copy in its collections, and the library receiving the copy must meet the section 108(a) requirements. This subsection permits the reproduction in facsimile form of any unpublished print or phonorecord work.[31]

There are two important limitations to this subsection. First, any copying done is also subject to any contractual obligations assumed by a library or archive in connection with the deposit of any manuscript collection.[32] Thus, if a library agreed at the time the unpublished work was deposited in its collection that no duplication whatsoever would be permitted, the library may not reproduce the manuscript even for preservation purposes.[33] Second, a library cannot convert an unpublished work into a different format. For example, a manuscript can be photocopied, but it cannot be entered into a database if it did not originally exist in that form.[34]

B. *Section 108(c): Reproduction to Replace Lost or Damaged Works*

Section 108(c) allows a library or archive to reproduce a lost, damaged, stolen or deteriorating work, but only after two important conditions are met: (a) the library must have made a reasonable effort to obtain an unused replacement, and if one is found, (b) it must be purchased unless a replacement is not available at a fair price.[35]

The library does not have to seek a used replacement; if the only copy available is a used copy, then the library is free to reproduce the work either in facsimile or digitized form[36] rather than purchase it. "Reasonable effort" is defined as a search that "will vary according to the circumstances of a particular situation."[37] It does require (a) recourse to commonly known U.S. trade sources, (b) that in most situations, the publisher or other copyright owner would be contacted or (c) that an authorized reproducing service would be used.[38] Librarians have conducted such searches for years when it is necessary to replace a work lost from the collection.

Fair price is somewhat more ambiguous, and it is impossible to define for all circumstances. A price at or close to the original list price, however, would certainly qualify. The Association of American Publishers (AAP) has defined fair price as follows: (a) the retail price if the work is available from book stores and vendors; (b) if not so available, the prevailing retail price; or, (c) if an authorized reproducing service is used, whatever it charges.[39] This published definition of fair price was developed by the publishers' trade group and is not necessarily how a court would determine fair price. The American Library Association (ALA) with the National Education Association proposed another definition of fair price. The fair price of a work in its original format is the most recent suggested retail price of an unused copy, and the price of a reproduction of a work is the manufacturing cost plus royalties. Frequently, a work is available only as a part of a larger work. The ALA definition states that if single volumes of a multivolume set are unavailable, the fair price should not be the cost of the full set.[40]

The Act is silent about how long a library must conduct a reasonable investigation. Since the House Report states that a reasonable effort varies according to the particular circumstances, presumably, in searching for an unused replacement, one may consider the demand for the item. A reasonable investigation for a high-demand item may be of considerably shorter duration than the search for which there is little demand. The ALA definition of fair price imposes a time limitation: the copy of the work should be available within 30 days.[41] The extent of the amount of the original copy that is damaged or lost also should factor into the determination. If a 20-page article has been removed from a bound volume of a journal, must the library purchase a new bound journal? Probably not since fair price also relates to the portion of the work to be replaced.

A library may even reproduce a copy for another library, assuming the above requirements have been satisfied, if the requesting library had a copy of the material before the request was made. The lending library, however, should require the requesting library to certify that the copy has been requested in compliance with section 108(c).[42]

C. *Tables of Contents*

Many libraries routinely copy title pages of new journal issues and route copies of tables of contents to users in their institutions or companies. Although journal titles cannot be copyrighted, title pages and tables of contents pages are protected by the copyright in the entire journal issue. Further, many tables of contents, particularly those which include abstracts of articles that appear in the issue, might be eligible for copyright protection as a compilation. Moreover, abstracts can be a condensation of a work which is a derivative work under section 106(2); if they are derivative works, they also are copyrighted. Since tables of contents are copyrighted it is unclear whether the practice of duplicating and routing these pages is permissible; on the other hand, there have been no suits against libraries in this area. When publishers are asked about this practice, some say routing tables of contents pages is no problem while others assert it infringes their copyrights. Libraries should be aware of the possible problem with such routing practices before imitating a tables of contents service. One thing is clear, however: libraries should limit copies to one article per journal issue made in response to requests generated by routed copies of tables of contents.

IV. COPYING FOR USERS

A. *Section 108(d): Articles or Small Excerpts*

Section 108(d) of the Act permits a library to make a copy for a patron of one article in a periodical issue or of a small portion of another copyrighted work, if three conditions are met: (a) the copy becomes the property of the user; (b) the library has no knowledge that the copy will be used for purposes other than private study, scholarship or research; and (c) the library prominently displays a copyright warning in the area where photocopy orders are received and on the photocopy order forms themselves.[43] Copies may be made from works in the library's own collection or via interlibrary loan. Section 108(d) must be read along with subsection 108(g)(2), which states that the library may not engage in systematic copying of single or multiple copies.[44]

Note that this section applies when the library itself makes the copy for the user, not when a user makes his own copy on unsupervised photocopy equipment. The wording for the warning was developed by the Register of Copyrights.

NOTICE: WARNING CONCERNING COPYRIGHT RESTRICTIONS
The Copyright law of the U.S. (Title 17, United States Code) governs the making of photocopies or other reproductions of copyrighted material.
Under certain conditions specified in the law, libraries and archives are authorized to furnish a photocopy or other reproduction. One of these specified conditions is that the photocopy or reproduction is not to be "used for any purpose other than private study, scholarship, or research." If a user makes a request for, or later uses, a photocopy or reproduction for purposes in excess of "fair use," that user may be liable for copyright infringement.
This institution reserves the right to refuse to accept a copying order if, in its judgement, fulfillment of the order would involve violation of copyright law.[45]

The Act was drafted in 1976 and it envisioned a library where a user physically presented herself to a desk and completed a form to request that the library photocopy an article for her. Thus, she would be confronted with the warning twice: once on the sign and once on the order form. In the years since 1976, however, library practice has changed significantly. True, some users still come to the library in person to request copies but requests also are received in a variety of ways today. For example, users may mail their requests for photocopies of articles, telephone their requests, fax them or send requests through electronic mail, etc. Must a library still comply with the warning?

Yes, in order to qualify for the section 108(d)-(e) exemptions, the library has no option except to give the mandated warning to users. Thus, the library must find the functional equivalent of placing the warning sign at the photocopy order desk and on the order form itself. Such equivalents might include reading the warning to users who telephone a request or faxing the warning to a user who submits a request by fax. In fact, many libraries require that fax requests for copies be submitted on forms that the library supplies, and the fax forms contain the warning. On an e-mail request system, the warning could appear as a screen when the user initiates the request. It could follow the screen after the user indicates that a request for a copy is to follow. Having a user sign a warning form once or even annually is not the functional equivalent of the user being confronted with the warning each time he requests a photocopy of a copyrighted work. The warning is to alert the user, and libraries should find ingenious methods for compliance with the statutory mandate to give the warning. What the library may not do is ignore the warning requirement and still incur no liability.

When the library has given the warning, then it may provide the copy to the user without doing more. No records need to be maintained nor royalties paid.

B. *Section 108(e)*

Section 108(e) expands on the exemption in subsection (d) which permits libraries to provide copies of *one* article from a journal issue. Subsection (e) applies to providing copies of a larger portion of a work. This could be even an entire work or a substantial part thereof from the collection of the library where the user made his request or from the collection of another library. Section 108(e) authorizes a library to copy an entire work (or substantial part of it) if the three requirements of section 108(d) are met and if the library has determined, by reasonable investigation, that a copy cannot be purchased at a fair price. This last requirement dictates that the same efforts would be necessary as under subsection (c); recourse to known trade sources, etc. Further, in the normal situation, the publisher or copyright owner would be contacted or an authorized reproducing service (such as University Microfilms) will be used. The word "authorized" means that the service either has permission from the copyright holder to reproduce and distribute the work or that the service will pay royalties to the copyright owner for doing so. Unlike 108(c), however, this subsection does not include the word "unused" in its language, which appears to mean that libraries must explore opportunities to obtain the requested work from various used book sources before it may copy the work for the user. Of considerable concern to librarians is the missing mention of the time element. How long must the library continue with the reasonable investigation? This apparently is subsumed in the House Report statement that what constitutes a reasonable investigation will vary according to the circumstances.[46]

As is the case with section 108(d), libraries may copy pursuant to this subsection either from their own collections or through an interlibrary loan arrangement.[46] Probably the most common request for larger portions of a work from a user are for symposia issues of a journal. The subsection applies, however, even to two articles from a periodical issue.

Because of the requirement to search for and purchase a copy of the work for the user under this subsection, many libraries have determined that this is just too difficult or time consuming. Therefore, as an administrative matter, a library may decide that it will limit copying for users to that permitted by the subsection (d), i.e., the limit of one article per issue of a journal. Although the Copyright Act permits more, staff and time shortages may dictate that a library not take advantage of the exemption in section 108(e).

The copyright issues are very clear under sections 108(d)-(e) when the user comes into the library to request a copy. Further, there is no problem with telephone or fax requests if all requirements discussed above are met. However, the Act permits only one copy be delivered to a user. The problem with current fax technology is that two copies are made: one when the library makes a photocopy of the article in order to transmit it and one when the transmission is received at user's end. Eventually all libraries will have telefax equipment that does not require the sender to make a photocopy as a prelude to transmission, but for the near future, this is not the case. What should the sending library do with that photocopy? It must destroy it. The Act does not permit the library to retain a photocopy of the article. In fact, subsections (d) and (e) specifically require that the copy must become the property of the user.

Some libraries fax a copy to the user and then follow by mailing the photocopy; this effectively delivers two copies to the user. Such practice does not comply with the Act, which specifies that a library may provide only one copy to a user. If concern about the quality of the transmission is the reason for the follow-up mailed photocopy, the library could hold the photocopy until it is sure a good copy has been received. Then the photocopy the sending library makes should be destroyed. If the library feels it is essential to follow the fax copy with a mailed copy, then the fax copy should contain a notice which says that under the Copyright Act the library is permitted to provide only one copy to a user; the library will send a mailed copy, but the recipient of the fax must agree to destroy the fax copy upon receipt of the photocopy. This should guarantee that the user never has more than one copy in hand. As plain paper fax machines increase in popularity, the need to follow fax transmission with a mailed copy decreases, so this should soon be an issue of the past.

For requests the library receives electronically but for which the response is a photocopy sent to the user, there are no unique problems, assuming the (d) and (e) requirements are met. For electronic delivery of copies, however, there are some additional considerations even after the warning is given. Just because a copy is in electronic format is not in itself problematic. Section 108 is not limited to photocopies; in fact, it uses the phrase "to reproduce no more than one *copy* or phonorecord of a work." Copies are defined as

> ... [M]aterial objects, other than phonorecords, in which a work is fixed by any method now known or later developed, and from which the work can be perceived, reproduced, or otherwise communicated, either directly or with the aid of a machine or device.[48]

If, upon the request of a user, the library scans a copy of a copyrighted work, sends the electronic copy to the user, meets the requirements of subsections (d) and (e), and the copy contains the notice of copyright as required under subsection (a), there is no problem unless the library retains the scanned copy. In other words, subsections (d) and (e) require that the copy made becomes the property of the user. When the library retains a scanned copy and creates a database of articles, it is the equivalent of retaining photocopies which clearly is not permitted under the Act. If the library wishes to retain the scanned copy, then it must seek permission from the copyright holder to do so and pay royalties if requested. This is precisely what commercial document delivery services such as CARL/UnCover are doing.

One area of debate between the Register of Copyrights and libraries concerns whether the copying privileges granted in subsections (d) and (e) apply to the reproduction of unpublished works. While the Register of Copyrights maintains that they do not contemplate such copying, some libraries hold that, in the absence of statutory language to the contrary, it is permissible to reproduce unpublished works under these provisions.[49] Given the close relationship between the fair use privileges in section 107 and the library photocopying rights in section 108, it seems likely that a similar rationale would apply to a library's copying of unpublished works. Since the passage of the Copyright Act, however, a series of court cases[50] stressed the more narrow scope of fair use with regard to unpublished works. Congress recently enacted legislation to clarify that fair use is applicable to unpublished works, however, and it is logical that the reasoning which permits fair use of unpublished works also applies to library copying under section 108.[51]

V. SECTION 108(f): GENERAL EXEMPTIONS

Section 108(f) has four clauses that provide exemptions for various practices; there is little relationship among these four clauses except for the first two. Section (f) seems oddly placed within the section since subsection (g) relates back to copying for users in (d) while this is a miscellaneous section.

A. *Unsupervised Reproduction Equipment*

Section 108(f)(1) frees a library from liability for unsupervised photocopy machine use by patrons if the library posts a notice of copyright near all unsupervised copiers. While libraries are free to choose their own wording for the notice, the best choice may be simply to use the language required to be displayed at the place where orders for copies are accepted (and on photocopying request forms).[52] Many libraries display a sign that says: "Notice: Making a Copy May Be Subject to the Copyright Law." Alternative wording for the notice on photocopy machines has been suggested by the ALA: "Notice: The copyright law of the United States (Title 17, U.S. Code) governs the making of photocopies or other reproductions of copyrighted material; the person using this equipment is liable for any infringement."[53] Failure to post a warning may result in the loss of immunity from liability which a library otherwise enjoys under section 108. An argument can be made to require that language similar to that mandated by section 108(f)(1) be posted adjacent to unsupervised microfilm or microfiche reader/printers and CD-ROM stations since such equipment also functions as reproduction equipment.

Section 108(f)(2) emphasizes that, although libraries will not be held responsible for patron copying which exceeds the law, the patron herself will be liable for any excessive copying. This is true regardless of whether the copying in excess of fair use is done on an unsupervised photocopy machine by the user himself or under section 108(d) where the library reproduces a work at the request of the user. These two subsections taken together illustrate congressional desire to relieve libraries of liability if they act in good faith in following the provisions of the statute. In other words, a user may misrepresent herself to the library employee about the nature of a request for a photocopy. The user would be liable for infringement but not the library if the unsupervised reproduction equipment contained some type of sign or notice concerning copyright compliance.

B. *Audiovisual News Program*

Section 108(f)(3) specifically allows libraries to copy and distribute audiovisual news programs, if the three conditions of section 108(a) are satisfied. An important question is what constitutes "the news"? According to the House Report, the exemption was intended to apply to daily newscasts of national television networks, but it was not intended to apply to documentary, magazine format or public affairs broadcasts.[54] The language suggests that programs such as *60 Minutes, Prime Time* and *20/20* do not qualify under this exemption. The exemption is intended to encourage efforts by universities and other repositories to build libraries of newscasts for use in current or future research.[55] It also represents recognition that there is something unique about the news. While news stories may be copyrighted, the facts included in the stories are not. Facts belong to everyone and are in the public domain. The reason for this strange exemption is found in the history of the revision of the Act.

During the revision period, a suit between the television network, the Columbia Broadcasting System (CBS), and Vanderbilt University was pending. The issue was whether the library at Vanderbilt infringed CBS's copyright in the *Evening News with Walter Cronkite*. For years CBS had not videotaped its evening news program. Recognizing Walter Cronkite as a national treasure, Vanderbilt University began to videotape the news and, upon request, distribute copies of a particular day's news to other libraries. CBS eventually began to tape its news programs and filed suit against Vanderbilt to stop this activity. The section 108(f)(3) exemption represents congressional opinion about this suit.[56] The case was dismissed after the Act was passed.

C. *Contractual Obligations*

Section 108(f)(4) states that nothing in section 108 affects the right of fair use or any prior or subsequent contractual obligations a library may have assumed as a condition upon its receipt of any material.[57] It is important to note that this section 108 exemption can be waived or expanded by contract.[58] In other words, a library can sign a license agreement that would give it even greater rights than those found in section 108. An excellent example would be a site license for software that gives the library the right to reproduce multiple copies of the printed manual that accompanies the software. Unfortunately, it is much more common for the library to sign license agreements that are more restrictive than section 108 requires. For example, in order to obtain a work for its collection, the library may have to sign an agreement that it will allow no copying of the work whatsoever.

VI. SECTION 108(g): SYSTEMATIC REPRODUCTION

A. *General*

As discussed earlier, subsection (1) of section 108(g) consists of two preliminary requirements. First, any copying done pursuant to section 108 must be "isolated and unrelated reproduction or distribution of a single copy."[59] Moreover, even a single, isolated copy may not be reproduced if the library or its employee knows or has serious reason to believe that it would be "engaging in the related or concerted reproduction or distribution of multiple copies."[60]

Subsection (2) states that the rights granted in subsection 108(d) do not extend to situations where the library or its employee engages in systematic copying.[61] More than one copy, even made on more than one occasion, may qualify as systematic reproduction; such infringement need not be intentional or in bad faith to constitute a violation of the law.[62] What constitutes systematic reproduction of single copies has been the subject of considerable debate. Perhaps the best example is a library's classic selective dissemination of information (SDI) service. As students, librarians were taught that good library service dictated learning what researchers in the organization were doing and then notifying the researcher when articles or books were received dealing with the topic of interest. After the advent of photocopying, many librarians simply photocopied such articles and sent them to the researcher without a request from him for the particular article. This has come to be standard SDI practice. Such copying appears to constitute systematic reproduction since it is done regularly for many researchers in the organization.[63] Another example which could be either single or multiple copying is the cover-to-cover reproduction of newsletters which has been held to constitute copyright infringement.[64] This rule appears to apply equally to the copying of newsletters by nonprofit associations as well as by libraries in the corporate environment.[65]

B. *Interlibrary Loan*

The Act includes an exception for certain interlibrary loan arrangements.

> (N)othing in this clause prevents a library or archives from participating in interlibrary loan arrangements that do not have, as their purpose or effect, that the library or archives ... does so in such aggregate quantities as to substitute for subscription to or purchase of such work.[66]

While the Act was still being debated, controversy erupted over the uncertain nature of the terms used in the subsection and specifically how its restrictions would affect interlibrary networks and other systems geared toward the exchange of photocopies.[67] To eliminate some of the confusion, the National Commission on New Technological Uses of Copyrighted Works (CONTU) was appointed by Congress, and one of its charges was to produce guidelines to cover interlibrary loan copying affected by section 108(g)(2).[68] These guidelines are not the law itself, nor do they provide explicit rules or directions that govern all situations.[69] Libraries that act within the limitations of the guidelines, however, are presumptively abiding by the law by not copying in such aggregate quantities as to substantiate for subscription to or purchase of the work.

Section (1)(a) of the Interlibrary Loan (sometimes called the CONTU guidelines) states that within any calendar year, copying in excess of five articles from a periodical title released within the past five years violates the section 108(g)(2) proscription of systematic copying.[70] Note the two different time limits: only five copies within a year, but if the periodical issue is over five years old, then the guidelines do not apply.

Assume it is *January 1994*. Within all of 1994, a library may request through interlibrary loan copies of five articles from a periodical *title* covering these years.

Volume years	1993> 1992> 1991> 1990> 1989>	Five times per calendar year within these volume years

All requests could be from the 1993 volume; the requests could be for one article from each of the five volume years; or three from 1991 and two from 1992. The requests could even be for the same article five separate times, in other words, for five different users. For periodical volumes older than 1989, the guidelines take no position.

For books, section (1)(b) of the guidelines restricts reproduction under section 108(d) of small portions of copyrightable work (such as a chapter from a book) to five per year for any calendar year. This limitation applies to all copyrighted works for the duration they remain protected by copyright, i.e., life of the author plus 50 years.[71]

Section (2) of the guidelines focuses on copies made pursuant to interlibrary loan. Such reproduction is limited to those instances in which "the requesting entity would have been able, under the provisions of section 108, to supply such copy from materials in its own collection."[72] Two additional statements in the guidelines relax the suggestion of five somewhat and assist a borrowing library. If a library has entered a subscription for a periodical or is unable to locate a work which it has in its own collection, reproduction may be done pursuant to interlibrary loan; this copying should be treated as if the library make the copy from its own collection for the purposes of the limitations in section (1)(a). Therefore, the library does not count this in the suggestion of five.

Section (3) imposes a duty on a supplying library in an interlibrary reproduction to receive a certification from the borrowing library to the effect that the request conforms with the CONTU guidelines. So, the requirements on lending libraries are very slight. Section (4) mandates that borrowing libraries maintain records of all copying done under the guidelines for three calendar years.[73] The records either must be maintained by title or else must provide easy access by title. A library can determine when it has reached its five copies per year for a particular title only if this information is kept by title. It is recommended that libraries retain interlibrary loan records no more than three calendar years as required by the Guidelines since such records could be subpoenaed if a library became involved in litigation over copying activity.

When a library has reached its five requests within a given calendar year, it has several choices of action when the sixth request is received. (1) The library may simply tell the user that the library has reached its five requests for the year and is unable to process the request until the next year. (This is possible in academic and public libraries but is not a likely solution for libraries in the for-profit sector.) (2) The library may request the copy from an authorized document delivery service that charges a fee and pays a royalty to the owner of the copyright. (3) The library may request the copy and pay royalty fees directly to the copyright owner or to the Copyright Clearance Center (CCC) if the publisher is a CCC member. (4) The library can immediately place a subscription to the journal (so the library then owns the journal for interlibrary loan purposes and may make a copy without counting against the five). (5) The library could determine that it will exceed the safe harbor guidelines this one time. This should be done only in very narrow circumstances, however, and can be thought of as the "once in a blue moon exception" to the suggestion of five.

C. *Fee-Based Services*

Libraries that provide copies but charge a fee to cover costs do not run afoul of section 108(a) so long as the copying is done without the purpose of commercial advantage. The commercial benefit refers to the copying itself and not to the enterprise in which the library is located.[74] But if a library creates a profit-center unit within the library which charges fees in excess of cost recovery for copies provided, it probably loses the benefits of the section 108 exception. A library may still have section 107 rights, but it may be difficult to meet the four fair use factors. Under section 108(d), a library may make a single copy for a user provided the library has no notice that the copy will be used for other than private study, scholarship, research, etc.[75] Providing single copies of articles to users is permissible if each request for a copy is isolated and unrelated. Section 108(g)(2), however, prohibits multiple or systematic copying.[76]

A library as an agent of the requestor may be able to make copies that the requestor could make; additionally, even if the requestor could not justify copying, a library may be able to do so under section 107. The library, of course, must undertake the four factor analysis if it relies on section 107. First, the purpose and character of the use is still education, research or scholarship. If a library charges more than overhead costs for this service, then the purpose becomes commercial and section 107 would not be available. Thus, the library would have to pay royalties for the copies made and distributed as a part of this service. Assuming a library is merely meeting its own in-house costs, however, providing copies to external nonprimary users for cost recovery is within the limits of the first fair use factor. Second, the nature of the work must be considered as in any other fair use overview. Factual works have greater fair use rights than creative works intended for entertainment. Most special library copying is of factual works.[77] Even when motion pictures or music is copied for special library patrons, the purpose is not entertainment but rather serious study and scholarship. Therefore, even these works should be eligible for copying under the nature of the copyrighted work.

Third, the amount of copying weighs against a library when a significant amount is copied. Much of what is copied under the auspices of document delivery service is single journal articles. As part of its analysis of the amount and substantiality of copying, the court in the *Texaco* case discussed copying of entire articles.[78] Generally, it is the issue of a journal that is copyrighted, not each individual article within a journal issue.[79] Texaco argued that the copying of an eight to ten page article from approximately 200 pages of an issue amounted to approximately four percent of the copyrighted work.[80] The argument did not prevail, however, and the judge held that each article is a separately authored work protected by copyright. According to the judge, because of expense and convenience, issues rather than articles are registered. For purposes of fair use analysis the court held that the copyrighted work consists of the entire issue rather than the individually authored articles.[81] If this interpretation prevails on appeal, arguably this factor could weigh against a finding of fair use when a library copies an entire article.[82] Yet few are likely to want to copy only a portion of a scientific or technical article where accuracy is essential.

The fourth factor, generally considered the most important, focuses on the effect of the copying on the market for the original work. An individual may copy an article assuming the user will not profit financially from this act. Publishers argue that they are harmed by this kind of copying, that every article has some commercial value and that every copy of a valuable article made without compensation harms the copyright holder. Publishers assert that the marketing of reprints supports this contention.

The Association of American Publishers (AAP) issued a controversial position paper in mid-1992 which argued that libraries which provide copies as a fee-based service infringe copyright. Moreover, the AAP states that libraries which provide such services must pay or get permission from the copyright holder.[83] Yet this is not the critical determination. The issue is not cost recovery but whether the library makes a profit on the sale of copies.

There has been a proliferation of commercial document delivery services in the past few years. As noted earlier, these centers reproduce documents and then deliver the copies to clients for a fee. They differ in this regard from the CCC, which merely collects royalties from users, who then must copy the documents themselves.[84] Many commercial document delivery services physically pay the royalties for their clients as a service; of course, the requesting entity pays through fees to the document delivery service. Universities have thrown their hats into the document delivery game as well; examples of the academic sources are Columbia, Purdue and Rice universities, which offer customers quick access to academic information. Finally, the British Library has its own Document Supply Centre, which provides copies from its wealth of materials. The Centre offers a journal contents service for more than 55,000 serials and a document retrieval and supply from more than 200,000 journals and three million books.[85] Recently, the Centre announced that a royalty payment would be due together with the bill for copying.[86] For a flat fee of $2.25 the Copyright Clearance option provides compliance with copyright laws.[87]

All of these profit-making services assume the risk of copyright infringement in exchange for their services and fees. Given the still uncertain nature of how much and what kind of copying constitutes a violation of the law, document delivery services are more attractive than they might otherwise be.

VII. SECTIONS 108(h) AND (i)

A. *Section 108(h)*

Section 108 deals specifically with library reproduction of literary works and dramatic works. Unfortunately, there is no provision anywhere in the Act that permits libraries to reproduce audiovisual, musical or pictorial, graphic and sculptural works; and section 108(h) states that the section does not apply to audiovisual works.[88] Thus, section 108 is primarily a print section.

Libraries are permitted under subsections (b) and (c) to reproduce all types of works, but these are the situations where the library is permitted to copy for itself. The other exception to the section 108(h) limitation is for pictorial or graphic works that appear as a part of an article; those works may be reproduced along with the article as if they are text.

B. *Section 108(i)*

As enacted in 1976, section 108 included a final subsection (i) which required the Register of Copyrights to submit a report to Congress every five years, "after consulting with representatives of authors, book and periodical publishers, and other owners of copyrighted materials, and with representatives of library users and librarians."[89] This report was to focus on "the extent to which this section has achieved the intended statutory balancing of the rights of creators, and the needs of users."[90] Because there was general agreement among all concerned with the success of section 108 in achieving a rough balance of interests,[91] Congress was urged from the early 1980s to either amend section 108(i) to require less frequent reports or to abolish it altogether. In June 1992 section 108(i) was repealed.[92]

Despite the consensus on the elimination of the five-year reports regarding library photocopying, there has been a growing call for a similar investigation on the part of the Register into new technological means of reproduction such as electronic copying, laser disks, and the like.[93] Congress has not responded to these appeals to date.

VIII. COPYING FOR LIBRARY RESERVES

Many libraries view the reserve room as an extension of the classroom and, therefore, copying for reserves is covered by the *Classroom Guidelines*.[94] The guidelines represent a compromise among authors, publishers and users and may be workable in some situations but are entirely too restrictive for colleges and universities and graduate education.[95] Copying under the guidelines falls within a safe harbor and they impose no liability for unauthorized copying.[96]

Additionally, the ALA has promulgated guidelines on library reserve collections for colleges and universities. See Chapter 7 for a detailed discussion of library reserves.

IX. RELATIONSHIP OF SECTION 107 TO SECTION 108

Whether libraries have section 107 rights in addition to section 108 rights has engendered much debate. Section 108(f)(4) reads in part "Nothing in this section ... in any way affects the right of fair use as provided by section 107." This clause has raised two important questions: may libraries copy pursuant to section 107, and if so, how much may be copied under that subsection?

A. *Are Libraries Permitted to Copy Under Section 107?*

The Special Libraries Association (SLA) has stated that this question generally is an academic one, since much of the copying a library would otherwise do under fair use is covered by section 108.[97] A library may be beyond the provisions of section 108, however, but within fair use. SLA notes further that "the rights granted in section 108 are *in addition* to the fair use rights in section 107."[98] Given this construction of the statute, the Association, along with the American Association of Law Libraries (AALL) and other library organizations hold that libraries may copy under section

107 as well as under section 108.[99] The Authors League of America and the Association of American Publishers contest this view of section 107 and maintain that library copying is restricted to section 108.[100]

In his 1983 report, the Register of Copyrights writes that while the circumstances for library reproduction outside section 108 may be rare, there may be occasions of fair use photocopying beyond section 108.[101] Several noted copyright law experts including Paul Goldstein and Melville Nimmer concur with the library associations. Goldstein states that section 108(f)(4) provides a supplemental role exemption,[102] and Nimmer holds that a library or archive whose photocopying practices exceed the scope of the section 108 exemption may still rely on the defense of fair use.[103] This interpretation is supported by a plain reading of the text at issue; if "nothing in this section ... affects the right of fair use,"[104] then the right of fair use continues as it stood before the statute when libraries had no choice but to copy under the fair use privilege. If Congress had intended to limit library photocopying to section 108, surely it would have said so and drafted more stringent language than that found in section 108(f)(4). In the words of the House Report,

> Nothing in section 108 impairs the applicability of the fair use doctrine to a wide variety of situations involving photocopying or other reproduction by a library of copyrighted material in its collections where the user requests the reproduction for legitimate scholarly or research purposes.[105]

B. When May Libraries Copy Under Section 107?

There are two situations that frame this question. The first scenario is a situation in which a library copies beyond the scope of section 108, in the sense that section 108 does not cover the desired copying at all. It seems clear that if section 108 does not contemplate the copying but section 107 does, then if a library is indeed covered by section 107, it may make the copy. There is little debate on this question.

The controversy arises in another context stated neatly by the Register of Copyrights:

> The issue is ... whether a librarian who has made all of the photocopies permitted in a given type of transaction may thereafter make one or more additional photocopies under the fair use provisions of section 107, or whether such copying is infringing unless authorized by the copyright owner.[106]

The library and publishing communities are predictably at odds over this question. As mentioned earlier, the Authors League of America and the Association of American Publishers resist the notion that libraries may copy at all under section 107, while the American Association of Law Libraries believes that a library may copy for a researcher, scholar or educator if that person herself could copy under the fair use doctrine.[107]

The Register appears to take a position somewhere in between those two positions, a stance endorsed by Paul Goldstein.[108] On the one hand, the Register notes, some copying may be done by libraries under section 107; on the other hand, "(t)o read section 108(f)(4) as permitting 'post-108' reliance on fair use as if no section 108 copying had occurred is to come dangerously close to reading section 108 out of the statute."[109] To determine if a library may copy under section 107, the Register proposed a two-part test.

(L) ibrary photocopying "beyond" 108 may be fair use if both:
a) the transaction is of a *type* which could be fair use in the absence of section 108, *and*
b) the fair use analysis ... of *this* transaction takes into account the "108" copying with has already occurred.[110]

When weighing the effects of the beyond section 108 copying, one must take into account the section 108 copying already done since a different result may be achieved by counting the section 108 photocopying previously done rather than counting later copying as if no section 108 copying had occurred.[111]

XI. CONCLUSION

A. *Fair Use Cases*

The *Williams & Wilkins Co. v. National Library of Medicine*[112] case was decided in 1973, three years before the library reproduction provisions in the revised Copyright Act were enacted. Thus, it was decided as a fair use case, and it remains a significant case in the field of library reproduction. In essence, it restated the common law. For that reason, the decision continues to hold the force of precedent, and it must be taken into account in any analysis of fair use and library copying.

The plaintiff, Williams & Wilkins Company, published 37 medical journals. Most of its revenue came in the form of subscriptions; since the publications covered rather specialized fields, subscribers were relatively few. The defendants were two government organizations, National Library of Medicine (NLM) and the National Institutes of Health (NIH); because of their government agency status, the suit was filed in the Court of Claims. Williams & Wilkins sued NLM and NIH charging that the two agencies had infringed copyrighted journals by making unauthorized photocopies of articles featured within those publications and distributing them to medical researchers.[113]

The facts of the case were not disputed by either party. In 1969, NIH copied 93,000 articles which were then distributed free of charge to staff members. NLM received 127,000 photocopy requests for articles from various sources (mainly private drug companies) in 1968, and all but 7,000 of these requests were honored. While both defendants had internal policies designed to limit reproduction to a single article per week per requester and officially refused to copy any one article more than once, these restrictions were seldom observed.[114]

When heard by a single commissioner on the Court of Claims, the commissioner held for the publisher. On appeal, the Court of Claims held that the copying was fair use.[115] The court summarized its holding with three assertions. First, the court was not convinced that Williams & Wilkins was being harmed by the practices of NLM and NIH since the publisher was still making a profit. Second, the court believed that medicine and medical research would be harmed by a finding of infringement. Third, since Congress was debating the new Act but had not yet enacted the revision, it was more prudent to allow the status quo to continue in the interim. Other factors included the nonprofit nature of the copying, the character of copying (aimed at scientific advancement and medical research), and the restrictions the two agencies placed on their copying.[116]

This decision has been harshly criticized, nowhere more so than in the dissenting opinions, one of which labeled the majority's holding "the *Dred Scott* decision of copyright law."[117] Among the alleged mistakes and obfuscations made by the majority, one writer has pointed out that the concern with the advancement of science evinced in the opinion does not dictate a finding of fair use. "By publishing the research and by splitting its (meager) profits with medical societies, plaintiff was promoting the progress of those endeavors."[118] Moreover, Williams & Wilkins (and other similarly situated publishers) could easily be driven out of business if other potential customers resorted to photocopies. Should that happen, and the company cease to publish, the negative impact on medical knowledge and advancement would be great indeed.[119]

Several of the bases for free mass photocopying cited by the majority in *Williams & Wilkins* disappeared in the ensuing years. In the revised Copyright Act of 1976, Congress clearly sought to rein in liberal applications of the fair use doctrine; some commentators even see the 1976 provisions as statutorily reversing that decision.[120] Additionally, licensing organizations were formed which put to rest any fears that a narrow view of fair use would frighten individuals and institutions away from making copies and thereby hamper advances in knowledge. Finally, beyond the justifications of the *Williams & Wilkins* court, both the quality and the quantity of photocopying technology advanced swiftly, making the threat posed to publishers (and, by extension, knowledge) by the 1980's far more immediate than it was even ten years earlier.[121]

Given these changes, new legal challenges to unlicensed photocopying were inevitable. The one that has attracted the most recent attention is *American Geophysical Union v. Texaco, Inc.*[122] which posed the question of whether the making of single copies from copyrighted journals by an employee in a commercial business is fair use under section 107 of the Copyright Act of 1976.[123]

The plaintiff, American Geophysical Union, and five other publishers, all of which are members of the Association of American Publishers, sued Texaco in 1985. While Texaco was a member of the CCC and paid royalties to that organization on a per-transaction basis, the plaintiffs alleged that these payments represented only a fragment of the actual copying done at Texaco.[124] In fact, publishers were alerted to the problem by the wide disparity in the royalties paid by other major oil companies compared to those paid by Texaco. Since the issue of fair use was central to the dispute, the parties agreed to a mini-trial on that issue. Accordingly, the plaintiffs selected a Texaco scientist at random, and the case subsequently focused on photocopies of eight articles found in that employee's own files.[125]

In a lengthy decision, Judge Pierre Leval of the Southern District of New York rejected Texaco's defense of fair use, despite its claim that the contested photocopying was both customary in the industry and essential to technological progress. Applying the four fair use factors, Judge Leval found that only one, the nature of the work, favored Texaco, since "the scope of fair use is greater with respect to factual than non-factual works,"[126] and the articles copied were of a factual nature. The first factor, purpose and character of the use, favors the defendant only if the use is transformative and nonsuperseding or noncommercial; since Texaco certainly aimed to profit from the knowledge gained from the copies, and since the act of making a photocopy adds nothing to the original work, the plaintiffs won on this factor. Left unnoticed by Judge Leval was the fact that the copies themselves were not of a commercial nature. The third factor, the amount and substantiality of the use, also went against Texaco, as entire articles were copied and the judge found that each article was a separately copyrightable item. Arguably, the copyright applies to the entire issue of the journal instead of to individual articles. The fourth factor, the effect of the use on the potential market, also was held to favor the plaintiffs because, by photocopying the articles, Texaco eliminated the need to buy either more subscriptions or pay royalties through the CCC, thus depriving the copyright owners of income.[127]

Williams & Wilkins was distinguished on several grounds. First, the institutions doing the copying in that case were nonprofit research organizations, while Texaco's photocopying was done for the purpose of enabling the company to make a profit.[128] Second, no mechanism such as the CCC existed at the time of the *Williams & Wilkins* decision which made concerns about the stifling effect on science real. By the 1980s, however, the advent of the CCC and other such entities meant that a court looking at copying practices was no longer faced with a Hobson's choice of possible infringement and scientific progress on the one hand, or no infringement and stymied research, on the other.[129]

The impact of *Texaco* will be felt primarily at profit-making institutions; moreover, since the outcome was predicated on fair use, libraries' section 108 rights are not affected at all. The decision definitely narrows the plausible scope of fair use, however.

In an *amicus* brief filed by several library associations, including the Special Libraries Association, the groups have urged the federal appeals court to reverse the *Texaco* decision on the grounds that Judge Leval misapplied the four fair use factors in holding Texaco liable for copyright infringement.[130] With regard to the first factor, the purpose and character of the use, the brief argues that the District Court's focus on the commercial nature of the copying was too narrow and unsupported by law. The brief maintains that Texaco's use had a socially beneficial purpose — the furtherance of scientific research — in addition to the goal of improving Texaco's commercial prospects. Further, established precedent holds the former to outweigh the latter when both are present.[131] In addition, the associations assert that the fourth factor, the effect of the use on the market for or value of a work, goes against an alleged infringer only when the use eliminates the demand for the original work. Since Texaco's use did not have such an effect, the brief argues, the fourth factor should have supported Texaco's use. Finally, the brief argues that the court misapplied section 108 in its holding that Texaco's copying was for commercial advantage, since section 108 allows copying by for-profit entities so long as there is no immediate commercial motivation behind the reproduction or distribution itself.[132]

As noted earlier, the case is on appeal to the Second Circuit U.S. Court of Appeals. Both parties have indicated that should they lose at the circuit court level, they will appeal to the U.S. Supreme Court. Not only is this the most significant library photocopying case since *Williams & Wilkins*, it is the only one to come to trial.[133]

Despite publishers' protests to the contrary, a final decision in favor of the publishers will have a great impact on practices in all types of libraries as well as on research methods throughout both the profit and not-for-profit worlds. University libraries support faculty research efforts and supply photocopies of articles to faculty. Much faculty research is grant funded, mostly by the federal government, but some from state governments and private companies. Will academic libraries be able to supply photocopies to faculty only for unfunded research or for research that is government funded, but be unable to do so if the funding comes from a software or pharmaceutical company? What about public/private research partnerships? Will libraries at government laboratories be able to provide photocopies to their own scientists but not to those from the private company with which they maintain a partnership for a particular project? The line between nonprofit and for-profit is not so clear as publishers believe.[134]

Public libraries also fear the consequences of a holding against *Texaco*. Many public libraries, especially in urban areas, have large sections that serve business and industry. Will they now have to inquire as to the reason a patron requests a photocopy? Will a public library be able to provide the copy if it is wanted to support a user's hobby but be unable to supply the copy if the patron will use the photocopy in his work?[135] Thus, reproducing articles for entertainment would be permitted while copying for research, which ultimately might have a commercial value, would be prohibited without paying royalties. This seems a peculiar turn of events given the constitutional purpose of copyright, i.e., to promote the progress of science and the useful arts.[136]

Another serious concern about the *Texaco* case is what will happen to journal prices. Many corporate librarians say that if their companies pay for every copy made, then there is no reason to subscribe to journals. Companies can order huge numbers of copies from authorized document delivery services and pay royalties for the price of subscriptions. If companies cancel subscriptions then prices will increase, which will be felt by university and other nonprofit libraries.

B. The Future

What will happen in the library copying area? No one can predict precisely, but much of what happens regarding library reproduction of copyrighted works hinges on the ultimate outcome of *Texaco*. Should the publishers be successful, additional litigation against libraries and companies is likely. More companies join the CCC every day and pay for copies made. These entities have made a business decision that it is easier to pay for all copies made than to worry about whether a particular activity is infringing or not. See Chapter 4 for a discussion of the CCC and other licensing.

ENDNOTES - CHAPTER 3

1   17 U.S.C. §§ 1-219 (1970) (repealed 1978).
2   17 U.S.C. § 108(a) (1988).
3   S. Rep. No. 94-473, 94th Cong., 1st Sess. (1975) *reprinted in* 13 *Omnibus Copyright Revision Legislative History* 67 (1977) [hereinafter Senate Report].
4   H.R. Rep. No. 1476, 94th Cong., 2d Sess. (1976) *reprinted in* 17 *Omnibus Copyright Revision Legislative History* 72 (1977) [hereinafter House Report].
5   U.S. Copyright Office, *Report to the Register of Copyrights: Library Reproduction of Copyrighted Works (17 U.S.C. 108)* 66 (1983) [hereinafter Register's Report].
6   17 U.S.C. § 101 (1988).
7   See Chapter 6 for a discussion of computer programs and computer databases and Chapter 7 for electronic reserves.
8   17 U.S.C. § 108(a) (1988).
9   *See* House Report, *supra* note 4, at 74.
10 *Id.* at 74-75.
11 Senate Report, *supra* note 3, at 67. Although the Senate Report was published first, the House Report has been far more influential.
12 House Report, *supra* note 4, at 75. *See also* James S. Heller & Sarah K. Wiant, *Copyright Handbook* 15 (1984) [hereinafter Heller & Wiant].
13 Conf. Rep. No. 1733, 94th Cong., 2d Sess. (1976) *reprinted in* 17 *Omnibus Copyright Revision Legislative History* 74 (1977) [hereinafter Conference Report].
14 American Geophysical Union v. Texaco, Inc., 802 F. Supp. 1 (S.D.N.Y. 1992).
15 *Id.*
16 *Id.* at 7-9. For a discussion of the CCC, see Chapter 4.
17 *Id.* at 15-16.
18 *Id.* at 28.
19 *Amicus Curiae* Brief of Association of Research Libraries, American Association of Law Libraries, Special Libraries Association, Medical Library Association, American Council of Learned Societies, National Humanities Alliance, and the Association of Academic Health Science Library Directors, *American Geophysical Union v. Texaco* (No. 92-9341) (2d Cir., filed March 4, 1993).
20 Conference Report, *supra* note 13.
21 *See* text accompanying notes 97-111.
22 *See* text accompanying notes 120-36 for additional discussion of *Texaco.*
23 Register's Report, *supra* note 5.
24 *Id.* at 78.
25 17 U.S.C. § 401(b) (1988).
26 Heller & Wiant, *supra* note 12.
27 17 U.S.C. § 108(g)(1) (1988).
28 *Id.*
29 *Id.*
30 House Report, *supra* note 4, at 77.
31 17 U.S.C. § 108(b) (1988), which reads in full:

The rights of reproduction and distribution under this section apply to a copy or phonorecord of an unpublished work duplicated in facsimile form solely for purposes of preservation and security or for deposit for research in use in another library or archives of the type described by clause (2) of subsection (a), if a copy or phonorecord reproduced is currently in the collections of the library or archives.

32 17 U.S.C. § 108(f)(4) (1988).

33 This limitation should be taken into consideration when libraries accept donations of unpublished works. It might be useful to add a provision to the standard agreement that the donor agrees to preservation in any format so that the library could make an electronic copy.

34 17 U.S.C. § 108(b) (1988).

35 *Id.* § 108(c).

36 Although neither subsections 108(d) nor (e) mentions digitized images, given the dictionary meaning of the term, it is probably an acceptable means of reproduction as a facsimile.

37 House Report, *supra* note 4, at 75-76.

38 *Id.* at 76.

39 Association of American Publishers (AAP) and Authors League of America, *Photocopying by Academic, Public, and Non-profit Research Libraries* 14 (1978).

40 Mary Hutchins Reed, *The Copyright Primer for Librarians and Educators* 10 (1987) [hereinafter Reed].

41 *Id.*

42 *See* text accompanying note 73.

43 17 U.S.C. § 108(d) (1988). The copy also must contain the notice of copyright as required by § 108(a)(3).

44 For a discussion of interlibrary loan *see* text accompanying notes 59-73.

45 37 C.F.R. § 201.14(b) (1991).

46 House Report, *supra* note 4, at 75-76. *See* text accompanying notes 35-41.

47 17 U.S.C. § 108(e) (1988).

48 *Id.* § 101.

49 Heller & Wiant, *supra* note 12, at 21.

50 Salinger v. Random House Inc., 811 F.2d 90 (2d Cir. 1987); New Era Publications v. Henry Holt & Co., Inc., 873 F.2d 576 (2d Cir. 1989); Wright v. Warner Books, Inc., 953 F.2d 731 (2d Cir. 1989). For further discussion of these cases, see Chapter 7.

51 17 U.S.C. § 107 (Supp. IV 1992).

52 *See* text accompanying note 45 for the language of the warning.

53 Reed, *supra* note 40, at 13.

54 House Report, *supra* note 4, at 77.

55 *Id.*

56 Cosette Kies, *Copyright Versus Free Access: CBS and Vanderbilt University Square Off*, 50 Wilson Libr. Bull. 242, 246 (1975).

57 For a discussion of the controversy about whether libraries have section 107 rights in addition to section 108 rights, *see* text accompanying notes 97-111.

58 17 U.S.C. § 108(f)(4) (1988).

59 *Id.* § 108(g)(1).

60 *Id.*

61 Special Libraries Association, *Library Photocopying and the U.S. Copyright Law of 1976* 8 (1977) [hereinafter SLA].

62 Pasha Publications, Inc. v. Enmark Gas Corp., 22 U.S.P.Q.2d (BNA) 1076 (N.D. Tex. 1992).

63 A better alternative for selective dissemination of information is to notify the researcher of a relevenat article but to refrain from copying it absent a request for that particular article from the researcher. For many libraries in the for-profit sector, it may be more desirable to provide this useful service and simply pay royalties for the copies made as a part of any SDI effort.

64 *Pasha*, 22 U.S.P.Q.2d at 1077.

65 Television Digest, Inc. v. U.S. Telephone Ass'n, 47 Pat. Trademark & Copyright J. (BNA) 32 (1993).

66 17 U.S.C. § 108(g)(2) (1988).

67   SLA, *supra* note 61, at 9.

68   Conference Report, *supra* note 13, at 71.

69   *Id.*

70   *Id.* at 72-73.

71   *Id.* at 73.

72   *Id.*

73   *Id.*

74   House Report, *supra* note 4, at 75. But the district court in *Texaco* suggests otherwise.

75   17 U.S.C. § 108(g)(1) (1988).

76   *Id.* § 108(g)(2).

77   802 F. Supp. at 16.

78   *Id.* at 17.

79   *Id.*

80   *Id.*

81   *Id.*

82   *Texaco* is on appeal to the U.S. Court of Appeals, Second Circuit. The case was argued in May 1993. Both parties have stated that they will appeal to the U.S. Supreme Court if they lose.

83   Statement of the Association of American Publishers on Commercial and Fee Based Document Delivery (1992).

84   For a discussion of the CCC and other licensing organizations, see Chapter 4.

85   Pamphlet, The British Library Document Supply Centre.

86   *British Library to Pay Copyright Fees on All Photocopies Sent to USA*, The British Library Press Information 93/49, Aug. 3, 1993; *A.A.P. Applauds Decision of British Library*, AAP News, Aug. 3, 1993; CANCOPY and the Canadian Institute for Scientific and Technical Information (CISTI) are negotiating a similar agreement which would enable CISTI to supply copies of copyrighted information to U.S. customers without infringing copyright.

87   Pamphlet, *Discover the Fastest Most Powerful Information Source in the World, Right Here in America.* British Library Document Supply Centre in America (1993).

88   17 U.S.C. § 108(h). See Chapter 5 for a discussion of audiovisual works.

89   *Id.* § 108(i) (1988) (repealed 1992).

90   *Id.*

91   Letter from David Bender, Executive Director, Special Libraries Association, to Ralph Oman, Register of Copyrights (June 18, 1992) (on file with authors).

92   Copyright Report to Congress, 17 U.S.C. § 108(i), *repealed by* Copyright Amendments Act of 1992, § 301, 17 U.S.C. § 108 (Supp. IV 1992).

93   *Id.*

94   House Report, *supra* note 4, at 68-71. See Chapter 7 for a further discussion of the Classroom Guidelines.

95   *Id.* at 72.

96   See Chapter 7 for a further discussion of reserve copying.

97   The example given is one of a library reproduction made at the request of a researcher. While this scenario is certainly within the scope of fair use, it also is explicitly permitted by § 108(d). SLA, *supra* note 61, at 4.

98   *Id.*

99   Tape of America Association of Law Libraries meeting, New Orleans, July, 1991 (on file with authors).

100   Association of American Publishers & the Authors League of America, *Photocopying by Academic, Public and Nonprofit Libraries* 4, 16 (1978).

101   Register's Report, *supra* note 5, at 96.

102   1 Paul Goldstein, *Copyright: Principles, Law and Practice* § 5.2.2.1(c) (1989) [hereinafter Goldstein].

103   Melville B. Nimmer & David D. Nimmer, 3 *Nimmer on Copyright* § 13.05[E][2] (1992) [hereinafter Nimmer].

104   17 U.S.C. § 108(f)(4) (1988).

105   House Report, *supra* note 4, at 78-79.

106   Register's Report, *supra* note 5, at 95.

107   Heller & Wiant, *supra* note 12, at 30.

108   Goldstein, *supra* note 99, at 559.

109   Register's Report, *supra* note 5, at 98.

110   *Id.*

111   *Id.*

112   487 F.2d 1345 (Ct. Cl. 1973), *aff'd by an equally divided Court*, 420 U.S. 376 (1975).

113   *Id.*

114   *Id.*

115   *Id.*

116   *Id.*

117   *Id.* at 1387.

118   William F. Patry, *The Fair Use Privilege in Copyright Law* 182 (1985).

119   Nimmer, *supra* note 103, § 13.05[E][1].

120   William F. Patry, *Latman's the Copyright Law* 245 (6th ed. 1986).

121   Nimmer, *supra* note 103, § 13.05[E][1].

122   *Texaco*, 802 F. Supp. at 1.

123   *Id.* at 4.

124   *Id.*

125   *Id.* at 5.

126   *Id.* at 16, quoting New Era Publications Int'l, ApS v. Carol Publishing Group, 904 F.2d 152, 157 (2d Cir. 1990).

127   *Id.* at 9-21.

128   *Id.* at 23.

129   *Id.* at 25.

130   *Amicus Curiae* Brief of Association of Research Libraries, American Association of Law Libraries, Special Libraries Association, Medical Library Association, American Council of Learned Societies, National Humanities Alliance, and the Association of Academic Health Science Library Directors, *American Geophysical Union v. Texaco*, (No. 92-9341) (2d Cir., filed March 4, 1993).

131   *Id.* at 9-18.

132   *Id.* at 27-34.

133   In the early 1980s two pharmaceutical companies, American Cynamid and Squibb, were sued by publisher members of the AAP for copyright infringement for internal copying of journal articles. Neither case came to trial; both were settled.

134   Laura N. Gasaway, *Wide Impact Seen for Photocopying Case*, 15 Nat'l. L. J., Aug. 16, 1993, at 21, 24.

135   The American Library Association also filed an *amicus* brief to the Second Circuit.

136   U.S. Const. art. I, § 8, cl. 8.

CHAPTER 4

LICENSING AGENCIES AND COLLECTIVES

I. INTRODUCTION

Before the technological revolution began in the middle of this century, unauthorized reproduction of entire documents was fairly rare and was fully ignored as causing any significant problems for copyright holders. Such copying was so infrequent and copyright holders were so indifferent to it that there are no reported cases on the issue.[1] The dearth of litigation is due likely to two reasons. First, would-be copiers had the choice of copying by hand (or by typewriter) or by investing in the equipment and skilled labor necessary for mass mechanical reproduction such as a printing press. The former method is dull and time consuming, which presumably discouraged most people from copying more than a fragment of any publication,[2] while the latter alternative is prohibitively expensive and easily detected.[3] In addition, before World War II, scientific progress was steady, but not spectacular; these advances were increasingly announced and published in the form of journal articles, which, while often difficult for researchers to obtain and libraries to index, nevertheless managed the flow of new information tolerably well.[4]

The innovations of what has been termed the "information age" changed everything. With the introduction of photocopiers, an institution or even an individual could quickly and easily copy volumes of information.[5] Thus, the means for widespread infringement were available. Further, there has been a veritable explosion of knowledge in the years since World War II; as new discoveries in every field led to still newer discoveries, at an almost exponential rate, the volume of scientific work increased enormously and published reports of research data overflowed all prior systems.[6] Thousands of new journals sprang up to report these findings; bibliographic services, overwhelmed at first, eventually were able to catalog these publications through the use of computers.[7] Despite these improvements, access to information in these journals remained frustratingly limited. So many journals were published that even the largest libraries could acquire only a proportion of them, as there was not enough money to pay for all the subscriptions.[8] In order to fulfill their patrons' needs, libraries formed networks to supply one another with articles; however, since subscriptions were few, lending libraries were reluctant to send the easily damaged originals, and thus resorted to photocopying the information requested.[9]

By the early 1970s, the situation was in dire need of reform. Journal publishers, who usually gained nothing from photocopies made of their articles, were faced with financial ruin;[10] with every silenced publication, the pool of knowledge accordingly decreased.[11] Researchers found the system cumbersome and inadequate to their growing needs.[12] Finally, larger libraries, which paid for the bulk of journal subscriptions, received nothing from the smaller libraries they supplied with copies, and such larger institutions had to shoulder the burden of these increasing costs alone.[13]

The passage of the revised Copyright Act in 1976 narrowed the scope of unauthorized reproduction, and at the same time, bolstered the legal foundation of the practice. While the Act explicitly reserved the right to reproduce the copyrighted work in copies for the owner of copyright,[14] it also limited this right by codifying the doctrine of fair use[15] and by granting to libraries and archives the right to reproduce one copy without the permission of the copyright holder,[16] Congress made it clear, however, that any copying done outside these limitations on the exclusive rights of the copyright holder would constitute infringement. The long controversy over the use of copyrighted works by educators led to the specific wording of section 107. Teachers felt there was a need for greater certainty and protection.[17]

In the course of deliberations on the revised Act, lawmakers noted the paucity of procedures to assist libraries that wished to engage in unprivileged reproduction. As the law took shape, it became increasingly clear that as a practical matter large users of copyrighted information needed a way to access that material easily and pay for it when appropriate. Seeing this gap as likely to frustrate efforts to abide by the law, the Senate called for the development of mechanisms to facilitate royalty payments to copyright holders. As the Senate Report notes, "Concerning library photocopying practices not authorized by this legislation, the committee recommends that workable clearance and licensing procedures be developed."[18]

Collective licensing organizations are predicated on three assumptions: (a) that users have concern for the value and development of intellectual property and abuse of intellectual property is harmful; (b) that rights holders recognize that information does not exist if no one uses or buys the information; and (c) that the rights and values of information need to be easily recognized.[19] Publishers of technical books, authors and librarians responded to this congressional call to action in 1977 with the establishment of the Copyright Clearance Center (CCC).[20]

II. COPYRIGHT CLEARANCE CENTER

A. *General*

Since 1978, the CCC (Copyright Clearance Center) "has provided a service through which libraries (public and private), commercial organizations, and others may centrally pay license fees for photocopying of certain copyrighted publications."[21] The CCC is the incarnation of a licensing and collecting system for unauthorized copying of copyrighted works. Users and publishers realized that to make this work as a private nonprofit organization, they had to stay away from another governmental regulatory scheme, and they had to develop common procedures. In 1990, over 600,000 domestic and foreign journals, primarily in the scientific, medical and technical fields, were registered with the CCC.[22] By 1992, however, only a small percentage of other publishers in fields such as law were registered.[23]

The CCC, a not-for-profit corporation created at the suggestion of Congress under the sponsorship of publisher and author organizations, serves as an organization for photocopy and royalty remittance fees.[24] To date, the CCC has not been licensed by publishers to collect royalties for electronic copying; thus, it is solely an agency to collect and distribute photocopy royalty fees. Initially, each journal publisher established a standardized code published on the first page of copyrighted articles which identified both the article and the individually designated per-copy fee, set by each individual publisher.[25] Publishers established different fees based on the purpose of the use of the copy. For example, some publishers charge nothing or only a few cents per article for course packets but several dollars per copy per article in the corporate world.

The CCC code is being phased out and replaced with the SIID (Serial Issue - Level Identifier) developed by SISAC (Serials Industry Systems Advisory Committee).[26] The SIID is a unique issue identifier (enumeration and chronology) much like the ISSN serves as a unique title identifier. The SIID symbol can be transferred into machine-readable or bar code form. Together the SIID and the bar code form the SISAC symbol for a serial issue.[27]

User organizations including libraries electing to use the services of the CCC also register with the CCC. Registered organizations receive identification numbers to use when reporting and paying royalties associated with copying.[28] Many corporations and other libraries in the for-profit sector joined the CCC in order to reduce administrative time and effort to determine what was a fair use copy and what was a copy on which royalties should be paid. These libraries determined as an administrative matter that it was simpler to pay for all internal copying than to separate the types of copying and monitor it. AT&T Bell Laboratories is an example of such a library which early joined the CCC.[29]

B. *Types of Licenses*

1. Transaction Reporting Service

The CCC currently operates in two different ways. The first and oldest approach calls for subscribers to report, on a regular basis (generally monthly), any photocopies made from CCC-registered publications. This is called the Transaction Reporting Service (TRS). The CCC then calculates what each publisher participant is owed and bills the subscriber.[30] Early in the history of the TRS, libraries submitted to the CCC the first page of the article copied as the method of reporting. Concern over protecting trade secret information lead to the use of ISSN numbers instead of article-specific copying. This method requires the subscriber to keep a record of the photocopying done at each photocopier within the organization; continuous monitoring of copying is mandatory. Larger companies must identify and report all copying of copyrighted material in excess of fair use, whatever the location, along with a comprehensive implementation plan to assure total compliance. Many larger corporate members with multiple locations sought an easier method for paying royalties and thus the Annual Authorization Service was developed in the early 1980s.

2. Annual Authorization Service

The CCC offers so-called blanket licenses, the Annual Authorization Service (AAS). This method permits licensed for-profit commercial users or industrial corporations for internal use to make unlimited copies of works registered with the CCC for an annual fee. Licensees then do not have to report photocopying transactions, and the collective distributes royalties to its members in proportion to the value they assigned to their works.[31] Users may select one of the following methods for determining the license fee: the wall-to-wall survey, the limited sample of copying within a company which is then extrapolated to yield the payment due or the pooled industry model.

Initially the wall-to-wall survey was based on a ninety-day survey of copying activity at all U.S. branches of a corporation.[32] The results of the survey were multiplied by four to project an annual copying level. Like the TRS publishers set rates for copying fees for individual titles.

Not every company must conduct a survey any longer. Over the years the CCC has collected sufficient data from individual surveys or econometric models based on an intricate system of sampling the type of business, the number of employees, etc., to project fees. Copies company-wide were extrapolated from the data, an average price per copy based on publisher-set fees was developed and the CCC rate was generated.[33] This second approach appears to be the preferred method now for the CCC and reproductive rights organizations in foreign countries.[34]

In the late 1980s the more streamlined pooled industry model was implemented. Because the statistical model is industry based and not company-specific, a license can be priced before surveys are conducted.[35] For all three models an annual license based on the surveys may be renewed at the same fee for a second year.

CCC surveys are done in a confidential fashion, so the subjects of a subscriber's photocopying are not made known to publishers or the public. It should be noted that the AAS covers only in-house copying and does not cover interlibrary loan copies for libraries that do not qualify for the section 108(g)(2) exception. These libraries must pay for interlibrary loan copies on a per transaction basis or else obtain copies through an authorized document delivery service.

The CCC offers several advantages. First, since the fee paid to the CCC is based on a survey of copying done over a given amount of time (generally sixty days), the subscriber does not have to maintain its own records detailing the amount and subjects of copying. Second, to encourage use of the CCC, many members of the CCC charge less for copying done through the CCC than they do for other copying requests. Third, an annual license based on the surveys may be renewed at the same fee for a second year. The CCC collects an administrative charge approximating one-third to one-quarter of the annual fee in addition to distributing the royalties to registered copyright holders. Fourth, CCC surveys are done in a confidential fashion, so the subjects of a subscriber's photocopying are not made known to publishers or the public.

Since its establishment the CCC has collected and distributed royalty payments based on the fees schedules set by each publisher. Recently, in an attempt to develop a more efficient service, the CCC sought and received from the Department of Justice a business review letter which permits the CCC to negotiate directly with users in setting rates.[36] Under this proposal the CCC would discontinue its AAS blanket license.[37] Negotiations for license fees permitting unlimited copying for internal use of copyrighted works in the CCC library may result in price levels that more accurately reflect the market.[38] Nonetheless the CCC will continue to license corporations for photocopying large repertories of copyrighted materials.

C. *Other CCC Activities*

In 1989, as part of a move to serve universities in the same fashion as corporate copiers, several universities were selected by the CCC to enter a new program aimed ultimately at collecting royalties from educational institutions for copying in excess of fair use.[39] In exchange for authorization from participating publishers to copy freely for internal educational use, the schools agreed to monitor their own copying practices over a two year period.[40] At the outset the CCC was looking to develop a university licensing system. The schools participating in the pilot project were Northeastern University, Columbia University, Stanford University, the Cornell School of Hotel Management, Utah State University and Principia College. The CCC recognized that the size and decentralization of most universities would require a variation of the corporate licensing schemes.[41] In a follow-up seminar of publishers and university officials, there seemed to be as much difference of opinion among the publishers as between publishers and university users.[42] At the conclusion of the pilot project nothing further has been done. There is no blanket license available for universities. Some academic libraries use the TRS but others simply do no copying beyond that permitted by sections 107 and 108. It is apparent that universities are looking to find ways to respect copying of copyrighted information. Perhaps there may be a way to develop a program for a phased university licensing system especially for electronic copying should the CCC be licensed to cover reproductions in electronic format.

One point bears repeated emphasis. The CCC does not make copies itself; rather, "it operates as a clearinghouse for licensing of copying and as a conduit for the transfer of license fees to copyright holders."[43] Library staffs or corporate employees still must make the copies themselves.

A recent entry into the copying and distribution field is CARL/UnCover. This service available from CARL (Colorado Association of Research Libraries) offers the option of searching online the UnCover database of tables of contents of more than 15,000 periodicals. With an appropriate password and profile, UnCover offers quick delivery by fax of articles selected from the service. Subscribers pay a per article fee plus copyright royalties and fax surcharge where applicable.[44] In other words, CARL/Uncover is a commercial document delivery system which pays the royalty on every copy made regardless of whether that copy would have been a fair use copy or not. It pays the royalty through the CCC if the publisher is registered with the CCC.

A relatively limited number of publishers participate in Pubnet, an electronic mail system developed by publishers as a book order and inventory system.[45] Following the decision in *Basic Books, Inc. v. Kinko's*,[46] Pubnet built a permissions module linked to some book stores. More recently, the CCC has entered into a partnership with National Association of College Stores in which the CCC has taken on responsibility for processing permissions formerly handled by member stores.[47]

D. *Licensing for Educational Copying*

Publishers believed the decision in *Basic Books, Inc. v. Kinko's*[48] would stop the widespread copying done by college professors to produce anthologies and course packs. Instead, a dramatic increase in permission to copy has occurred.[49] As the economics of publishing and copying continue to change, copyright holders and users are finding appropriate ways to be clear about owner's expectations and to put that information in an unequivocal form. Further, because of high royalty fees, faculty have had to reevaluate the use of course packs or at least to limit the contents of such packets.

The *Kinko's* decision has especially important ramifications for users of educational information. Eight New York publishing houses, including Basic Books, sued Kinko's Graphics Corporation over the practice of reproducing copyrighted material to be included in anthologies and course packs prepared by college professors for use in their courses. Kinko's had offered this as the university "Professor Publishing" service since the mid 1980s which is aggressively marketed to college faculty.[50] Professors submitted collections of material related to a given course (for example, articles and scholarly pieces on energy generated by wind turbines for an engineering class) to Kinko's. Kinko's would duplicate the material, combine the articles, book chapters, etc., according to the professor's instructions, and then sell the resulting anthology or course packs to students at a profit.[51] Many faculty members at universities across America used this service because it allowed them to tailor class assignments to material they felt was important, rather than simply rely on published textbooks, which provide far less flexibility and currency. Kinko's neither sought nor obtained permission from the copyright holders before using the excerpts;[52] for that matter, neither did professors make a general practice of seeking permission.

The court rejected Kinko's various defenses to the charge of infringement; the most important of these defenses was the doctrine of fair use,[53] which the court rejected after considering the doctrine's four statutory factors.[54] The first factor, the purpose and character of the use, weighed against Kinko's for several reasons. First, while nonprofit educational uses are presumptively fair, the court found that Kinko's copying was not educational in nature, since Kinko's is neither a nonprofit entity nor an educational institution.[55] Its goal in reproducing the copyrighted material, the court held, was not education; rather, Kinko's had "the intended purpose of supplanting the copyright holder's commercially valuable right."[56] Second, although so-called transformative uses, which enhance or build upon the original, are arguably enough to win this factor, Kinko's use added nothing to the copyrighted works. In the words of the court, it was a "mere repackaging."[57]

Fair use is generally maintained to be broader in scope with respect to factual works than it is regarding nonfactual ones, as the former "are believed to have a greater public value and, therefore, uses of them may be better tolerated by the copyright law."[58] The second factor, the nature of the use, is generally won by an alleged infringer if the works at issue are primarily factual. Since all the works at issue in the case were factual, the court found Kinko's to have prevailed on the second factor.[59]

The third factor, the amount and substantiality of the work, focuses on both the qualitative and quantitative aspects of the use. Since Kinko's copied between five percent and 25 percent of the works at issue, and often took whole chapters, the court judged them to have crossed the quantitative line. On the qualitative side, the court concluded "that the portions copied were critical parts of the books copied, since that is the likely reason the college professors selected them for use in their classes."[60] Thus, Kinko's lost the third factor.

The fourth and final factor, the effect of the use on the market for the work, is the most important fair use factor.[61] "To negate fair use one need only show that if the challenged use `should become widespread, it would adversely affect the potential market for the copyrighted work.'"[62] The court found that Kinko's copying hurt the market for both the sales of plaintiff publishers' books and the market for royalty fees to those publishers.[63] With the most important factor tallied against them, along with two of the three other factors, Kinko's failed the fair use test.

In addition to their fair use argument, Kinko's also claimed that its actions fell within the *Agreement on Guidelines for Classroom Copying in Not-For-Profit Educational Institutions*. The guidelines are part of the legislative history of the Copyright Act of 1976,[64] and establish the minimum reach of fair use with respect to educational copying.[65] The court held that the guidelines cover only copying done by educational institutions without profit, a fact which renders Kinko's, a corporation engaged in the pursuit of profit, beyond the shield of the guidelines.[66] Even if Kinko's copying was to fall under the guidelines, the court found their copying was in excess of what the guidelines permit.

As the court noted, the guidelines set up a three-part test to determine if a given instance of photocopying falls within its limits: (a) brevity and spontaneity, (b) cumulative effect and (c) notice of copyright.[67] Since the articles copied exceeded the specific word-count limitations imposed by the brevity requirement, this test went against Kinko's. Also, as the copying of the collections was done at the beginning of the semester, yet intended to last until the end of the term, the court held that the copying did not meet the spontaneity requirement; thus, Kinko's lost the first factor. The court said as to cumulative effect Kinko's copying encompasses more than the nine instances of multiple copies for one course during one class term. Furthermore, copying should be limited to one course only and to no more than one piece of work per author.[68] Since four of the five collections at issue exceeded the nine instances allowed by the guidelines, the court weighed the second factor against Kinko's.

The third test requires a notice of copyright to be placed on each copy. Kinko's included no notice on any of the course pack collections, so they failed the final factor as well, and by virtue of this failure, placed itself outside the guidelines. Finally, even beyond exceeding the limitations of the guidelines, since Kinko's copying led to the creation of new anthologies and garnered a profit for Kinko's, its actions violated two of the express prohibitions contained in the guidelines.[69] One of the more amusing argument Kinko's raised in support of its fair use defense was that the copying actually whetted the students' appetite to read more from the author.[70]

In the wake of the *Kinko's* decision, some practical suggestions have been made concerning how academic institutions can bring their photocopying practices into line with the law. The following comments also serve as a primer for continuing education programs and conferences, both of which often provide similar anthologies for the use of participants. While most of these recommendations merely restate existing law, they serve as a helpful reminder. The suggestions include:

1. Students should be charged no more than the price necessary to cover the cost of copying any material sold to them.
2. Place a notice of copyright on every copy distributed.
3. Do not copy solely to create anthologies, without the permission of at least most of the copyright holders.
4. Copies done under the guidelines must be spontaneous.
5. Consumable material, such as workbooks and standardized tests, should not be copied.[71]

With the growing demand for permission services and for sophisticated electronic services, Kinko's announced it would discontinue its course works program at the end of 1993.[72] The CCC has already expanded its services in this area providing its Anthologies Permission Service. Other groups such as the National Association of College Stores now have joined with the CCC in handling permissions but college stores produce the course packs for faculty in addition to obtaining the necessary permissions, payment or royalties, etc.[73]

III. INTERNATIONAL MUSIC AND FILM LICENSING

A. *International Collectives*

The United States is not alone in having licensing organizations. In fact, there are reproductive rights organizations throughout the developed world, nineteen of which are members of the International Federation of Reproductive Rights Organizations (IFRRO).[74] The Copyright Licensing Agency (CLA) is the equivalent of the CCC in the United Kingdom. Others include BONUS in Sweden, CANCOPY in Canada, Kopinor in Norway, Literar-Mechana in Austria, Centre Francais du Copyright in France, Pro Litteris in Switzerland and VG Wort in Germany.[75] The CCC has reciprocal arrangements with the United Kingdom, Canada, France, Germany, Australia and New Zealand, and with VAP, the collective licensing agency of the former Soviet Union. Thus, these agencies collect royalties for the photocopying of materials done abroad and send them to the CCC and vice versa.[76] In a move that recognizes a growing exchange of information internationally by users, the British Library announced a decision to work with the CLA to develop an appropriate method to pay individual publisher-set fees for copies of U.S. copyrighted material supplied to its customers in the United States. This is in connection with the British Library's document delivery service.[77] The CCC is coordinating with the CLA to expedite the collection and distribution of royalties. Further, the British Library is discussing with certain members of the Association of American Publishers ways to participate in a similar collective licensing arrangement involving electronic publishing copying and distribution;[78] however, since the CCC currently does not handle collection of royalties related to electronic rights further investigation is needed. CANCOPY is close to finalizing an agreement with the Canadian Institute of Scientific and Technical Information (CISTI) which would enable CISTI to supply copies of copyrighted materials to U.S. customers under an arrangement similar to the one established between the British Library Document Supply Service and the CCC. As is the case with the CCC, these organizations now appear to be moving toward the use of blanket licenses, although some (including the CLA) continue to base their fees on the amount of copying actually done.[79]

B. *Music Collectives*

Organizations to collect royalties for the public performances of music have existed for years. The two main groups in the United States are the American Society of Composers, Artists and Publishers (ASCAP) and Broadcast Music, Inc. (BMI). Once a copyright owner assigns her right of public performance[80] to one of these organizations, it issues blanket licenses entitling users to perform works in the societies' repertories.[81] Fees are generally based upon a percentage of the licensees' gross receipts, not the popularity of the works.[82] Bars, restaurants, dance studios, etc., as well as radio or television studios and cable companies have blanket licenses covering music performances. Even melodies played on an elevator or a telephone may require royalty payments. (Muzak includes the royalty payment in the subscription fee.) Colleges and universities must pay fees for the music performances outside the classroom if there is any payment of fees to performers or promoters.[83]

ASCAP has been particularly vigilant in enforcing the rights of copyright holders. Not only has there been litigation of national scope involving blanket licenses to television networks, but a large number of suits have been filed and won against local restaurants and bars that perform ASCAP compositions yet fail to take an ASCAP license.[84] Establishments that have a license from ASCAP and BMI display a decal, usually near the door, to indicate that it is licensed for the music performances.

In 1993, a new organization was formed by the Recording Industry Association of America "to distribute the royalty shares allocated to featured performers and record companies."[85] The Alliance of Artists and Record Companies (AARC) hopes to collect digital home-taping royalties due member recording artists under the Audio Home Recording Act of 1992,[86] which requires makers and importers of digital audio recorders and blank media to pay fees to various parties responsible for a given piece of recorded music (including the record company, performer, nonfeatured musicians, writer and publisher).[87] AARC will concentrate on assuring adequate compensation for record companies and performers.[88]

C. Film Collectives

Since the Copyright Act grants to the copyright owner of a motion picture the exclusive right to perform that work publicly, most public showings of movies require authorization. While there is the so-called Face-to-Face Teaching Exemption,[89] this covers only performance or display by an instructor or pupils, in each other's presence, as part of a teaching exercise. This exemption does not protect students, teachers or schools from liability for the display of films for entertainment purposes, or any purpose that does not qualify as strictly educational (such as assemblies or graduation ceremonies).

Films, Inc. is an example of a licensing agency that offers schools the right to perform its members' motion pictures.[90] Its roster includes most major Hollywood studios, as well as established independent filmmakers, and the films available range from current box-office hits to classics to foreign films. Most films purchased through Films, Inc. have the performance rights included in the purchase price of the film. If the performance right is not included in the purchase price, it usually is available separately from Films, Inc.[91]

Film, Inc. generally charges a flat fee determined by each particular film's commercial worth. If admission is charged, it requires a guaranteed minimum fee in addition to a pre-set percentage of the admission receipts. Certain films with less established box-office potential have a negotiable price.[92]

A newer licensing agency is the Motion Picture Licensing Corporation (MPLC), which was created by the major Hollywood movie studios to license public performance of "home use only" videos. The MPLC offers annual blanket licenses, called an umbrella license, to nonprofit organizations, private companies and government agencies for public performance of these videotapes. Licenses are for a one-year period and are automatically renewable. The MPLC does not provide copies of the videotape or videodisk. Instead, licensees purchase or rent them from their local video store or public library or other distributor. The licensee then enables an organization to exhibit home videos in public performance.[93]

IV. MISCELLANEOUS LICENSES

Information is increasingly available in digital format, either on CD-ROMs or through online data systems. Since any entry of such information into a computer's memory, even if it is later erased, constitutes the making of a copy, owners of copyrighted material reproduced in that manner in excess of fair use are owed compensation unless a license agreement permits reproduction.[94] To facilitate payments, another consortium of several publishers of research and scientific periodicals called ADONIS, delivers regularly updated CD-ROMs containing the texts of the periodicals to online services to which users may subscribe.[95]

Subscribers to ADONIS can quickly retrieve and print articles from more than 500 journals published by over 40 publishers.[96] Currently the journals are primarily from biomedical disciplines but related areas such as chemistry, biochemistry, bioengineering and biotechnology also are included.

ADONIS plans to expand into other fields. The ADONIS search retrieval software can operate on a variety of workstations and is available in a network version. The annual subscription rate is based on hard-copy subscriptions, software, CD-ROM disks and technical support. Each week subscribers receive additional CD-ROM disks with the latest journal issues. The system keeps records of prints made whether they are printed according to codes defined by the subscriber. Each time a print is made or the article is downloaded to a disk, a royalty, set by the publishers, is due. When the disk with the record of prints is returned to the publisher, users receive a management report detailing copies made and a bill for the royalties.[97]

V. CONCLUSION

While reproductive rights organizations like the CCC have streamlined the procedure for reporting copying and paying royalties for making photocopy in excess of fair use within the limits of the law, they are an imperfect solution at best. They do nothing to alleviate the physical burden of making the copy currently resting on libraries or individual library users.[98] Such organizations offer publishers only an approximation of how often their work is used, however. Among the proposals put forward for a less burdensome, more accurate system, Paul Goldstein[99] has suggested a computerized licensing solution in which researchers could log on to a database and scan titles and abstracts of articles in a chosen disciplines.[100] Once the researcher found the information he needed, the system would provide information on whether, and at what price, the publisher was willing to license the reproduction of a copy.[101] If the price was acceptable, the researcher could just punch "Yes," make a copy, and have the system bill him automatically.[102] The electronic environment will provide many opportunities to experiment with "pay as you go" systems. Unfortunately, such systems are unlikely to be able to differentiate fair use copies from those for which a user should pay. This means that users probably will pay for all uses of works, whether reproduced as a print copy or downloaded to disk.

Collectives are designed to meet the broad needs of many, but they are not suited to individual needs. The development of collectives confirms the importance of clearinghouses, but more work on clearance procedures is needed, especially given rapidly changing information needs.[103] Additionally, the actual behavior of information users is often different from perspectives on what that use will be and the scope and breadth of copying are changing. The CCC does not handle permissions or collection and distribution of royalties for electronic rights or electronic copying. At present there is no mechanism for handling this type of copying other than bilateral agreements. Many libraries in the for-profit sector want to deliver copies to users electronically, but without a central licensing organization, establishing such bilateral agreements is difficult and paying royalties to individual publishers is just too cumbersome.

ENDNOTES - CHAPTER 4

1   Melville B. Nimmer & David Nimmer, 3 *Nimmer on Copyright* § 13.05[E][4] (1992) [hereinafter Nimmer].

2   Sony Corp. of Am. v. Universal City Studios, Inc., 464 U.S. 417, 467 n.16 (1984) (Blackmun, J., dissenting).

3   Mary L. Mills, Note and Comment, *New Technology and the Limitations Of Copyright Law: An Argument For Finding Alternatives To Copyright Legislation in an Era of Rapid Technological Change*, 65 Chi.-Kent L. Rev. 307, 313 (1989) [hereinafter Mills].

4   Daniel Lacy, *The Photocopy Issue: Past, Present, and Future, in Essential Elements of a Copyright Clearinghouse* 5-6 (Proceedings of a Conference, Feb. 11-12, 1976, Washington, D.C.) (1977) [hereinafter Lacy].

5   Mills, *supra* note 3, at 310.

6   Lacy, *supra* note 4, at 6.

7   *Id.*

8   *Id.* at 7. Storage space also began to be in short supply for these large libraries as bound journal collections threatened already tight space.

9   *Id.* at 7-8.

10 *Id.* at 7.

11 Nimmer, *supra* note 1, § 13.05[E][1].

12 Lacy, *supra* note 4, at 9.

13 *Id.* at 8.

14 17 U.S.C. § 106(1) (1988).

15 *Id.* § 107.

16 *Id.* § 108.

17 H.R. Rep. No. 1476, 94th Cong., 2d Sess. (1976) *reprinted in* 17 *Omnibus Copyright Revision Legislative History* 67 (1977) [hereinafter House Report].

18 S. Rep. No. 94-473, 94th Cong., 1st Sess. (1975) *reprinted in* 13 *Omnibus Copyright Revision Legislative History* 71 (1977).

19 Isabella Hinds, Copyright Workshop, July, 1992 (tape available from the AALL) [hereinafter Hinds].

20 Edwin McDowell, *The Media Business; Royalties from Photocopying Grow*, N.Y. Times, June 13, 1988, at D9. The CCC may be contacted at 222 Rosewood Drive, Danvers, MA 01923; 508-750-8400.

21 Copyright Clearance Ctr., Inc. v. Comm'r, 79 T.C. 793 (1982).

22 Pamphlet, Copyright Clearance Center. Recently, the CCC was endorsed as a means of adhering to copyright law in *American Geophysical Union v. Texaco*. For information on the *Texaco* case, see Chapter 3.

23 *Id.*

24 National Commission on New Technological Uses of Copyrighted Works, *Final Report* 154 (1978).

25 *Handbook for Serial Publishers: Procedures for Using the Programs of the Copyright Clearance Center, Inc.*, prepared for the Copyright Clearance Center, Inc. by the AAP/TSM Copyright Clearance Center Task Force 2-3 (1977).

26 Betty Landesman & Judith M. Brugger, *Everything You Always Wanted to Know about SISAC (Serials Industry Systems Advisory Committee)*, 21 Serials Librn. 211 (1991).

27 *Id.*

28 *Handbook for Libraries and Other Organizations Users Which Copy From Serials and Separates: Procedures for Using the Programs of the Copyright Clearance Center, Inc.* 3 (1977).

29 Louise Levy Schaper & Alicja T. Kawecki, *Towards Compliance: How One Global Corporation Complies with Copyright Law*, Online, Mar. 1991, at 15.

30   Pamphlet, Copyright Clearance Center.

31   Paul Goldstein, *Commentary on "An Economic Analysis of Copyright Collectives,"* 78 Va. L. Rev. 413 (1992) [hereinafter Goldstein Commentary].

32   Virginia Riordan, *Copyright Clearance Center, 1988: A Progress Report*, 15 Serials Libr. 43, 44 (1988) [hereinafter Riordan].

33   *Id.* at 45.

34   Stanley M. Besen, Sheila N. Kirby, & Steven C. Salop, *An Economic Analysis of Copyright Collectives*, 78 Va. L. Rev. 383, 387 (1992) [hereinafter Besen].

35   Riordan, *supra* note 32, at 45.

36   Letter from Anne K. Bingaman, Assistant Attorney General, Antitrust Division, Department of Justice, to R. Bruce Rich, Weil Gotshal & Manges, on behalf of the Copyright Clearance Center, Inc. (Aug. 2, 1993) (on file with authors).

37   *Justice Does Not Plan to Challenge Copyright Clearance Center Proposal*, 46 Pat. Trademark and Copyright J. (BNA) 298-99 (1993).

38   *Id.*

39   *Copyright Clearance Center to Document College Photocopying*, Pub. Wkly., Sept. 29, 1989, at 10.

40   *Id.*

41   *Photocopying in the Academic Setting: Final Report of the CCC University Licensing Pilot Program* (1991).

42   *Id.*

43   Copyright Clearance Ctr. v. Comm'r, 79 T.C. at 794-95 (1982).

44   Pamphlet, *A New Look for UnCover* (1993).

45   *Pubnet Permissions [Electronic System to Facilitate Copyright] for Course Anthologies Set for Test Launch*, 238 Pub. Wkly. Mar. 1, 1991, at 24.

46   758 F. Supp. 1522 (S.D.N.Y. 1991).

47   National Campus Marketplace (July 1993).

48   758 F. Supp. 1522 (S.D.N.Y. 1991). Marian Kent, *From Sony to Kinko's: Dismantling the Fair Use Doctrine*, 12 J. L. & Com. 133 (1992).

49   Hinds, *supra* note 19.

50   *Kinko's*, 758 F. Supp. at 1529.

51   *Id.*

52   David J. Bianchi, Grant H. Brenna & James P. Shannon, Comment, *Basic Books, Inc v. Kinko's Graphic Corp.: Potential Liability for Classroom Anthologies*, 18 J. of C. & U.L. 595-97 (1992) [hereinafter Bianchi].

53   For a full explanation of this doctrine, see Chapter 2.

54   17 U.S.C. § 107 (1988).

55   *Kinko's*, 758 F. Supp. at 1531-32.

56   *Id.* at 1531 (quoting Harper & Row Publishers, Inc. v. Nation Enters., 471 U.S. 539, 562 (1985)).

57   *Kinko's*, 758 F. Supp. at 1530.

58   *Id.* at 1533.

59   To some extent this was a peculiar finding since some of the works Kinko's is known to have copied were frequently poetry, short stories and other fiction. However, the copying was not for entertainment but for serious academic study. The collections duplicated by Kinko's contained only factual works.

60   *Id.*

61   *Harper & Row*, 471 U.S. at 566.

62   *Id.* at 568 (quoting Sony Corp. of Am. v. Universal City Studios, Inc., 464 U.S. 417, 451 (1984)).

63   *Kinko's*, 758 F. Supp. at 1534.

64   House Report, *supra* note 17, at 68. See Appendix C.

65   *Id.*

66   *Kinko's*, 758 F. Supp. at 1535-36.

67   House Report, *supra* note 17, at 68.

68   *Kinko's*, 758 F. Supp. 1537.

69   *Id.* Readers interested in the definitions and elements of the preceding terms of art should consult Chapters 2 and 7 for a detailed treatment of the Guidelines.

70   *Id.* at 1534. Clearly, Kinko's attorneys must never have been exposed to course packs when they were students. The likelihood of students being inspired to read more from the author is slight indeed.

71   Bianchi, *supra* note 52, at 619.

72   *Kinko's Drops Course Packet to Focus on Electronic Services*, Pub. Wkly. June 14, 1993, at 21.

73   National Campus Marketplace, July, 1993. *See* text accompanying note 7.

74   Besen, *supra* note 34, at 386.

75   *Id.*

76   *Id.* at 387

77   *British Library to Pay Copyright Fees on All Photocopies Sent to USA*, Press Information 93/94, Aug. 3, 1993.

78   Association of American Publishers, AAP News, Aug. 1993.

79   Besen, *supra* note 34, at 386.

80   17 U.S.C. § 106(4) (1988).

81   Jay M. Fujitani, Comment, *Controlling the Market Power of Performing Rights Societies: an Administrative Substitute for Antitrust Regulation*, 72 Cal. L. Rev. 103 (1984). It is also possible to make arrangements with ASCAP and BMI to pay for a single performance.

82   *Id.*

83   17 U.S.C. § 110(4) (1988).

84   *See, e.g.*, Columbia Broadcasting Sys., Inc. v. American Soc'y of Composers, Authors and Publishers, 620 F.2d 930 (2d Cir. 1980), *cert. denied*, 450 U.S. 970 (1981); Broadcast Music, Inc. v. Niro's Place, Inc., 619 F. Supp. 9589 (N.D. Ill. 1985).

85   Bill Holland & Ken Terry, *RIAA Spearheads New Royalty Group; Alliance Will Collect Artists' Digital Fees*, Billboard, Feb. 13, 1993, at 1 [hereinafter Holland].

86   Audio Home Recording Act of 1992, 17 U.S.C. § 1001 (Supp. IV 1992).

87   *Id.*

88   Holland, *supra* note 85.

89   17 U.S.C. § 110(1) (1988).

90   Pamphlet, Films Inc. (1992). To contact Films, Inc. call 1-800-323-4222.

91   *Id.*

92   For further treatment of this topic, see Chapter 5.

93   Pamphlet, Motion Picture Licensing Corporation, *Do You Show Home Videocassettes?*. To contact the MPLC, call 1-800-462-8855.

94   Alan Latman, Robert Gorman & Jane C. Ginsburg, *Copyright for the Ninties* 466-67 (3d ed. 1989).

95   *Id.* at 467.

96   Pamphlet, *Adonis: Quality Document Delivery Services* (1992).

97   *Id.*

98   Lacy, *supra* note 4, at 8.

99   Professor Paul Goldstein is on the faculty at Stanford School of Law; he is a recognized expert in copyright law and is a prolific author.

100   Goldstein Commentary, *supra* note 31, at 415.

101   *Id.*

102   *Id.*

103   *Id.* at 414-15.

CHAPTER 5

AUDIOVISUAL AND NONPRINT WORKS

I. INTRODUCTION

Audiovisual and nonprint materials have become increasingly important in special library collections in recent years. Many libraries house extensive collections of films, videotapes, laser disks, audiotapes, phonorecords, compact disks, photographs, prints, slides and transparencies. In fact, the collections of some special libraries consists primarily of these materials rather than works in traditional print format. New formats of audiovisual and nonprint works will continue to be developed and special libraries will collect them as they collected materials in formats developed earlier. Audiovisual and nonprint works are protected by the Copyright Act, but their status relative to library users is less clear than for works in traditional print format.

A. Definitions

Audiovisual materials and nonprint works are subject to protection under the Copyright Act within the following categories: (a) musical works, (b) pictorial, graphic and sculptural works, (c) motion pictures and other audiovisual works, (d) sound recordings and (e) architectural works.[1] Musical works are not defined in the statute. The pictorial, graphic, and sculptural works category is defined as including "... two-dimensional and three-dimensional works of fine, graphic and applied art, photographs, prints and art reproductions, maps, globes, charts, diagrams, models, and technical drawings including architectural plans."[2] Works of visual art are defined more specifically than other categories because of special rights attached to them. The Act defines a work of visual art as a work of fine art such as a painting or sculpture that exists in a single copy or a limited edition of 200 or fewer copies which are numbered consecutively and signed by the author. A similar limitation concerning 200 signed and numbered copies exits for sculpture and photographic prints.[3] Audiovisual works are defined in the Act as " ... works that consist of a series of related images which are intrinsically intended to be shown by the use of machines or devices such as projectors, viewers, or electronic equipment, together with accompanying sounds, if any, regardless of the nature of the material objects, such as films or tapes, in which the works are embodied."[4] Motion pictures clearly are a type of audiovisual work and are defined as " ... audiovisual works consisting of series of related images which, when shown in succession, impart an impression of motion, together with any accompanying sounds, if any."[5] In 1990, architectural works were added as a protected category.[6] An architectural work is defined as "the design of a building as embodied in any tangible medium of expression, including a building, architectural plans, or drawings."[7]

B. Library Practices

The owner of a copyright in one of these works receives the same five rights as the holder of a copyright in a print work. The most likely type of infringement in which a library is likely to be involved is unauthorized reproduction and distribution of copies of a literary work. With nonprint and audiovisual works, however, libraries may be involved in practices that could infringe any of the five rights since audiovisual and nonprint works are meant to be performed but also may be reproduced. Further, musical works, motion pictures and audiovisual works are easily adapted to another format. Finally, works of art, graphic works and the like usually are displayed. Thus, problems may arise whenever a library duplicates a copyrighted work, prepares a derivative work based on a copyrighted work, or performs or displays it. There are standard library practices which use pictorial works such as transparencies and photographs. Some of the materials may be original while others are duplicated in some manner from the original sources. Libraries also are involved in the creation of derivative visual works such as producing slides from photographs in a textbook and preparing transparencies from published charts. Some libraries produce in-house audiovisual works designed to communicate information to various user groups.

Occasionally, libraries are asked to duplicate audiovisual material for users. It still is somewhat unclear under what circumstances a library that owns such duplication equipment can reproduce audiovisual material from its collection for users without infringing copyright. Libraries may be asked to loan such materials to other libraries through interlibrary loan. Because of the high probability of damage to these materials from mailing and handling, many libraries simply refuse to loan audiovisual and nonprint works. With the advent of inexpensive duplicators, however, it is easy to make a copy for interlibrary loan and thus make the transaction similar to the exchange of photocopies to satisfy interlibrary loan requests. The difficulty with all of these practices, however, is that there is very little indication in the Act that such reproduction is either permissible or risk free.

II. SECTION 108: LIBRARY EXEMPTIONS

The Copyright Act includes a section specifically for libraries and archives which has as its purpose to exempt certain copying which otherwise would constitute infringement.[8] Unfortunately, the reproduction of audiovisual and pictorial, graphic and sculptural works is excluded from the general section 108 exemption. "The rights of reproduction and distribution under this section do not apply to a musical work, a pictoral, graphic or sculptural work, or a motion picture or other audiovisual work other than an audiovisual work dealing with the news."[9] This section is limited, however, and the copying of audiovisual material is permitted under the following circumstances:

1. Whenever pictorial and graphic works are published as illustrations, charts, diagrams, etc., to accompany textual material, they may be copied under the same conditions as the textual portion.[10]
2. For purposes of preservation, security or deposit for research use in another library, an unpublished work may be copied.[11] This includes pictorial, graphic and audiovisual works.
3. Copying to replace damaged, deteriorating or lost copies or phonorecords is permissible if the library first made a reasonable effort to obtain an unused replacement copy but found that one could not be obtained at a reasonable price.[12]

Outside of these specific circumstances, sections 108 is limited to printed materials and nonmusical sound recordings, other than soundtracks from motion pictures or other audiovisual works.[13]

III. SECTION 107: FAIR USE

A. General

Fair use likely has some applicability to library uses of audiovisual and nonprint works. In fact, section 108 states that nothing in the library section affects the right of fair use as provided in section 107.[14] The Senate Report that accompanied the Act goes further and states that not only is nothing in section 108 intended to take away from any rights that exist under the fair use doctrine but that section 108 actually authorizes certain photocopy practices that might not otherwise qualify as a fair use.[15] In discussing the reproduction of works for classroom uses, the Senate Report states that it is expected that the fair use doctrine would be applied strictly to the reproduction of entire works such as a musical composition, dramas and audiovisual works, including motion pictures. The reason is because these works, by their very nature, are intended for performance and public exhibition.[16]

The *Guidelines for Educational Uses of Music*[17] pertain to nonprofit educational institutions and provide that a single copy of a sound recording of copyrighted music may be made from a sound recording owned by the school or the individual teacher but only for the purpose of constructing aural exercises or examinations. The copy made may be retained by the educational institution or by the individual teacher. This exemption applies only to the copyright in the music and not to any copyright that may exist in the sound recording.[18] The guidelines also allow some copying of sheet music, but only in three instances. First, when performance is imminent, one may replace purchased copies that are unavailable for that performance. Even after the performance, replacement copies must be purchased. The second instance in which sheet music may be reproduced is to make multiple copies, up to one copy per student; this is limited only to excerpts of a work so long as the excerpt comprises less than 10 percent of the work. Third, a single copy of a performable unit may be made for a teacher for research or preparation to teach a class if one of two of the following conditions are met: (a) the library obtains confirmation from the copyright holder that the unit is out of print or (b) the unit is available only as a part of a larger work. Any copy made by a nonprofit educational institution must contain the copyright notice that appears on the printed copy.[19] See Appendix D for the text of the guidelines.

Thus, there is little mention of nonprint and audiovisual works in the sections of the Copyright Act that are specifically applicable to libraries. This, despite the tremendous increase in the availability of such material and the change in the nature of library collections.

B. Sony Case

The U.S. Supreme Court first applied the fair use test to the reproduction of audiovisual works in *Sony Corporation of America v. Universal City Studios, Inc.*[20] Although the Ninth Circuit Court of Appeals held otherwise, the Supreme Court held that home videotaping of television programs off the air did not constitute infringement of the copyrights on these television programs.[21] The Supreme Court opinion dealt only fleetingly with the four elements of fair use; instead, the decision focused primarily on the issue of whether a video recording was capable of noninfringing uses sufficient to relieve the manufacturer of the video recorder of vicarious liability, rather than on whether the taping of television shows was per se an infringement.[22] The case, however, contains an extremely important statement regarding the fair use factors: the test is based on an "equitable rule of reason," which is to be achieved by weighing the four factors articulated in section 107 against the particular facts of a case.[23] The House Report also expressly states that the fair use doctrine is an equitable rule of reason.[24] This statement may be used to justify some fair use reproduction of audiovisual works.

The Court also discussed the issue of infringement but not define its boundaries. Instead, it found that there were certain types of videotaping for home use that were not infringing, most importantly time-shifting, in which a viewer tapes a program to be watched at a later time and then erases that tape. Time-shifting is not an infringing activity, according to the Court, because the taping is not-commercial (that is, the consumer does not anticipate selling the finished tape) and because the program originally was offered free for the viewers' consumption.[25]

As it stands, the *Sony* decision fails to offer a solid model by which to judge whether the reproduction of an audiovisual work is a fair use under section 107 outside of this home exemption. It does, however, establish some general considerations to examine. If the intent behind a use is commercial gain, this is of supreme importance to the Court. The opinion states that there is a presumption of infringement if the use is for profit.[26] Since most library reproductions tend to be for nonprofit educational and research purposes, one might argue that the *Sony* opinion supports a fair use in these situations. Whether the argument is equally valid for libraries in for-profit companies is less clear, especially in light of the current *Texaco* litigation.[27]

If the use is noncommercial, then the nature of the work and the amount and sustantially of the portion taped may be held to be of small consequence. The programs that entered into the discussion were works intended for entertainment purposes and most were taped in their entirety, yet, because they were recorded merely for the private viewing of the copier, those factors were not fatal to a fair use defence. Further, *Sony* is limited to home use.

Unfortunately, for libraries, there are many circumstances not covered by *Sony*. There is no discussion of library-building, the very opposite of time-shifting. A person who creates a tape of a copyrighted work in order to save the copy for repeated future viewing is building a library. If a work of entertainment is copied for this purpose in lieu of purchasing a lawfully prepared copy, it is logical to presume this activity infringes the copyright in the work. If the work copied is for research and is to be retained for extended study, it might be permitted under *Sony* depending on the facts of the case.

Also, the taping in *Sony* was held to be noncommercial, in part, because it was from broadcast television, shown free to the viewer. The assumption is that the owner of the copyright will receive compensation for the display of these works in the form of advertising revenue.[28] The Court makes no reference to taping from another audiovisual work. If a copy is made of a tape owned by a special library, either for distribution to a client or to another library, that situation clearly is beyond the *Sony* exemption.

The circumstances surrounding the copying of an audiovisual work for a library user are beyond the specified limitations of the majority opinion of *Sony*. Thus, one must rely on the "equitable rule of reason" language which dictates that the fair use of audiovisual works will have to be decided on a case-by-case basis.

There is, however, a strong dissenting opinion in *Sony*, joined by four justices.[29] The dissent strictly follows and analyzes the four-pronged test in much the same manner and to a similar effect as the Ninth Circuit had done.[30] Although the dissenting opinion is not law, it could be persuasive when a future case concerning infringement of an audiovisual work is next heard by the Supreme Court. The dissent agreed with the lower court that only a productive use, i.e., one that increases the benefits the public derives from the copied work, is evidence of fair use. Productive uses were defined as teaching, scholarship, research and the traditional fair use exceptions.[31]

An examination of the typical copying of works done in libraries using the *Sony* reasoning supports the fair use of much of this activity as defined by the four fair use factors.

1. *The purpose and character of the use.* Library copying generally is noncommercial and the works reproduced usually are informational in nature, "productive uses" under the dissent's definition. Use in a special library copying situation is most likely to be research and scholarship, clearly a productive use unlike either entertainment or commercial use. Most library copying, since it passes the majority requirement of noncommerciality and the dissent's test of productivity, should satisfy this requirement.[32] The later *Texaco* court, however, says that research is not the kind of use productive use envisioned. Instead, it means that something new should be produced from the copy.[33] Thus, whether research and scholarship uses are a productive use is unclear although the Supreme Court's definition of "productive use" should prevail.

2. *The nature of the copyrighted work.* The *Sony* majority disagrees that more copyright protection is required for "original" works, such as fiction, than for informational works, such as news or biography although this factor appears to be of minor importance in *Sony*. A library might be asked to reproduce copies of works of both informational works and those originally intended for entertainment. While special libraries are more likely to be interested in works that have information as their purpose, this is not an absolute. There are many special libraries and information centers that have large collection of television and radio programs, recorded music, photographs and motion pictures. Although the original purpose of these copyrighted works may have been entertainment, they now are used for serious research, study, and criticism. The nature of the copyrighted work alone cannot define fair use for library reproduction of copyrighted audiovisual works.[34]

3. *The amount and substantiality of the portion used.* The *Sony* majority held that taping the entire work did not prevent a finding of fair use if the use was noncommercial. The dissent believed that the fair use doctrine allowed only the reproduction of small portions of a work, if that copy is to be used for a productive purpose.[35] Seldom is a library asked to reproduce only a small portion of a copyrighted audiovisual work for research purposes, although this would be satisfactory under the dissent's more restrictive view. Much more frequently a user requests a reproduction of the entire audiovisual work. As libraries are asked to reproduce works to be used in the new multimedia world, requests for copies of small portions of audiovisual works probably will increase.

   The extent of the portion requested by the user remains a critical factor, although in view of the majority holding in *Sony*, if the use is noncommercial, that fact will militate in favor of fair use despite the amount of substantiality used. One might inquire whether serious television study should permit the development of visual exercises for students from copyrighted audiovisual works much as is permitted under the *Guidelines on Educational Uses of Music*[36] permits for the construction of aural exercises for use in a nonprofit educational institution.

4. *The effect of the use on the potential market.* The dissent allows a broad definition of harm to the potential market, by stating that the respondents in *Sony* "can show harm from the VCR use simply by showing that the value of their copyrights would increase if they were compensated for the copies that are used in the market."[37] Commerciality of the use is the critical factor in the majority opinion as well. The majority treats the issue of commerciality as that of the intent of the user to profit economically from the copying of a work; the dissent considers any potential loss of revenue evidence of market harm. Using this definition, any copying by a library in a for-profit company ultimately aims at some economic gain for the company. The legislative history, however, indicates that a library may respond to a spontaneous user request even in the for-profit corporate library.[38] The *Texaco* decision appears to adopt the minority view on market effect. There the court held that publishers lose royalty income from Texaco's photocopying activities not that Texaco profits commercially from the scientist's reproduction of scientific articles.[39]

C. General Issues

Obviously, home videotaping of telecasts affects potential sales of commercially produced tapes. When a library reproduces a copyrighted audiovisual program, whether from off-the-air, from a commercially purchased tape or from a copy, the effect on the potential market may be equally significant. If the market for informational programs is libraries and educational institutions, any unauthorized duplication may directly harm the copyright proprietor. Some argue that the market for such works is finite, and that libraries purchase the number of copies needed as permitted by their budgets and thus number copies duplicated by libraries would not negatively impact on the owner's market.

The Senate Report cautions that under section 107, the attention paid to uses of copyrighted material by nonprofit educational institutions - such as multiple copying for classroom use - should not be construed as limiting fair use only for such copying; the same general principles are applicable to situations other than education, but the weight given to them may vary from case to case.[40] Copying done by various organizations including the Library of Congress and the American Film Institute to preserve pre-1942 motion pictures as well as the copying of literary works to produce talking books for the blind are given as other examples of fair use. Even though the copyright term may not have expired on such works, copying for these purposes would be within fair use exception.[41]

The recent Audio Home Recording Act[42] has considerable ramification for audiovisual works. The law allows consumers a fair use of digital audiotape technology in home recording for personal consumption, in exchange for which manufacturers of the technology would pay royalties on every machine built or imported to this country.[43] The importance of this legislative development is that it allows phonorecords, cassettes, and CD recordings to be copied or transferred into a high-quality, compact, digital format without risk of infringement. It expressly applies only to home taping; although library reproduction usually will be noncommercial, it is not immediately apparent whether library reproduction will be permitted under the law.[44]

One of the difficulties in this area is that the 1909 Copyright Act[45] was a reaction to the end of the Guttenberg era and the beginning of the electronic age. The demise of this era was heralded by the advent of radio and television.[46] The fair use test was developed for printed literary material and functions well for those works. With the advent of electronic media, some scholars believe the concept of fair use will require redefinition.[47] Older fair use cases primarily involved instances of plagiarism in which one author borrowed from the work of another to improve her own reputation. The potential injustice behind the reproduction of copyrighted works today is almost purely commercial; the worry is hardly that a user will attempt to pass off sections of *Howard's End* as his own work, but rather that he will obtain a copy of the film or perform it publicly without rendering royalties to the owner of the copyright. The "equitable rule of reason" opens the door for the courts to weigh and balance the justifications for the alleged infringement instead of simply conforming the facts of a case to four narrow categories. The problem, of course, is the lack of standards by which easily to judge the likelihood that its use will be held to be infringing. As one commentator stated,

> A definite standard would champion predictability at the expense of justification and would stifle intellectual activity to the detriment of the copyright objectives. We should not adopt a bright-line standard unless it were a good one - and we do not have a good one.[48]

The difficulties embodied in applying the fair use standard, particularly in the audiovisual arena, is evidenced by the dissention between and among the various levels of courts in the *Sony* case.

It is said that examining the purpose of a use for audiovisual works under the first fair use criterion calls for arbitrary distinctions.[49] Perhaps the nature of the user rather than the nature of the work should be the test for audiovisual works. Thus, researchers in all fields should be granted broad exemptions because they are engaged in "promoting the progress of science and the useful arts" in accordance with the constitutional copyright provision.[50] While copyright owners certainly should reap economic rewards for their creations, the rewards should not be such as to frustrate the constitutional purpose to advance science, art, and industry.[51]

IV. LIBRARY DUPLICATION OF AUDIOVISUAL AND NONPRINT MATERIALS

A library might want to duplicate musical works, audiovisual or pictorial and graphic works for many reasons. Preservation clearly is an important motive. The medium in which the work is stored makes handling a problem, and audiovisual works particularly are more easily damaged than printed, bound materials. Users may request copies of a portion of a work or the entire work for scholarship and research. A library may wish to duplicate a work if it is out of print and otherwise unavailable at a fair price. Also, a library may receive interlibrary loan requests for copies of photographs, music or audiovisual material and may prefer loaning a copy rather than the original.

Performance of audiovisual or musical works within a library presents entirely different issues from the reproduction of those works. A discussion of these issues follows this section.

A. *Nonprint Works*

1. Musical Works

In any discussion of audiovisual and nonprint works in special libraries, some attention must be paid to music. Although music fixed in musical notation format is subject to photocopying much as literary works, some aspects of copyrighted music clearly are relevant in this chapter especially when the musical work is fixed in a sound recording and available on phonorecord, audiotape, compact disc (CD), etc. Special libraries may deal with music in this format in two ways: as a type of work housed in the library's collection and as a music performance in a library either as background music or for a special performance event.

Musical works on tapes or phonorecords as a part of the library's collection are treated as other works in that format. In special libraries, such musical works generally are circulated to users for serious study and examination whether it be for research on blues musicians or various arrangements of a Beethoven sonata. The reproduction of musical works generally is prohibited under section 108.[52]

Under section 107 fair use, however, some reproduction of music is permitted. In fact, the *Guidelines for Educational Uses of Music* published in the House Report permit even multiple copying of music for performance in emergency situations. For academic purposes, other multiple copying of excerpts of musical works is permitted but limited to one copy per student.[53] Further, printed copies of purchased music may be edited or simplified.[54] Single copies of recorded music may be made in two circumstances: (a) to record the performance of students for the purposes of evaluation or rehearsal, and (b) to construct aural exercises for students.[55] The Senate Report admonishes that even in a nonprofit educational institution the fair use doctrine would be applied strictly to works such as musical compositions which are meant by their nature to be performed.[56]

The only mention of musical works in section 108 appears in subsection (h) which states that the section 108 rights of reproduction do not apply to a musical work. In discussing this subsection, however, the House Report states that although musical works generally are excluded from section 108, the fair use doctrine remains applicable.[57] "In the case of music, for example, it would be fair use for a scholar doing musicological research to have a library supply a portion of a score or to reproduce portions of a phonorecord of a work."[58] Thus, a special library can reproduce portions of a musical works for a scholar to use for her research, but there are no specific guidelines on how much may be copied or what other purposes other than scholarship might qualify for such copying.

2. Illustrations, Photographs, Drawings and Maps

Graphic works are subject to copyright protection, as are photographs, drawings, and the like. Many special libraries have extensive collections of pictures, maps, drawings and models. Clearly, under the 1976 Act, a library which purchases an original work of art or a photograph does not own the copyright to that work[59] unless it was produced as a work for hire, i.e., the library specifically hired the person to produce the work, or unless the work was reproduced by a library employee within the course of his employment.[60] When a library does not own the copyright, any reproduction of the work is subject to the requirements of sections 107 and 108. Under the provisions of section 107, fair use might permit the duplication of such material if the use meets the four fair use criteria.

In the *Agreement on Guidelines for Classroom Copying in Not-For-Profit Educational Institutions*, an example is used that includes an illustration being copied in multiples.[61] In the nonprofit educational setting, therefore, the library has some flexibility in copying these materials for classroom use. There also are other instances, however, in which fair use may allow a library to make a single copy.

Under section 108, a work may be reproduced only for preservation if it is an unpublished work,[62] or if it is necessary to replace a damaged, deteriorating, lost or stolen work.[63] Generally, libraries are prohibited from copying audiovisual and nonprint works for users with one narrow exception — pictorial or graphic works that are illustrations or diagrams which accompany a text may be copied along with the text if reproducing the text meets the section 108 requirements.[64]

The Act exempts certain displays offered in the course of instruction if specific criteria are met. The primary criterion is that the display take place within a nonprofit educational institution.[65] The display must involve face-to-face teaching activity and take place within a classroom. Classroom is broadly defined to include library, laboratory, gymnasium, and the like. There is no limitation on the works included in the exemption; thus, a teacher may display text or pictorial works to the class by means of a projector so long as there is no projection beyond the place where the copy is located.[66] This exemption apparently carries with it the right to produce slides and transparencies in order to permit the projection so that the illustration may be seen. Section 107 may permit other copying not included in these examples, but there is no specific language in the legislative history to define these situations with the degree of certainty most librarians prefer.

Many archival collections own the only known copy of a photograph although the collection does not own the copyright. The practice of such libraries has been to sell the rights to use the photograph to publishers and authors. The right of these libraries to license the use of the photograph is not based in copyright law but rather is because the library has the only copy of the work. There is an interesting copyright issue of note, however: photographs, as well as other works that were unpublished as of the effective date of the 1976 Copyright Act, i.e., January 1, 1978, will pass into the public domain on December 31, 2002. There is an exception for previously unpublished works that are published between January 1, 1978, and December 31, 2002. They will pass into the public domain in 2027.[67]

Archival collections of photographs will lose an important source of revenue for older photographs published before 2002 since anyone who wants to reproduce them after that date may do so by optically scanning the published photograph rather than obtaining a print and a license from the archives. For one-of-a-kind photographs that remain unpublished after 2002, the archive will be able to sell the right once, to one user, again because it has the only copy of the photograph. After the photograph is published, however, anyone may use the photo from the published work — it will be in the public domain.

3. Slides and Transparencies

Slides and transparencies produced as original works are subject to copyright just as are photographs, illustrations and diagrams. The copyright considerations relative to the production of original slides and transparencies have not been a problem of much consequence to libraries and information centers since original works clearly are defined as subject to copyright.[68] The library responsible for their creation may claim copyright for works produced in these media. The copyright difficulties libraries traditionally encounter deal with duplication of copyrighted photographs, diagrams and illustrations in the form of slides and transparencies. Libraries may wish to reproduce graphics from copyrighted textual material for various reasons, such as to illustrate a lecture by the librarian or for use in a multimedia presentation. Moreover, in a nonprofit educational institution, the Act permits projection of copyrighted material in a classroom through a transparency or slide.[69]

If paper-to-paper or microform photocopying of a chart or illustration either for internal library use or for a client is permissible under the Act, then preparation of the slide or transparency is permissible. Throughout section 108, the phrase used is "reproduce no more than one copy or phonorecord." Clearly, a slide prepared from a copyrighted illustration, chart or graph is a reproduction of the work. Publishers do not agree with one another whether permission should be sought prior to duplicating such a work in slide or transparency format.[70]

A different problem is raised when a library duplicates, in either slide or transparency format, either a substantial number of or every illustration from a copyrighted work which is primarily comprised of photographs or illustrations and adds that group of slides or transparencies to the library's collection. In this situation, the library has created a second copy of the entire work in another format. This is comparable to photocopying an entire work, binding it, and then adding it to the collection. Absent the section 107 and/or section 108 exemption, this constitutes an infringement of copyright. In creating a duplicate copy of a work in slide format, an additional infringement may be found in the conversion from one format to another (printed photograph to slide), which abridges the owner's exclusive right of adaption.[71]

Should the library have a collection of copyrighted slides or transparencies (i.e., not copies made by the library of copyrighted photographs in slide or transparency format), the reproduction of those slides is governed by section 108. This subsection provides that the rights of reproduction and distribution detailed in section 108 do not apply to pictorial, graphic or sculptural works, that is, works subject to copyright protection under the category "pictorial, graphic, and sculptural works." Libraries that duplicate such slides or transparencies infringe the copyright on the slides. Under fair use, however, the House Report seems to indicate that libraries may duplicate works under situations not detailed in section 108.[72] Many libraries with large slide collections have begun to experiment with digitizing these slides for preservation, improved storage and better indexing and access. Whether reproduction in digital format and storage for later repeated use is a fair use is unclear. Publishers who produce slides for sale are not likely to look kindly on further duplication, and by digitizing the slide, the library technically makes another copy every time the slide is displayed on a screen. If the slides being digitized were made from photographs in a book, the situation is even more complicated. Even if the library wanted to seek permission for the digital storing and reuse of the slide, the publisher of the book in which the photograph appeared and from which the slide was made does not likely hold copyright in the photograph. The publisher probably had permission to use the photograph in the book but has no rights to use it further or to license anyone else's use. Tracking down the copyright holder for photographs is particularly difficult although sometimes stock photograph companies can help.

B. *Audiovisual Works*

1. Audiotapes and Phonorecords

Audiotapes, phonorecords and CDs merely represent a storage media for a performance of types of works that exist in print format. Just as literary works most often are preserved in printed book format, musical works frequently are stored on audiotapes and phonorecords. Other works also might be stored in this format, however, such as literary or dramatic works being read aloud. Sound recordings also may be preserved in this format.

Some libraries have huge collections of audiotapes. There are three basic reasons a library may duplicate audiocassette tapes and phonorecords: (a) for preservation, (b) in response to patron demand and (c) to convert the format of the material, i.e., from phonorecord to cassette. Audiocassette tapes present a unique problem relative to copying. Because of the nature of tapes and the hard use they receive in some libraries, tapes wear out and become unusable; also, they are subject to accidental erasure. While books also may be destroyed, audiotapes pose a more serious problem. Libraries that serve young users may have a more difficult time with tape destruction than do special libraries, but accidental erasure of tapes remains a significant problem for all libraries.

Some librarians, in an effort to solve this problem, created a master tape file. Employees made one or more copies of the audiocassette tape and retained the original as a master. Should the copy be damaged, another copy was produced from the master and the copy was circulated to users. Absent permission from the copyright owner, this duplication constitutes copyright infringement. Unfortunately, many libraries purchased only one copy of the tape and did not seek permission to duplicate it and may be involved in practices that clearly constitute infringement. The only permission to make an archival copy of a work is found in section 117(1) and relates only to computer programs; there is no similar exemption for audiovisual works.

Except for music libraries, most libraries prefer audiocassette format to phonorecord when it is available. Many libraries still have extensive collections of phonorecords, however, and they also are subject to damage from being scratched, warped and otherwise damaged. In order to preserve their phonorecord collections, some libraries converted them to audiocassette tape and circulated only the tape copy, thereby making the original phonorecord an archival copy. Absent permission from the copyright holder, this also is infringement. Under section 108(c) a librarian is permitted to replace a lost, damaged, stolen or deteriorating work but only after the destruction has occurred. The above described practice anticipated the loss, which is not envisioned by the Act. Further, under section 108, audiovisual works generally are excluded. Even under section 107, it is debatable whether the above described practices qualify as fair use. Another difficulty with this practice is that it converts the format of a work from record to cassette, and, absent permission from the copyright holder, this represents an infringement of the owner's adaptation right or the right to prepare derivative works.[73]

While the goal of preserving expensive library materials is a laudable one, the method selected may have a negative effect on publishers' markets for their materials.[74] If the material in question happens to be a published book, the library would not photocopy it for archival purposes in case the book might be destroyed accidentally. By analogy then, this practice of duplicating phonorecords onto tape is an infringement of copyright without the express permission of the copyright owner and the payment of royalties if the owner so demands. Record producers are very willing to sell multiple copies of tapes to libraries so that they have more than one copy in case of accidental erasure or damage.

Libraries may contract with individual producers of audiocassette tapes to purchase one copy and create a master tape file through duplication of the tape. Since the owner of the copyright on an audiocassette tape has the reproduction right, he can sell or give that right to libraries. Some producers are willing to allow such duplication, but prior permission should be requested. There frequently is an additional charge for the right to duplicate. Some libraries have sought to force audiotape producers into granting blanket permission to make a master copy. This can be done, but only with an agreement signed both by the librarian and by an employee of the tape producer who has sufficient authority to commit the publisher to the terms requested. This likely is not a visiting sales representative. Some libraries have attempted unilaterally to force producers to permit the creation of a mater tape with a statement on the order form to the effect that filling this order represents an agreement that the library may maintain a master tape and reproduce copies to circulate as needed. This is a useless and ineffective practice which has little chance of being enforced by a court. The employee filling orders for the tape producer undoubtedly lacks authority to commit the producer to such an agreement. A much better alternative is to write a separate letter to send with the order form asking for signed permission with a statement in the letter that if the tape producer is unable to grant such a request, the order is to be canceled.[75]

Another reason some libraries duplicate phonorecords and audiotapes is to convert from one format to another for ease of use. If the library has good tape equipment but an old, poorly functioning record player, a librarian may think an ideal solution is to duplicate the phonorecord onto audiocassette tape or digital audiotape. This practice is an infringement since the copyright owner has the right to determine the format in which the work will be reproduced and distributed. Conversion of format infringes the owner's adaption right in addition to the rights of reproduction and distribution. Most producers of phonorecords also have works available in audiocassette format and will sell libraries either or both formats or license the library to do its own conversion. The license to convert the format may be free but most likely there is a charge attached.[76]

In some school libraries, a tape duplicator was placed alongside the photocopier. Users were free to make copies of tapes whether the tape existed in the library's own collection, was owned by the user or was borrowed from another individual. Under fair use, some copying of audiovisual, pictorial and graphic works probably is allowed, but the copying of an entire work as described above probably cannot meet the fair use test. The library might even be seen as encouraging reproduction of an entire work.

In 1973, a similar practice was involved in a nonlibrary case. In *Electra Records Co. v. Gem Electronics Distributors*[77] a federal district court ruled that the duplication of copyrighted sound recordings by a record store for its customers was an infringement of copyright.[78] Defendant Gem Electronics owned fifteen retail outlets for electronics supplies and equipment. It installed ten Make-A-Tape systems at these outlets which enabled customers to duplicate commercially produced 8-track tapes for the price of a blank tape which the defendant also sold. Additionally, the defendant maintained a library of musical recordings on 8-track tape. The user could select a recording from the library and duplicate it using the Make-A-Tape system for $1.49 to $1.99 each, whereas the commercially produced tapes were sold for $6.00 each.[79] Plaintiff copyright owners sued for a preliminary injunction which the court granted. The defendant claimed that the Make-A-Tape system was equivalent to a photocopier in a library. The court disagreed, stating that instead of copying one article from a journal, the entire sound recording was copied. The court further held that while photocopies of printed materials generally are less desirable than the original, a duplicated tape is identical to the original and, therefore, is just as desirable.[80] Many media specialists disagree with the court, claiming that a tape copy made from a phonorecord under such circumstances is always of poorer quality.[81]

The court also pointed to a further difference. While a public library makes materials available to users, the purpose is altruistic and the photocopier is a device merely to assist in the dissemination of information. The defendant in this case, in contrast, derived a source of income from the sale of blank tapes even though the loan of the tape from the store's library was free.[82] Thus, the use was a commercial use. Special libraries may be public in nature or they may exist in a for-profit organization. The reasoning used by the court concerning the altruistic purpose for duplication of the tape would not necessarily be applicable to libraries in the for-profit sector.

A later case mirrored the opinion of the *Electra* court. In *RCA Records, et al. v. All-Fast Systems Inc.*,[83] a copy store, which kept a Resound machine, from which its customers could have audiotapes dubbed onto blank cassettes, was sued by the holders of the copyrights of several musical works for damages and for an injunction against the retention of the copying machine within the store. The defendant, in the aftermath of *Sony*, claimed in defense that the Resound was capable of noninfringing uses and that, therefore, an injunction was not warranted. The court made clear that a noninfringing capability was a viable defense only for a manufacturer; a user that employs technology for the purpose of unauthorized reproduction of copyrighted material for profit is a common pirate.[84] Despite the vehemence of the decision, the operative word once again is "profit"; the court did not address the potential liability of a noncommercial copier. A lack of economic motive for the reproduction may militate against infringement.

In all likelihood, a library that provides tape duplication equipment for its users would treat the equipment as if it were a photocopier. So long as there was no sale or profit made, the practice would comply with the above cases. The problem of duplicating the entire work indicates that libraries which offer tape duplication equipment should closely examine this practice to ensure compliance with the Act. The Audio Home Recording Act of 1992 has changed somewhat[85] the tenor of copyright law in the area of sound recording. By allowing an exemption for noncommercial home taping, the Act permits reproduction of copyrighted materials in exchange for royalties paid to the copyright holders, not when a tape or CD is purchased, but when a digital audiotape (DAT) record or recording is manufactured or imported. This royalty, originally paid by the distributor, eventually would be passed down to the consumer, so the user ultimately will pay.[86] A library that purchases a DAT recorder would pay into the royalty pool along with a person buying the machine for personal consumption. Although the bill makes no explicit mention of library use, because it tends to be as noncommercial as home use, there may be some applicability to libraries in the future. Librarians will simply have to watch for developments in this area.

As the law stands under section 107, many commentators believe that some copying of sound recordings for library users is permitted. The copy should not be used for purposes other than scholarship, research or teaching and probably does not extend to the duplication of an entire work. The librarian should consider the amount to be copied in relation to the work as a whole and the prohibition against selling the copy so made for profit.[87] There are neither guidelines nor specific examples from the legislative history dealing with library duplication of tapes and phonorecords, although the four fair use criteria would be applied to such a situation to determine whether the use qualifies as fair use.

An argument could be made that libraries which engage in the duplication of audiotapes and phonorecords for users are acting for the benefit of those users if the copyrighted work has been purchased directly, leased from the copyright owner or borrowed from another library. The library that duplicates the work is merely engaged in the dissemination of information while storing the original work for future use by patrons.[88]

On the other hand, the same could be said of traditional print materials. It certainly would save wear and tear on books to photocopy the entire work and circulate the copy rather than the purchased printed copy. Yet such practice clearly would infringe the reproduction and distribution rights afforded copyright owners. The application of the fair use test to such copying necessarily produces a negative result. The nature of the underlying work may be a musical composition, a literary work, a dramatic work or a sound recording; the amount copied is the entire work; the economic effect is to deprive the owner of a potential sale. Only the first factor would seem to favor such copying activity by libraries — the purpose and character of the use would be for scholarship or research. Taken as a whole, however, it is unlikely that duplication of an entire audiocassette tape or phonorecord for a user would satisfy the fair use test.

An entirely different situation is presented when a library merely circulates audiotapes and phonorecords to its users even if those users ultimately duplicate the tape at home. In 1984 the Copyright Act was amended to require royalty payments by commercial businesses that rent or loan phonorecords.[89] This amendment altered the "first sale" doctrine which held that royalties were paid to the copyright holder only for the first sale of the work.[90] For example, when someone buys a copy of a novel, royalties are paid to the author or copyright holder by the publisher based on the first sale of each copy. When the purchaser has read the novel and sells it to a friend or a used book store, no royalties are paid to the copyright owner for subsequent sales. The Act now states that owners of the copyright in a sound recording and the musical work embodied therein may dispose of that work through rental, lease, lending, or the like, for direct or indirect commercial advantage.[91] But record stores and other commercial concerns may not rent records without paying royalties for each such rental under a compulsory license.[92]

There is a specific exemption in the amendment for nonprofit libraries, however. "Nothing ... shall apply to the rental, lease or lending of a phonorecord for nonprofit purposes by a nonprofit library or nonprofit educational institution."[93] Thus, public, university and special libraries in the not-for-profit sector may lend or even rent phonorecords, tapes and CDs. Libraries in the for-profit sector may not. Some special libraries are particularly disadvantaged by the limitation of this exemption to nonprofit libraries. For example, a library in a recording company could lend or rent to employees copies of phonorecords on which it holds the copyright but not other recordings. The statute is clear, however, that the exemption is available only to nonprofit business; thus, there is little likelihood that a court would find that any internal loaning of records to users in a for-profit company is for a nonprofit purpose.

The Record Rental Amendment also represents the first time Congress separated libraries by whether they are in the nonprofit or for-profit sector.[94] It also is the first instance in which Congress made it clear that nonprofit libraries are not synonymous with nonprofit educational institutions. Until this amendment, many public libraries relied on a misunderstanding that any exemption available to nonprofit educational institutions also was available to public libraries.

2. Films, Filmstrips and Motion Pictures

The duplication of films and filmstrips presents problems for libraries similar to those presented for audiotapes. There are, however, some additional problems. For purposes of this article, films and filmstrips are used synonymously with the term motion picture in reel form but not with videotaped motion pictures. The potential damage to this material from users is somewhat reduced because films and filmstrips often are not be circulated directly to patrons. The library may arrange a showing, i.e., a performance of the film, or in the educational setting, the film may be loaned to an individual teacher for showing to a class. Unlike with audiotapes, the number of loan transactions is likely to be small. Most 16mm films now carry an FBI warning printed on the cannister or attached thereto indicating that duplication of the film is considered an act of piracy. Anyone who duplicates a film which carries this warning certainly cannot be considered an innocent infringer. The criminal penalties of the Act[95] can be evoked if there is any commercial purposes behind the duplication of a film. Whether a library in the for-profit sector would be held to have a commercial purpose if a film were duplicated is unknown and probably depends on the ultimate outcome of the *Texaco* decision.

Special libraries in organizations in which serious film study is conducted have additional problems regarding copyrighted films. One must assume that section 107 would allow some copying for teaching purposes. In examining filmmaking techniques, an instructor or scholar may want to copy portions of various copyrighted films or motion pictures in order to present or study them in a particular sequence. The library may be asked to copy portions of films for such purposes.[96] Although the creation of multiple copies or anthologies through photocopying is prohibited under the guise of fair use,[97] a film anthology is exactly what is needed for serious film study, much as is permitted for aural exercises in the *Music Guidelines*.[98] Should the creation of a visual anthology be allowed under fair use, the addition of that anthology to a library's collection raises additional problems related to the creation of audiovisual instruction packages and multimedia works.[99]

In applying the fair use test to library reproduction of films and filmstrips, the problem criteria are likely to be the nature of the work, the amount copied and the market effect. It has been argued that the nature of the copyrighted work factor should not weigh so heavily in evaluating the copying of films. Since the fair use test evolved in dealing with printed works, there is virtually no historical precedent for the fair use of motion pictures.[100] Film educators were not particularly active in the copyright revision process, and their influence was little felt. The Act reflects the film industry's restrictive distribution system, with little attention given to scholarship and research activities centering on motion pictures and film.[101] Because copyright owners are rigorously pursuing commercial film pirates,[102] it is unlikely that their immediate attention will be focused on libraries and educational institutions where serious film research is being conducted, but this does not insulate libraries from liability for copyright infringement.

Obviously, a library engaged in converting the format of a motion picture from 16mm to videotape is depriving the copyright owner of revenue;[103] moreover, the artistic integrity of the work may have been altered. For example, while Super-8 is a vast improvement over 8mm film, it is limited for use in small rooms due to image clarity and size problems. The impact of a motion picture originally offered in 16mm may be altered by the conversion to another form,[104] which changes the visual impact of the film. Thus, interference with artistic integrity has occurred in addition to copyright infringement.

The market effect of duplicating motion pictures and educational films is potentially quite serious. In describing the sales potential of film, an executive of a small production company described the normal costs and market for such films.[105] Recent estimates indicate that only about two-thirds of all films produced by members of the Motion Picture Association of America ever recoup their costs.[106] Of the revenues the average film accumulates, only about 10 percent derives from its theatrical release; the remainder comes from videocassette and cable sales (20 percent) and from broadcast television licensing (70 percent).[107] The latter two may be influenced by the copying of films or videotapes.

Libraries engaged in the duplication of films — whether the films are purchased, rented or borrowed — may have a serious impact on the producer's market. When the ultimate market is small, even one unauthorized copy may produce a negative impact. On the other hand, producers of educational films may take into account the potentially small market for their product and price the film accordingly. It can be argued, therefore, that few sales are lost through duplication since sales made have already exhausted the market.

Some libraries engage in the in-house production of audiovisual works for various purposes. In determining how much of a copyrighted motion picture or film can be used to create a new work, the following guidelines have been suggested: (a) the amount of copying should be small in relation to the overall length of the copyrighted motion picture; (b) any copied portion should not be the meritorious portion of the film, that is, it should not be exceptionally valuable or hard to reproduce; (c) if the portion copied from the film should represent only a small portion of the finished product, the copying then should be fair use if the other criteria are satisfied.[108]

The preservation of classic films has long been of concern. Films are now preserved by a number of organizations throughout the country. The first film archive in the United States was started by the Museum of Modern Art in New York.[109] In 1988, a National Film Registry was created by Congress with the National Film Preservation Act of 1988 (NFPA) to preserve films in their original form.[110] The collection at the Library of Congress houses approximately 100,000 titles, some 12,000 of which have been converted from perishable nitrate-based film to acetate-based "safety stock."[111] Other major film archives are in the University of California at Los Angeles and at the George Eastman House in Rochester, New York.[112]

3. Videotapes

For purposes of this chapter, "video" means videotapes, videocassettes and videodiscs. The Copyright Act differentiates between motion pictures and audiovisual works and makes it clear that motion pictures are a type of audiovisual work.[113] As far as copyright is concerned, videotapes are merely a storage media for motion pictures or other audiovisual works such as television programs. The technical difference between video and motion picture lies in the fixation process. In cinematography, the recording is made on film; in video, images and sounds are fixed in the tape or disc by a magnetic, mechanical or electronic process.[114] Another difference in the two media is the means of projection; a motion picture involves the use of a projector while video uses a television set. Also, it is far less expensive to reproduce video programs than to duplicate films.[115] Since the copyright exists in the underlying motion picture or other audiovisual work, it is technically incorrect to speak of a copyrighted videotape. Libraries purchase videotapes from a variety of producers, organizations and distributors many of which have a license from the copyright owner to further license the purchaser to duplicate the videotape and to perform it.[116]

In many library settings, the reproduction of video materials parallels the reproduction of audiotapes.[117] Libraries may have been tempted to purchase one copy of a videotape and duplicate sufficient copies to distribute to branch libraries without permission from the copyright owner. Occasionally libraries have made a duplicate copy of a videotape for circulation purposes while retaining the original purchased videotape as a "master." Should the circulating copy be damaged, then a new circulating copy would be produced from the master. Format conversion is another reason libraries want to duplicate videotape.

As with audiotapes, the purchase of a videocassette tape from the producer does not give one the right to reproduce additional copies of it. Such activity is direct infringement of the copyright holder's reproduction right. Some producers of videotapes offer their products for sale at two prices. The lower price is for a single copy of the tape. For a higher price, a purchaser may buy one tape along with the right to produce a specified number of duplicate copies.[118] In order to market their products, producers are devising various plans to induce purchase. It is likely that there will be more creative marketing techniques devised to permit library reproduction of videotapes after payment of some type of royalty that will be included in the purchase price. This is especially likely for video materials produced for the education market.

Video works also are easily destructible, and the more libraries circulate materials on videotape, the greater the chance of destruction or loss. The temptation to create a master file of videotapes thus increases. For libraries that tape in-house presentations or classes, a master tape file is advisable since the library or its parent organization owns the copyright in such in-house presentations, and there is no prohibition on the library in that organization's reproducing copies of the tape.[119] *Sony* permitted the home taping of television broadcasts only for time-shifting; creating a master file clearly presupposes the building of a library, which, though not expressly addressed by the *Sony* court, is at odds with any taping for short-term use. The other means of creating a master file is to reproduce copies of copyrighted videotapes either from lawfully purchased or a bootlegged tape. To be able to do this, one must have two video recorders (VCRs) and, for the many tapes that are scrambled to prevent reproduction, a decoding device commonly called a "black box."[120] Legislation against these devices has long been sought by the owners of the video copyrights. In 1990, bills were introduced in Congress to make it copyright infringement to "make, sell or distribute" such devices.[121] Opponents of this bill claimed that, if passed into law, the legislation would effectively reverse *Sony* by outlawing technology that has noninfringing capabilities.[122] There appears to be little interest on the part of Congress in enacting such legislation today.

The conversion of format from 3/4-inch Beta to 1/2-inch VHS videotape creates the same problem as converting audio formats, i.e., the reproduction of an additional copy which infringes the adaptation, reproduction and distribution rights of the copyright holder. In the past, video producers made their products available in both 3/4- and 1/2-inch formats, and conversion should not have been undertaken without the permission of the copyright holder. Today, however, VHS has become the standard and Beta equipment is no longer produced. Libraries that purchased videotapes in Beta format are now faced with a dilemma. They must convert to VHS format, but video producers vary in their willingness to permit format conversion free of charge. When contacted, some producers have agreed to permit a library purchaser to convert from Beta to VHS free of charge. Some have agreed to a no-charge conversion, but they have asked the library to return the Beta copy to the producer. A few producers have agreed simply to exchange the Beta copy for a VHS copy at no charge. Unfortunately, too many video producers have refused to permit format conversion and instead offered a new VHS copy at a reduced or even at full charge. Librarians must remember that the copyright holder owns the copyright as property, and the owner determines the format for copies of the work. Also, librarians should be aware that many video vendors do not have a license from the copyright holder to grant the library the right to convert the format so the owner must be contacted. Inconvenience or even the inability to use videotape because of old or malfunctioning equipment is not sufficient reason to excuse infringement of the rights of the copyright proprietor.[123]

Videotapes of classic television shows and children's and educational series are widely available for purchase. Public television stations frequently advertise videotapes of programs shown on PBS. Tapes are offered for sale for popular and important series that range from documentaries to artistic adaptations. Announcements are made on-the-air following programming; additionally, PBS has a mail order catalog through which tapes can be purchased. Increasingly, commercial television networks are advertising the sale of videotapes for programs ranging from documentaries and historical fiction to news magazine shows. Also, many popular early television series are available for sale on videotape.

Television programs preserved on videotape have the same characteristics as motion pictures relative to serious scholarship and research. The Act recognizes the importance of archival collections of television programs and grants libraries the right to videotape off-the-air television programs dealing with the news.[124] Additionally, the Radio and Television Archives was established at the Library of Congress[125] to house off-the-air recordings.

Various television archives have been created in academic institutions, as well as in private foundations and corporations; such collections are intended for the serious study of telecasts. Vanderbilt University Library houses an extensive television news archive and loans copies of news programs to users and other libraries on demand. Probably the largest of the university-based television archives is at the University of California at Los Angeles. There the Regents of the University of California and the Academy of Television Arts and Sciences established a television library in 1965. Material from that collection may not be copied and does not circulate off the premises. Additionally, the Museum of Broadcasting in New York is a heavily used research facility.[126] Libraries that want to create a television archive often must negotiate with networks to videotape or purchase their programs, with the exception of hard-core news programs. Normally, these materials may not be copied or removed from the premises; this restriction is likely to be contained in the contract between the network and the library relative to videotaping.

The ability to monitor and tape news broadcasts without infringing the copyright on the program may also be in the offing. A bill that would amend section 107 to add video news monitoring to the list of fair uses has been was reintroduced in Congress in 1991.[127] Should it eventually become law, independent monitoring services would have the right to tape and sell broadcasts to researchers, campaign workers and other interested parties. The parameters of fair use in the case of news monitoring have yet to be decided.

Missing from the legislative history of the 1976 Copyright Act is any recognition of serious television study. Television is the modern portrayer of culture, and students need the opportunity to view both the cultural and the social aspects of the development of this media.[128] Not only are there many high school and college courses that focus on American culture as portrayed on television, but universities offer degree programs in radio and television studies. In order to study this media, scholars and research libraries may need to create anthologies similar to those needed by film scholars. There is a strong public policy argument to allow copying to the extent that it encourages the serious study of television.

The need to provide materials for serious television study, however, does not give libraries free rein to videotape off-the-air. Libraries in nonprofit educational institutions are subject to the guidelines developed by a negotiating committee appointed by Representative Robert Kastenmeier, chair of the House Subcommittee on Courts, Civil Liberties and Administration of Justice. The committee was comprised of members who represented educational and library organizations, copyright proprietors, creative guilds and unions.[129] The guidelines, which took more than two years to negotiate,[130] provide more flexibility than did the Act as originally enacted in which off-the-air taping appeared to be limited to public broadcasting programs, even for nonprofit educational institutions and government bodies, with a limit of seven-day retention for those programs so copied.[131] Public broadcasters had made some voluntary expansions of the retention period and circulated lists of programs that might be taped by educational institutions and retained for a period of one year. Programs not included in the listing, however, were limited to the statutory seven-day retention.

The *Videotaping Guidelines* allow nonprofit educational institutions to copy from both commercial and public television with a 45-day retention, although the tape may be shown to students in the school only during the first ten days. During the remaining 35 days, the tape may be reviewed by teachers to determine whether a license should be sought for longer retention and subsequent showing.[132] The license may be negotiated by the educational institution directly with the network.[133] The guidelines contain some important limitations that also restrict how such videotapes can be used. For example, although such recordings need not be used in their entirety, no alteration from the original content is permitted. Moreover, a teacher cannot create or merge video teaching anthologies by physically or electronically combining or merging recordings.[134] These restrictions certainly do not take into account how advertising classes, television studies courses, etc., use video material. Although these guidelines, along with the others that accompany the Act, are not law, they have had considerable effect on universities and other nonprofit schools. The congressional endorsement all of the guidelines enjoy creates a powerful incentive to follow them.[135]

The guidelines do not cover libraries outside the nonprofit educational setting, and other libraries must either purchase tapes of television programs from producers or negotiate a license for off-the-air taping directly with the copyright holder. A network's primary concern is the illegal duplication of videotapes and loss of the rerun market. Such license agreements surely will require the payment of royalties to the copyright owner for the privilege of duplicating and retaining the program.

While the Act states that the provisions of section 108 do not permit the copying of audiovisual materials,[136] the House Report's language intimates that section 107 may be available for duplicating certain audiovisual programs.[137] These two provisions appear to conflict, but discussions of statutory interpretation and the weight to be given legislative history in interpreting a statute are not within the scope of this article. Nevertheless, while some exception to the owner's rights may exist under fair use, librarians should cautiously approach unauthorized reproduction of video materials. Videotaping for the purpose of building a tape library, while not specifically recognized in *Sony*, almost certainly indicates potential infringement. Whether taping for the purpose of research mitigates that potential is uncertain. One commentator has noted that while the *Sony* court speaks of productive use being an important factor, the taping it ultimately allows is not in itself productive; instead, it is for entertainment as opposed to scholarship and is merely the exact duplication of a complete copyrighted work.[138] Possible legislation suggested to deal with the problem of videotaping rights has ranged from compulsory licenses, which would be in the form of a tax on videotape recording equipment and blank tapes,[139] to royalty schemes.[140] Interest in enacting any such controls appears slight at the present time.

Because of the publicity this matter has received, few libraries would assume that videotaping anything for users outside of the news and that which is allowed under the videotaping guidelines for nonprofit educational institutions is permissible. Copyright owners have demonstrated through the *Sony* litigation the seriousness with which they view this issue. Under the 1909 Act, the situation regarding videotaping was much less clear. In *Encyclopedia Britannica Educational Corp. v. Crooks*,[141] a federal district court ruled that under the 1909 Copyright Act,[142] off-the-air videotaping conducted by BOCES, a school cooperative, which videotaped for 19 schools in the Buffalo, New York area was an infringement of copyright.[143] The cooperative neither obtained permission for such copying or retention of the programs nor paid royalties for the privilege of doing so.[144] After applying the fair use test, the court issued a permanent injunction to stop the copying but did not order destruction of the tapes, although this is a standard remedy provided under section 503(b) of the Act. The court declined to order erasure in order to give the parties an opportunity to meet and determine whether some type of licensing arrangement could be reached.[145] In light of the BOCES case, even libraries in nonprofit educational institutions should examine closely their collections of pre-1978 recorded videotapes to determine whether to pay for previously videotaped materials or to erase them. Clearly, post-1978 off-the-air videotaping should comply with the above discussed guidelines.

If producers of audiovisual programs are to continue to make such works available to libraries and schools, they must be afforded adequate financial compensation for the creation of copyrighted works.[146] There is a need for a fair and easily administered system to pay copyright royalties for the duplication of audiovisual material.[147] Even the expanded use of videodisc technology will not alleviate copyright problems. While videodisc is primarily a playback medium, discs now can be converted to videotape. Moreover, the possibility of central bank storage may eliminate the need to copy the material at all. Anyone with access to the central bank information system will be able to call up the material at any time. Presumably either the initial purchase price will be quite high or maintenance of the system will be financed by a per search and playback charge similar to that now in place for electronic database searches. On the other hand, some producers of videodiscs advertise them for use in multimedia works for which presumably needed portions would be reproduced.

Thus, current copyright problems surrounding videotapes will carry over into newer technology if they are not solved. More scholarly material is likely to become available in video format. For example, some universities now permit master's theses to include video presentation. Videotapes are becoming a common part of scholarly presentations at conferences, seminars and workshops. Even the most erudite and traditional special libraries eventually will come to recognize the contribution of scholarly works in video formats and include these in their collections.[148]

4. Audiovisual Packages

Increasingly, all types of libraries are engaging in the creation of in-house audiovisual packages and multimedia works. Slide and tape shows in which copyrighted slides (or slide reproductions of copyrighted photographs), poetry and music are presented simultaneously provide a good example traditional audiovisual package. An audiovisual work does more than tell a story. It is the simultaneous presentation of sights and sounds along with information that engages the senses of the audience.[149] In the production of such packages, libraries infringe all of the rights of the copyright owner — from reproduction, distribution and adaptation to performance and display. Once created, the audiovisual package itself is eligible for copyright protection, provided permission has been received from the owners of the portions used.

An audiovisual instruction program may consist of film clips, reproductions of pages from textbooks, workbooks, and so on, as well as original material along with background music. Originally of interest to schools and colleges as individualized learning "kits," their use and creation has seen a dramatic increase. Audiovisual packages have potential for library instruction as well. Instruction programs may be used to introduce users to the library, train staff members or instruct users in the use of particular research material in the library's collection. Librarians in the for-profit sector also are engaged in designing and using audiovisual packages. Since these programs are developed for repeated use, most are destined to form part of a library's permanent collection.

Section 108 contains no mention of the creation of audiovisual materials by libraries. One must, therefore, turn to other sections of the Act for guidance on the copyright status of these works. Normally, section 107 would apply and the fair use test may be used as a guide. Outside of fair use, permission should be sought from the copyright owner for copyrighted materials reproduced and included in audiovisual packages. The problem often is compounded if the library duplicates the program to allow simultaneous use by more than one user.[150] As audiovisual packages become increasingly common, libraries must learn to cope with the myriad of copyright problems that will in all probability attend their uses.

5. Multimedia Works

The advent of computer technology has made possible the creation of multimedia works, i.e., works that combine text, art works, clips from videotapes and sound all stored digitally. Such works often are embodied in compact discs. The software embedded in the CD can be used to manipulate the images as well as to search and retrieve information from the CD. Under the rules of the Copyright Office regarding the registration of audiovisual works, the software commissioned for the multimedia work can be registered as a part of the audiovisual work.[151] There also is software available on the market to assist in creating multimedia which is copyrighted separately but for which the purchaser of the software automatically receives a license to use to create multimedia works and to store the program on a disc or other format for the media. An example is Apple's Quicktime.

Just because there are software packages that can be purchased to use in developing multimedia does not mean that there are no copyright problems. In fact, despite the fact that technical problems with creating multimedia works continue to be solved, so far, the copyright issues have not.[152]

Corporate media producers, educational institutions and libraries of all types are poised to create multimedia works which will have a variety of training applications.[153] Students will create class projects in multimedia format, and even master's theses and dissertations are likely to consist entirely or in part of a student-created multimedia work. What is likely to slow the development of such works is the variety of permissions that must be obtained for any commercial use of a multimedia work. The principles of fair use will continue to apply, and should a librarian in a nonprofit library create a multimedia training program for use in her library, the use probably is a fair use and she would not have to obtain permission from each copyright holder to reproduce and display their works through a multimedia work for a nonprofit use. Should she start to market the multimedia training program, however, then the use has become commercial and would no longer be a fair use. At this point, permission must be obtained from each copyright holder whose work is reproduced and displayed in the resulting multimedia work.

Obtaining permission is difficult at best. Permissions for the various kinds of materials combined to create the multimedia work are negotiated individually. Further, they are negotiated differently for the various types of works such as photographs, music, video clips and text.[154] The easiest right to obtain probably is for text because librarians are more familiar with how to obtain such permission. The next easiest is for music, but traditional license fees may be so high as to preclude the use of the music in multimedia works. Moreover, different uses of music require different licenses, and as many as four separate music licenses could be required. First, a "mechanical license" is required if the user wants to produce and distribute physical objects in which the musical composition is embodied.[155] Second, a "synchronization license" is needed in order to synchronize the playing of the musical composition with visual images. The normal fees for a "sync license" are based on how much time the musical work will be played. Clearly, the traditional method for calculating sync fees will not work for multimedia works where the amount of time and the frequency with which a work is played will not be controlled by the producer of the multimedia work but by the user of that work. Third, a "performance license" is needed should the CD become a public performance such as by projecting the work onto a screen for a group outside of the classroom exemption for nonprofit educational institutions or the normal circle of family and friends. Lastly, a "master recording" license is used when someone wants to reproduce and distribute a specific performance of a musical composition by a particular recording artist.[156]

Obtaining permission to use photographs included in a multimedia work is likewise complicated. Presently, photographers charge $75 to $100 for the rights to use a single image in minor use (such as in a text to illustrate a particular point). In a multimedia work, a producer might well want to use 500 images which would result in $37,500 to $50,000 in fees. Even if the photographer reduced the fee to $10 per image, the fee would be $5,000.[157] Some of the difficulties in obtaining permission to use photographs can be obviated by using a stock photographic house. These businesses have obtained from photographers a large collection of photographs along with the right to license the use of those photographs to third parties.[158] At least one copyright collective has been planned. It is Benn's Picture Network International which will combine quality photographs, stock picture agencies and picture databases. It is a joint venture with a computer applications company and photographer Nathan Benn, a long-time contributor to *National Geographic*. Because of the recent interest in the use of photographs in multimedia works, the Society of Magazine Photographers is considering the possibility of creating a collective to handle rights for electronic media.[159] Needless to say, these are promising developments for multimedia producers.

The rights to use clips from video and movies also may be obtained from stock film companies and other institutions that maintain large collections of these works. Clips may contain music, voice or image of a performer who is a member of a union or guild, and additional permissions may be required from both the artist and the organization.[160] Film and television producers are not set up to deal with a huge number of permission requests, nor do they have any collective mechanism for handling rights such as the Copyright Clearance Center for photocopying. Fees for film and video clips normally are set for each minute of the clip used or fraction thereof, and the fees can quickly run into thousands of dollars.[161]

Clearly, before multimedia works can be developed and distributed widely, copyright holders must change their permissions and licensing practices. None of the above discussion takes into account the difficulties in and costs associated with locating copyright holders for some of these works. Using public domain materials certainly seems easier today! On the other hand, in order to capture the attention and imagination of likely audiences, more popular copyrighted music may be needed.

V. PERFORMANCE OF AUDIOVISUAL AND MUSICAL WORKS IN LIBRARIES

A. General

The most common copyright problem libraries encounter is the need to duplicate copyrighted works and this relates to the copyright holder's exclusive right of reproduction. Another of the five rights the statute affords copyright owners is the right to perform their works publicly.[162] Any public performance without the owner's permission may subject the library to copyright liability. Even though reproduction of an audiovisual or musical work by a library may be exempted under section 107 or 108, any exemption for reproduction does not carry with it an exemption for the performance right. For many performances within a library, royalties must be paid to the copyright owner.

Works that are performed privately for an individual library user is a private performance and does not infringe the owner's rights, however. Thus, a musical recording or videotape checked out by a user and played in a carrel is not a public performance. Nor is the circulation of records or videotapes to patrons who take them away from the library and play them at home for their own enjoyment or study. Works performed for very small groups may not be public performances, but clearly, as the size of the group increases, the greater the possibility the performance has become a public performance.

A public performance is defined as a performance outside the normal circle of one's family and friends.[163] Because a showing to a group in a library never really satisfies this limitation, the precise point at which a performance becomes a public one is not known.

B. The Classroom Exemption

Section 110 of the Act exempts certain performances and displays of copyrighted works if specific conditions are met.[164] The most common exemption is for performances and displays of a nondramatic literary or musical work in the regular course of instruction in a nonprofit educational institution. Even within a school or college, further requirements must be met. The performance must take place in a broadly defined classroom; libraries meet the definition if instruction routinely takes place in the library, and it does in most schools and universities. Teachers and students must be present in the same place, and the performance must be a part of the regular instruction.[165] In other words, it must not be a performance purely for entertainment purposes. Another important requirement is that the copy that is performed must be a lawfully obtained copy. In fact, the exemption is lost if the librarian knew or should have known that the copy was not lawfully obtained. When these requirements are met, students and teacher may sing a copyrighted song, read a copyrighted poem or perform a copyrighted play. Additionally, they may view the performance of a copyrighted motion picture or audiovisual work.[166]

Although many people believed that nonprofit libraries were nonprofit educational institutions, amendments since 1984 make it clear that they are not. Libraries in nonprofit educational institutions are covered by the exemption if the above mentioned conditions are met. Public libraries are not.[167]

As a general rule, a library which purchases a copyrighted motion picture does not obtain a performance right unless right is separately purchased. There are companies that sell copies of educational films and motion pictures to schools and libraries on which the performance right is either included in the purchase price or is available for an additional price. Films, Incorporated[168] is such an organization; using such a company is a particular benefit to libraries that are beset with requests by civic groups to loan films to them. The duplication right also may be purchased separately from Films, Incorporated.[169] It is clear that libraries which rent a videotape from the local video store do not have the right to hold a public performance since videos rented from a store are intended for home use only, i.e., a private performance.[170] In fact, most tapes rented at a retail video rental store carry a warning which appears on the screen that the tape is for home use only. Annual umbrella licenses are available to all types of libraries including those in for-profit companies for public performances in the library of home-use only videocassettes.[171]

For libraries in nonprofit educational institutions there is some confusion about whether purchased copies of motion pictures that carry a home-use only warning may be performed in the school if the section 110 requirements are met. The American Library Association takes the position that even a videotape which contains a home-use only warning may be performed in a nonprofit educational institution.[172] The Media Producers Association and other copyright holders disagree with ALA's interpretation.[173] Because this issue is not clear, librarians concerned about such performances should seek advice of legal counsel, but it appears today that the weight of authority from copyright scholars is that rented videotapes performed in a classroom as a part of instruction in a nonprofit educational institution is exempted and no performance royalties must be paid. Performances for entertainment or those in the evenings sponsored by student groups are not eligible for the section 110 classroom exemption.

Performances and displays of a nondramatic literary or musical work may be exempted even in the course of transmission if certain conditions are met.[174] First, the performance or display must be a "regular part of the systematic instructional activities of a government body or a nonprofit educational institution."[175] Second, the performance or display must be directly related to the teaching content of the transmission and be of material assistance to it. Third, the transmission must be made primarily for (a) reception in a classroom or other place usually devoted to instruction, (b) reception by disabled persons whose disabilities prevent their attendance in a classroom, or (c) reception by government employees as a normal part of their duties.[176] Note that this exemption does not include the performance of videotapes.

Libraries which have musical performances, whether by live musicians or by playing recordings for audiences, likely should seek permission for the performance or must pay royalties. Even so-called background music may be a public performance and subject the library to liability for the payment of royalties.[177] Libraries in nonprofit institutions may qualify for another exemption under section 110. If there is no payment of fees to either performers or promoters, and if the performance is free, it may be exempted. Even if there is an admission charge, if monies raised go back into nonprofit purposes, the performance may be exempted.[178] The owner of the copyright in the work to be performed may stop the performance, however, if he files notice on the library within seven days before the performance.[179]

Projection of sound or images beyond the place where the copy is maintained presents yet another problem. While there is a limited exemption for instructional broadcasting,[180] the playing of music through an intercom or other system of mass distribution loses the section 110 exemption for the library. Such performances clearly indicate that the library should consider an ASCAP license or subscribe to a service such as Musak through which royalties are automatically paid on copyrighted music. Few libraries are actually likely to use a "home-type receiving set" to receive radio or television broadcasts and play them in the library.[181]

VI. CONCLUSION

Prior to the advent of the printing press, there was no need for strictly enforced copyright laws, for it was not possible to produce copies of works in such numbers as to harm the copyright proprietor economically. When print became the medium for rapid production of copies of works expressing ideas, notions and aesthetic values, copyright and later fair use as an exception were developed to create a balance between the rights of the creator of the work and the rights of the public. The technological era has raised serious questions concerning the applicability of traditional fair use principles.[182]

A prime example of the different problems presented by print versus audiovisual material occurs in the application of one of the fair use factors — the substantiality of use. A scholar can excerpt portions of literary works and still get a feel for the author's style and taste. On the other hand, for materials such as paintings, photographs, drawings and audiovisual works, the user often must copy the entire work in order to sample the creator's style or taste.[183]

One scholar has proposed a new fair use test for audiovisual works. Instead of the traditional section 107 four-pronged analysis, the following should be substituted: (a) Does the reproducer belong to the class of persons engaged in the advancement of science, arts and industry or in the dissemination of information and ideas? (b) Is the purpose intended to advance science, arts and industry or the dissemination of information? (c) Will the intended use of the copy affect the potential market for or value of the copied work? (d) Is the use made of the material likely to produce substantial profits? If the answer to these four questions is yes, then royalties should be paid to the copyright holder. If the answer is no, then the reproducer may copy without incurring liability for royalties.[184]

In the 1983 report on section 108 by the Register of Copyrights the applicability of the fair use doctrine to libraries was of concern. The Report did not address the problems of reproduction and second-order technology or how fair use is applied in this modern era. Rather, it states that section 108 permits copying not otherwise permitted under section 107 and that fair use does not permit broad copying once the limits of section 108 have been reached.[185] The Register's Report five years later is silent on this issue.[186]

Does this mean that the Register currently considers section 107 inapplicable to *library* copying of copyrighted audiovisual, pictorial and graphic works? Are libraries to be strictly limited to section 108, which allows reproduction of such works only under the narrowest circumstances? The earlier Report states that a user's rights accrue under section 107.[187] Should a library circulate copyrighted audiovisual works to users and encourage them to make their own copies for section 107 purposes? This seems unrealistic given the cost of these works and the expense of duplication equipment as contrasted to the simplicity of the coin-operated photocopier for printed materials. It also seems naive in light of what is actually happening in the world of technology. More materials are becoming available in audiovisual formats, and demand for the information contained in these works does not necessarily change based on the format in which works are published.

All of this philosophy is predicated on user demand; libraries are not duplicating materials for their own purposes but in response to user requests. There is no commercial gain. Further, librarians tend to think of these works as information while Congress couples them with motion pictures. The difficulties that performance rights issues unnecessarily make libraries treat these works as separate and apart from other informational works.

The rights granted in section 108 of the Act apply to all types of libraries that meet the section 108(a) provisions. This includes special libraries, whether the parent organization is in the for-profit sector or not. All libraries engage in similar activities to meet the information needs of their respective user groups. The user of a public library is not required to state the purpose for information requests and any library may serve a user from a for-profit organization. Academic libraries, while primarily serving the academic community, also provide information and copies of copyrighted works to users for their work in the for-profit sector. The library charges nothing for disseminating the information, but the end user may be engaged in profit-making activities. The same is true of public libraries. The distinction whether a library is public in nature or attached to a for-profit organization is, therefore, not as clear as some might think. The *Texaco* litigation further highlights the difficulties of this distinction.

The reproduction of copyrighted audiovisual and nonprint media by libraries presents some unique problems. Clearly, more guidance is needed to assist librarians in applying the law to particular situations. Library associations and media producers should continue to work together to develop nonstatutory, voluntary guidelines aimed at assisting libraries to fulfill their missions while assuring fair returns to media producers. Some librarians believe that a new section should be added to Act to grant to owners of a copy of audiovisual works the same rights that exist for owners of a copy of a computer program.[188]

ENDNOTES - CHAPTER 5

1   17 U.S.C. § 101 (1988 & Supp. III 1990).
2   17 U.S.C. § 101 (1988).
3   *Id*. (Supp. III 1990).
4   17 U.S.C. § 101 (1988).
5   *Id*.
6   Architectural Works Copyright Protection Act, Pub. L. 101-650, 104 Stat. 5089, 5133 (1990) (codified at various sections 17 U.S.C.).
7   17 U.S.C. § 101 (Supp III 1990).
8   17 U.S.C. § 108 (1988).
9   *Id*. § 108(h).
10   *Id*.
11   *Id*. § 108(b).
12   *Id*. § 108(c).
13   See Chapter 3 for a detailed discussion of section 108.
14   Id. § 108(f)(4).
15   S. Rep. No. 473, 94th Cong., 1st Sess. (1975) *reprinted* in 13 *Omnibus Copyright Revision Legislative History* 67 (1977) [hereinafter Senate Report].
16   *Id*. at 64.
17   H.R. Rep. No. 1476, 94th Cong., 2d Sess. (1976) *reprinted* in 17 *Omnibus Copyright Revision Legislative History* 70-72 (1977) [hereinafter House Report]. The guidelines are reproduced in Appendix 4.
18   *Id*. at 71.
19   *Id*.
20   464 U.S. 417, *reh'g denied*, 465 U.S. 1112 (1984).
21   For the earlier holding, *see* Universal City Studios, Inc. v. Sony Corporation of Am., 659 F.2d 963, 974-77 (9th Cir. 1981).
22   464 U.S. at 434. The respondent, Universal Studio, did not seek relief against the individuals taping at home, but only against the manufacturer of the video recorders.
23   *Id*. at 448.
24   *Id*. at 448 quoting from the House Report, *supra* note 22, at 65. "Although the courts have considered and ruled upon the fair use doctrine over and over again, no real definition of the concept has ever emerged. Indeed, since the doctrine is an equitable rule of reason, no generally applicable definition is possible, and each case raising the question must be decided upon its own facts." *Id*. American Geophysical Union v. Texaco, Inc., 802 F. Supp. 1 (S.D.N.Y. 1992).
25   *Id*. at 443-56.
26   *Id*. at 449.
27   *Id*. at 446-47, n.28.
28   *Texaco* holds that a research scientist who photocopies articles for his own research for the company is not a fair use. One of the principal reasons given for the holding is the profit-seeking nature of Texaco. *Id*. at 15-16. The case currently is on appeal to the U.S. Court of Appeals for the Second Circuit. See Chapter 3 for a discussion of *Texaco*.
29   *Id*. at 457-500.
30   For an analysis of the fair use test as used in the Ninth Circuit holding in *Universal Studios v. Sony, see* Laura N. Gasaway, *Audiovisual Material and Copyright in Special Libraries*, 74 Special Libr. 222 (1983).
31   464 U.S. at 478-79.

32   *Id.* at 496.

33   The dissent acknowledge that scholarly use of entertainment media is a productive use. *Id.* at 478-79.

34   *Texaco*, 802 F. Supp. at 13-14.

35   *Sony*, 464, U.S. at 497.

36   House Report, *supra* note 17, at 71-72 (1975).

37   464 U.S. at 498.

38   House Report, *supra* note 17, at 75.

39   *Texaco*, 802 F. Supp. 17-21.

40   Senate Report, *supra* note 15, at 63.

41   *Id.* at 66.

42   Audio Home Recording Act of 1992, Pub. L. 102-563, 106 Stat. 4237 (codified at 17 U.S.C. §§ 1001-10 (Supp. IV 1992)).

43   *Id.*

44   *Id.*

45   17 U.S.C. §§ 1-216 (1970) (repealed 1978).

46   Ethan M. Katsh, *The Electronic Media and the Transformation of Law* 172 (1989).

47   *See id.* at 176.

48   Pierre N. Leval, *Toward a Fair Use Standard*, 103 Harv. L. Rev. 110, 140 (1990).

49   Note: *Toward a Unified Theory of Copyright Infringement for an Advanced Technological Era*, 96 Harv. L. Rev. 450, 458 (1982).

50   Sigmund Timburg, *A Modernized Fair Use Code for the Electronic Age As Well As the Guttenberg Age*, 75 Nw. U.L. Rev. 193, 221-22 (1980).

51   *Id.* at 223.

52   17 U.S.C. § 108(h) (1988).

53   House Report, *supra* note 17, at 71.

54   *Id.*

55   *Id.*

56   Senate Report, *supra* note 15, at 67.

57   House Report, *supra* note 17, at 78.

58   *Id.*

59   17 U.S.C. § 202 (1988).

60   *Id.* § 201(b). A work for hire is defined as "a work prepared by an employee within the scope of his or her employment, or a work specifically ordered or commissioned." *Id.* § 101.

61   *See* House Report, *supra* note 17, at 68.

62   17 U.S.C. § 108(b) (1988).

63   *Id.* § 108(c).

64   *Id.* § 108(h).

65   *Id.* § 110(1).

66   House Report, *supra* note 17, at 81-82.

67   17 U.S.C. § 303 (1988).

68   *Id.* § 102(5) (1988).

69   *Id.* § 110(1); Senate Report, *supra* note 15, at 75.

70   Jan Magnuson, *Duplicating AV Materials Legally*, 13 Media & Methods 52 (1977). The author reports a survey of journal publishers regarding duplication of illustrations in slide format.

71   Interestingly, the copyright holder determines the format in which a work may be reproduced and distributed. Conversion from one format (printed book) to slides infringes the right to prepare derivative works. 17 U.S.C. § 106(2) (1988).

72   House Report, *supra* note 17, at 72-73.

73   *See* text accompanying note 75.

74   Melinda V. Ogle, *Not by Books Alone: Library Copying of Nonprint Copyrighted Material*, 70 Law Libr. J. 153, 159 (1977) [hereinafter Ogle].

75   See Appendix H(2) for a sample permission letter concerning video or audiocassette tapes.

76   *Id.*

77   360 F. Supp. 821 (E.D.N.Y. 1973).

78   *Id.* at 825.

79   *Id.* at 821-23.

80   *Id.* at 824.

81   Telephone interview with Michael E. Rayburn, Media Consultant, Norman Independent Schools (December 6, 1982).

82   360 F. Supp. 821, 824 (E.D.N.Y. 1973).

83   594 F. Supp. 335 (S.D.N.Y. 1984).

84   *Id.*

85   Audio Home Recording Act of 1992, Pub. L. 102-563, 106 Stat. 4237 (codified at 17 U.S.C. §§ 1001-10 (Supp. IV 1992)).

86   *Id.* §§ 1003-04. For a discussion of DAT issues *see* Michael Plumleigh, *Digital Audio Tape: New Fuel Stokes the Smoldering Home Taping Fires*, 37 U.C.L.A. L. Rev. 733, 745-48 (1990) [hereinafter Plumleigh].

87   American Association of School Libraries and American Library Association, *Copyright, Media and the School Librarian*, at D (1978).

88   Ogle, *supra* note 74, at 158.

89   *See* Record Rental Amendment of 1984, Pub. L. No. 98-450, 98 Stat. 1717 (codified at 17 U.S.C. § 109(b)(1)(A) (1988)).

90   17 U.S.C. § 109(a) (1988).

91   *Id.* § 109(b)(1)(A).

92   *Id.* § 115(c)(3).

93   *Id.* § 109(b)(1)(A).

94   The Special Librarics Association pressed Congress to use the section 108(a) definition of libraries in the Record Rental Amendment rather than adopting the language that differentiates the types of libraries on the basis of profit or nonprofit status of the parent institution. The effort gained support among library associations but Congress was not persuaded.

95   17 U.S.C. § 506(a) (1988).

96   *See* Gerald Mast, *Film Study and Copyright Law*, in John Sheldon Lawrence & Bernard Timberg, *Fair Use and Free Inquiry* 72 (1980) for a good explanation of film study and copyright problems [hereinafter Mast].

97   U.S. Copyright Office, *Circular R21: Reproduction of Copyrighted Materials by Educators and Librarians* (1992).

98   House Report, *supra* note 17, at 71.

99   *See* text accompanying notes 148-60.

100   Mark Johnson, *The New Copyright Law: Its Impact on Film and Video Education*, 32 J. Univ. Film Ass'n. 67, 68 (1980) [hereinafter Johnson]. In the furor over artistic control of films, however, filmakers are becoming increasingly involved in legislation. *See* Report of the Register of Copyrights, *Technological Alterations to Motion Pictures and Other Audiovisual Works...* 10 Loyola Ent. L.J. 1, 90-95 (1990).

101   Johnson, *supra* note 100.

102   *See e.g.*, Ronald J. Ostorw, *FBI Official to Take on Film Pirates*, L.A. Times, Oct. 2, 1991, Part F, at 2(3); Robert Trautman, "U.S. Cites China, India, Thailand ...," Reuters Business Report, April 26, 1991.

103   Jerome K. Miller, *Applying the New Copyright Law: A Guide for Educators and Librarians* 55 (1979) [hereinafter Miller].

104 Mast, *supra* note 96, at 85.

105 Robert Churchill, *Golden Egg Production: The Goose Cries "Foul,"* in John Sheldon Lawrence & Bernard Timberg, *Fair Use and Free Inquiry* 169, 169-70 (1980).

106 *Report of Register of Copyrights, Technological Alterations to Motion Pictures and Other Audiovisual Works...* 10 Loyola Ent. L.J. 1, 46-47 (1990) [hereinafter Loyola].

107 *Id.* at 47.

108 Miller, *supra* note 103, at 52.

109 Loyola, *supra* note 106, at 111.

110 *See* National Film Preservation Act of 1988, Pub. L. No. 100-446, 1988 U.S.C.C.A.N. (102 Stat. 1782)). This act was a negative response to the colorization of films, indicating a desire to preserve films in their original forms as national treasures.

111 Loyola, *supra* note 106, at 111.

112 *Id.*

113 17 U.S.C. § 101 (1988).

114 Franca Klaver, *The Legal Problems of Video Cassettes and Audiovisual Discs*, 23 Bull. Copyright Soc'y 152, 155 (1976).

115 *Id.* at 155, 167.

116 *See* Beda Johnson, *How To Acquire Legal Copies of Video Programs; Resource Information* (6th rev. ed. 1993). *See also* text accompanying notes 162-81 for a discussion of performance rights.

117 Plumleigh, *supra* note 86, at 737.

118 Arnold J. Holland, *The Audiovisual Package: Handle with Care*, 23 Bull. Copyright Soc'y 104, 124 (1974).

119 If the company has contracted with an outside party to make an in-house presentation, there should be a written agreement that does two things: (a) grants permission from the speaker to the organization to tape the presentation and (b) specifies who will own the rights to the videotape.

120 *Joint Senate Panels Hear Support for Penalizing Makers of Video Code Breakers*, Bureau of National Affairs Daily Report for Executives, July 25, 1991 [hereinafter Joint Panels].

121 S. 1096, 102d Cong., 1st Sess.; H.R. 2369, 102d Cong., 1st Sess. (1991).

122 *See* Joint Panels, *supra* note 120.

123 See Appendix H for a sample letter that includes a format conversion request.

124 17 U.S.C. § 108(h) (1988). For a history of the development of this exception, *see* Ogle, *supra* note 74, at 55.

125 *Id.* § 407(e).

126 Douglas Kellner, *Television Research and Fair Use*, in John Sheldon Lawrence & Bernard Timberg, *Fair Use and Free Inquiry* 90, 96 (1980).

127 S. 1805, 102d Cong., 1st Sess. (1991).

128 Jon T. Powell & Wall Gair, *Public Interest and the Business of Broadcasting* 10 (1988).

129 *Federal Guidelines for Off-The-Air Recording Broadcast Programming for Educational Purposes*, Cong. Rec. E4750-52 (1981) [hereinafter Videotaping Guidelines].

130 Esther R. Sinofshy, *Off-Air Videotaping in Education* 99 (1984).

131 17 U.S.C. § 118(d)(3) (1988).

132 Videotaping Guidelines, *supra* note 129.

133 The Television Licensing Center (TLC) was a national clearinghouse designed to provide educators with information about off-the-air videotaping and with licenses to record, duplicate and retain television programs. It closed in the late 1980's probably because of the availability on videotape so many quality television programs and the fact that the cost of such programs becoming lower.

134 Videotape Guidelines, *supra* note 129.

135 Kenneth D. Crews, *Copyright, Fair Use and the Challenge for Universities; Promoting the Progress of Higher Education* 180 (1993).

136 17 U.S.C. § 108(h) (1988).

137 House Report, *supra* note 22, at 78-79.

138 Vincent F. Aiello, *Note: Educating Sony: Requiem for a "Fair Use,"* 22 Cal. W.L. Rev. 159, 172 (1985).

139 For example, H.R. 5705, 97th Cong., 2d Sess. (1982).

140 Compulsory licenses already exist for cable television systems, 17 U.S.C. § 111(d) (1988); juke box royalties, *id* § 116(b); mechanical license fees for phonographs, *id.* § 115(a); for public broadcasting, *id.* § 118(b); and for satellite transmission, *id.* § 119(a).

141 542 F. Supp. 1156 (W.D.N.Y. 1982).

142 17 U.S.C. §§ 1-215 (1970) (repealed 1978).

143 542 F. Supp. 1156, 1185-87 (W.D.N.Y. 1982).

144 *Id.* at 1159.

145 *Id.* at 1187-88.

146 There is no organization similar to the Copyright Clearance Center (CCC) for collecting and distributing royalties for the copying of audiovisual works. The CCC was established for the purpose of collecting and distributing royalties for photocopies made in excess of fair use. See Chapter 4.

147 Daniel Callison, *Fair Payment for Fair Use in Future Information Technology Systems*, Educ. Tech., Jan. 1981, at 20, 24.

148 Bernard Timberg, *New Forms of Media and the Challenge to Copyright Law*, in John Sheldon Lawrence & Bernard Timberg, *Fair Use and Free Inquiry* 247, 256 (1980).

149 Bayard F. Berman & Joel E. Boxer, *Copyright Infringement of Audiovisual Works and Characters*, 52 S. Cal. L. Rev. 315, 324 (1979).

150 Jerome K. Miller, *The Duplication of Audiovisual Materials in Libraries*, in John Sheldon Lawrence & Bernard Timberg, *Fair Use and Free Inquiry* 128, 135 (1980).

151 William A. Tanenbaum & William K. Wells, Jr., *Multimedia Works Require Broad Protection*, 16 Nat'l L.J., Nov. 1, 1991, at S11 [hereinafter Tanenbaum]. Registration regulations regarding audiovisual works that embody a computer program are found at 37 C.F.R. § 202.20(c)(2)(xix) (1993).

152 Barbara Zimmerman, *The Trouble with Multimedia: Copyright Clearance and the Uncertain Future*, AV-Video, Jan., 1993, at 46 [hereinafter Zimmerman].

153 *Id.*

154 *Id.*

155 Tanenbaum, *supra* note 151, at S12. The mechanical license is a compulsory license authorized in 17 U.S.C. § 115 (1988). The mechanical license provides that if a composer either makes a sound recording of his copyrighted musical composition or permits someone else to do so, then he must permit everyone else to make a recording of the song. Along with the right to record the song, a performer also gets a limited adaptation right to prepare a musical arrangement of the work for use on that record. The recording artist then must pay a compulsory license fee to the composer based on the number of phonorecords sold. *Id.* § 115(3)-(4).

156 Tanenbaum, *supra* note 151, at S12. *See also* text accompanying notes 161-80 and chapter 4 for a discussion of performance rights and licenses.

157 Zimmerman, *supra* note 152, at 47.

158 Tanenbaum, *supra* note 151, at S12.

159 Zimmerman, *supra* note 152, at 48.

160 Tanenbaum, *supra* note 151, at S12.

161 Zimmerman, *supra* note 152, at 48.

162 17 U.S.C. § 106(4) (1988).

163 *Id.* § 101.

164 *Id.* § 110.

165 *Id.* § 110(1).

166 *Id.*

167 *See id.* § 109(b)(1)(A) (Supp. III 1990).

168 Films, Incorporated, 5547 N. Ravenswood Avenue, Chicago, IL, 60640-1199. Telephone: 800-323-4222, Est. 43.

169 *Films, Inc. Corporated Video: 1990 Comprehensive Film and Video Catalog* 176 (1990).

170 Mary Hutchings Reed, *The Copyright Primer for Librarians and Educators* 32-35 (1987) [hereinafter Reed]; and Ohio Attorney General Opinion in Copyright L. Rep. (CCH) 26, 240 (1987).

171 For information on an umbrella license for performances contact the Motion Picture Licensing Corporation, 209 Dunn Avenue, PO Box 3838, Stamford, CT, 06905-0838 (1-800-968-8855) or *Films Inc., supra* note 167.

172 Reed, *supra* note 170.

173 *Action Exchange,* 23 Am. Libr. 47 (1992).

174 17 U.S.C. § 110(2) (1988).

175 *Id.* § 110(2)(B).

176 *Id.* § 110(2)(A)-(C).

177 Robert A. Wyunbrandt, *Musical Performances in Libraries: Is a License from ASCAP Required?,* 24 Pub. Libr. 224, 224-25 (1990).

178 17 U.S.C. § 110 (1988).

179 *Id.* § 110(4)(B).

180 *Id.* § 110(2).

181 *Id.* § 100(5).

182 Sigmund Timberg, *A Modernized Fair Use Code for Visual, Auditory and Audiovisual Copyrights: Economic Context, Legal issues and the Laocoon Shortfall,* in John Sheldon Lawrence & Bernard Timberg, *Fair Use and Free Inquiry* 311, 316 (1980).

183 *Id.* at 318-19.

184 *Id.* at 324.

185 Report of the Register of Copyrights, *Library Reproduction of Copyrighted Works* (17 U.S.C. § 108) 97-102 (1983) [hereinafter 1983 Report].

186 Report of the Register of Copyrights, *Library Reproduction of Copyrighted Works* (17 U.S.C. § 108) 97-102 (1988).

187 1983 Report, *supra* note 185, at xii.

188 See Chapter 6 for a discussion of these rights for software.

CHAPTER 6

COMPUTERS, SOFTWARE, DATABASES AND COPYRIGHT

I. INTRODUCTION

A. General

Soon after the introduction of computers in industry, librarians recognized the benefits of automation to improve library operations. Because libraries are also businesses, functions such as accounting, personnel record maintenance, and the like were natural for migration to the computer. Technical services operations, indexing and abstracting were the first library-specific functions that were extensively automated. In the early days, library software products tended to be developed in-house since existing commercial application software was expensive and required tremendous computer capacity. Some of these products were developed for one specific library, but the market demanded software with broader application than one particular library. So library application software was developed and marketed commercially. Today, most library application software is available commercially. Librarians know the sources for these products and are willing to pay the high annual software maintenance fees in order to benefit from software upgrades, organized users' groups, etc. All librarians probably realize that such library application software is copyrighted and may not be duplicated unless so permitted by the terms of the license agreement. Any need to reproduce, alter or transfer the software should be handled directly with the vendor. This chapter does not discuss library software that manages circulation, cataloging, etc., separately from other types of software.

With the advent of personal computers, a plethora of software for business and home applications developed. Libraries purchase a variety of software for their own uses (such as word processing and spreadsheet software) and for circulation to library users. Libraries may lend software packages directly or may make them available to users through a local area network within the library.

Librarians must also be concerned with the relatively recent development of books that contain computer diskettes within the binder. Frequently, advertisements for the book title do not indicate that it will be accompanied by a disk, which may contain a program application such as forms or a database, or both.

Libraries also access a huge number of electronic databases to obtain abstracts or full-text copies of works, to answer reference inquiries and for access to bibliographic records. How this information may be used and by whom are important concerns in this area.

A final issue dealing with the use of computers is electronic publishing. After nearly two decades of predictions that electronic publishing was the wave of the future, there finally are some true electronically published journals, i.e., ones that meet the general requirements to qualify as a scholarly journal (refereed, etc.) and which are available only electronically. Although now there are such journals published, many questions remain unanswered about their copyright status, use by nonsubscribers, availability to fill interlibrary loan requests, etc.

License agreements govern many of the activities that a user of one of these computer-related works may conduct. License agreements are contracts between the owner of a copyrighted work or a vendor such as Dialog and a user of the work. Contracts are governed by state law. The terms of the license agreement may either broaden or narrow the rights a user has under the Copyright Act. Such agreements usually specify restrictions on the user's rights to copy the software, to access the database, and to download information, or it may specify what constitutes legitimate uses of information obtained from the database. Rights at the expiration of the agreement also are normally included in the terms of the contract.

B. The Problem

Computer have driven the cost of making copies virtually to nothing or at least to the minimal price of a blank diskette. A person who wishes to make a copy of a diskette will spend a few seconds and less than a dollar to obtain a perfect copy of the original disk. In this sense, computer technology is the latest development in the trend that has driven down the price of making copies: photocopying is cheaper than typesetting, which in turn is superior to actual transcription.

As it became easier to copy original works, the legal protection afforded to authors expanded as the term of copyright was lengthened. It is thought that the costs of illicit reproduction, activity that is likely to escalate as it becomes cheaper and less tedious, act as a disincentive to producers of original works. To counterbalance this effect, Congress has periodically increased the incentive to produce by increasing the duration and extent of copyright protection.[1]

In pushing the process of copying to a kind of economic limit, computers have created major problems for the legal protection of intellectual labors. In fact, it is questionable whether copyright protection is even adequate to the task. Various alternative forms of legal protection have been proposed, including patents, trade secrets, contracts and license agreements as well as hybrid forms of protection that adopt portions of the existing protection regimes to create a new type of protection for computer software.

The dominant form of protection remains copyright, however, and this creates special problems for libraries. Unlike European law, which has long recognized moral rights for authors, American copyright law ultimately aims to serve the public. The Constitution is explicit on this point: authors have the "exclusive right" to their labors so that "science and useful arts" will advance.[2] The tension between the limited monopoly the author receives and the interest of the public in her work is clear: if the monopoly were total the public would be legally blocked from making use of the work and thus would derive no benefit from it. On the other hand, if a prospective author stood to gain no rights in a work, he would lack the incentive to produce it.[3] Libraries occupy a curious place in this scheme. As repositories and disseminators of knowledge and information, libraries may attract those who would pirate the intellectual fruits of others. Because libraries now use and loan easily reproducible software, their roles are even more delicate in the computer arena than in the more traditional photocopying areas.

This chapter examines copyright law as it applies to software, electronic databases and electronically published information in libraries. The chapter examines the issue of software "shrinkwrap" or "boxtop" licenses, which today are all but ubiquitous. Throughout the chapter there is practical advice concerning methods libraries can employ to minimize their exposure to legal liability. The copyright status of software, electronic databases and electronic journals, license agreements and the rights of libraries as licensees to use, reproduce and distribute these works is discussed in this chapter along with the rights a library patron has to make of these materials.[4]

II. LIBRARY PRACTICES

A. General

Because computer technology is expanding so rapidly, any description of current library practices is outdated almost as soon as it is written. Nonetheless, it is possible to highlight basic practices relating to electronic media and to pose a series of questions relating to the use of computers and computer-related materials by libraries.

Libraries utilize some computer programs as an end user, for example budget management software. They also purchase software to circulate to users just as they acquire books, journals, videotapes and other materials. Libraries access these databases and utilize the information obtained to answer reference questions, for collection development, for cataloging and to create bibliographies for users. Electronic journal subscriptions are used just as other journal subscriptions. Users can print individual articles and other data located in the database or download them to a disk.

B. Why Is This Area Important To Special Librarians?

Consider these scenarios:

■ A library purchases a network version of Microsoft Word to load on the library's local area network (LAN) for in-house library use. May a library patron use the word processing program on the network?

■ A library has Quattro Pro loaded on the LAN available to library users. May a patron bring in a blank disk and copy the program for his own personal use?

■ A one-volume book arrived in the library with a computer disk to accompany it. May the library make a backup copy of the disk?

■ In order to utilize a piece of expensive software purchased by the library, a librarian must load it onto the hard disk of her personal computer. Is this permitted under the law?

■ Must the library invest in the more expensive network version of software in order to load it onto the computer laboratory's network?

■ The library has purchased two copies of a software upgrade. The Acquisitions Department asks to use the old version since the administrative office will be using the upgrade. Should the library permit this use?

■ A library patron wants to donate several computer disks that contain copies of freeware. May the library accept the gift?

■ May a reference librarian download bibliographic citations from a variety of databases to create a bibliography which will become a section of the library's monthly newsletter?

■ Users have access to the library's LAN and can use all programs loaded on the LAN. Should the library check the network periodically to ensure that no patron has added other software to the network?

■ After a complicated computer database search, the librarian retains copies of both the results and his search strategy. May he reuse either or both?

■ The library purchases a copy of software. May it modify the program for its own internal use?

■ A library staff member has requested permission to duplicate a copy of a program owned by the library to use on her home computer so she may work at home. Should the library approve this request?

■ Instead of producing a printout to record the results of a computer search, a librarian downloads the results to a disk. May he give the results to the user in this format?

For answers to these questions, see Appendix B.

III. COMPUTER SOFTWARE USE BY LIBRARIES

A. What Is Computer Software?

The Copyright Act defines a computer program as "a set of statements or instructions to be used directly or indirectly in a computer in order to bring about a certain result."[5] Another definition is that a computer program is "a collection of instructions that tells a computer to count, sort, print or perform other operations in a known sequence."[6] Computer programs are often also called software, as distinguished from the computer's circuits, or hardware, because they can be easily changed to perform different functions. In fact, the hardware of the same computer can do almost infinite number of jobs when different computer software is loaded onto it.

In deciding whether an action for copyright infringement will lie, the label "computer program" is of little help. What is a computer program and how is it copied? With books the answer is straightforward; the pages of a book contain text, usually a literary work. Copying the text entails reproducing a substantial sequence of that text elsewhere, e.g., by using a photocopier to copy a page or by scanning the page and storing the text electronically. With software the mode of expression is not so clear. To understand what it means to copy software, one must be familiar with certain basic working concepts and definitions.

Hardware is the computer itself and the devices connected to the computer. These devices are often called peripherals and include printers, video monitors, keyboards, cables and mice. The computer itself comprises mostly chips,[7] though devices such as the power supply, the disk drives and the hard drive are integral parts of the computer.

Software, on the other hand, is the information that the computer uses to perform tasks. The most common medium of software is no longer floppy disks but is a diskette. Indeed, the term software has its genesis in the physical flexibility of computer diskettes, called floppy disks. But hard drives contain the same type of information as floppy disks, only in greater quantities, yet hard drives are considered hardware. The inconsistency reflects sloppy usage of the term "software" as well as the history of the development of software in various formats. In one sense, software suggests floppy disks; in another sense a user may say the Quattro Pro is a good piece of software. Clearly, the latter use refers to the easy and efficient way in which Quattro Pro operates and not to the disks on which it is contained. In other words, Quattro Pro is a good intellectual creation. This is the better use of the term, but by no means is it the only one. Software may depend on a physical machine for its execution but not for its existence. If this page contains a written series of programs, routines and instructions that effectively analyze financial information to predict market conditions, a piece of software has been written. For example, WordPerfect contains hundreds of data files and programs. Data files contain information (e.g., books in a library and their classification numbers); programs execute instructions (such as how to arrange a series of classification numbers into a coherent system). Again, however, usage is not strict, and the terms are often used interchangeably.

Perhaps the best way to understand the distinction between hardware and software is to think of the hard drive. A hard drive is a ceramic cylinder, a piece of equipment. In this connection it is a piece of hardware. The hard drive can store many pieces of software, however.

Without software, a computer is only worth its weight in scrap metal. Each computer program is different; it depends on what the program will do, the kind of data that it uses and how it will work.[8] One classification is as an application program or a system program. Systems programs are programs that manage the resources of the computer, such as memory, disk space and processing time for any application programs running on the computer.[9] They perform diagnostic checks and allocate memory in an efficient manner. Examples of systems programs are the operating systems, such as MS-DOS, UNIX and MVS. These systems provide the programs necessary to carry out such functions as creating and reading disk files, starting and ending programs on the computer and allocating memory.[10]

Application programs such as WordPerfect, Lotus 1-2-3 and PageMaker perform real-world tasks for the computer user such as inventory, payroll, spreadsheet manipulation or word processing and permit users to search files of information.[11] Each application program is tailored to the individual type of use, i.e., WordPerfect is primarily for word processing. Other programs such as various Windows programs are neither completely a system nor an application program; instead they are combinations of both referred to as a graphical user interface, or GUI.

Programmers create programs by entering a sequence of instructions into a computer where they can be executed on command.[12] After a programmer has identified what the program must do, how it must do it and what data it must use to do it, she normally will formulate a high-level description of the program's structure. This description can take various forms: it may be a series of flow charts, pseudocode or a set of structure diagrams.[13] The programmer then writes source code to describe the program's operation to the computer. Source code is the form of one of the various computer programming languages that are relatively standard across the computer industry.[14] Since a computer does not understand the words of the language as a person would, the source code must be converted into object code that the computer can read.

Object code is a translated version of the source code that instructs the computer how to manipulate its internal circuits.[15] Source code is converted into object code by the computer through the use of a compiler or assembler, and the computer can directly execute the resulting object code.[16] There are two things to note about object code. First, it is easily copied. Second, to the naked eye it makes no sense, unlike the text in a book, which the human mind easily assimilates. Despite the second consideration, object code is copyrightable.[17] Legally, the first consideration carries more weight with Congress. Because software is so easy to copy, legal protection is necessary. Another important point about object code is that it is not the *lingua franca* of computer programmers. Like most people, a programmer thinks in concepts first and worries about details later.

Both the source code and the object code can be stored on a computer disk for later use. To modify a program, the source code must be changed first and then "recompiled" into new object code.[18]

A computer program can be thought of as layers of ideas and expressions created by the programmer who develops the program.[19] Because of the intellectual and intangible nature of computer programs, it is inherently more difficult to visualize the physical aspects of a computer program than it is in other engineering areas.[20] However, programs are usually developed in stages where the program evolves from vague ideas into something more tangible that the computer can directly execute.[21] Thus, it can be said that there are different layers of abstraction in a computer program as the programmer goes through the series of steps from defining the problem to be solved to a finished program that solves the program.[22]

B. Copyrightability of Software

Early in the development of computers there was concern about whether a computer program was a copyrightable work since it simply provides a series of instructions to a machine.[23] Today there is no debate. The legislative history of the 1976 Copyright Act simply declared that computer programs are literary works, and, as such, they are copyrightable.[24] The fact that software may require the aid of a machine to be perceived is not a problem. The Act states that copyright exists for any work of authorship

> fixed in any tangible means of expression, now known or later developed,
> from which they can be perceived, reproduced, or otherwise communicated,
> either directly or with the aid of a machine or device.[25]

While the Copyright Act was being debated it became clear to Congress that the entire area dealing with computers and copyright was very complex. In order to complete work on the Act and get it passed, it was enacted without dealing with the copyright of software. Section 117 as enacted froze the current status of the law until additional work could be done to clarify and define these issues. To this end, Congress created the Commission on the New Technological Uses of Copyrighted Works (CONTU) and appointed its members. CONTU had two tasks assigned to it: develop the interlibrary loan guidelines and determine what to do about computer software and electronic databases.[26] The first amendment to the Copyright Revision Act of 1976 was a new section 117 to replace the existing one which had merely preserved the status quo.[27]

Not all authors of computer software claim copyright in the programs they create. Freeware is a type of software in which the developer makes no claim of copyright and everyone is free to use the program without fear of infringement. Shareware generally is authored software which is distributed free of charge initially but with the understanding that a user will compensate the author if the user decides to adopt that program for a particular application. The remainder of this chapter deals with software in which the author or the author's employer claims copyright.

C. Section 117

Section 106 of the Act defines the exclusive rights of a copyright owner. Those rights include reproduction, distribution, adaptation, performance and display. Sections 107 to 118 limit those rights and give some leeway to the owners of the copies to use them in certain ways. Section 117 applies specifically to computer programs and grants the "owner of a copy of a computer program"[28] the right to make a copy or adaptation of the program in two circumstances. The owner of a copy of a computer program and the owner of the program (the copyright owner) are distinct. For example, Microsoft owns the copyright on DOS while a user owns a copy of the DOS program.

Section 117 limits the rights of the copyright owner by expanding the rights the lawful possessor of a copy of the program has. Distinctions in forms of possession (owner of copyright versus the owner of a copy, possessor of a copy, etc.) appear frequently in copyright law as applied to computer programs, especially in the area of licenses.

The first instance in which a copy owner may reproduce the program when doing so is "an essential step in the utilization of the computer program in conjunction with a machine."[29] This provision stems from the recognition that the use of any software automatically entails making a copy. A computer executes programs from its memory.[30] For the computer to do this, a user must copy the program from its magnetic storage medium, e.g., a floppy diskette, into the computer's memory. This is done automatically by instructions on the software that interact with instructions built into the computer.[31] For example, someone using WordPerfect may type in the command "wp," whereupon the WordPerfect program is copied into the computer's memory and executed. WordPerfect then is up and running on that machine. But two copies now exist, one on the hard drive (or diskette) and a second in the internal memory of the computer. Because the second copy is fixed in a tangible medium of expression (RAM) and is retrievable, it fits the definition of a copy under copyright law.[32] This internal copy, however, does not infringe WordPerfect's exclusive right of reproduction,[33] because the copy is essential to using the program. Section 117(1) codifies what users know by common sense: the routine use of a lawfully purchased program should not be a violation of the law.

Another example of when a copy must be made in order to utilize the program is when it is necessary to convert if from one disk size to another. When the new section 117 was adopted, the original thought was that the copy would need to be made in order to convert from one computer language to another such as from FORTRAN to COBAL. Today, conversion of disks from 5 1/2-inch to 3 1/2-inch size is likely to be more common.

The second instance in which an owner of a copy of a program may make a copy is for archival purposes.[34] The intent of this provision is to protect the copy owner from the risk of electronic destruction[35] of the software, which can occur as the result of power surges, computer gliches or garden-variety human error.[36] It is important to note that the right to have an archival copy rests only in the rightful owner of the program copy. If X makes a backup copy of TurboTax and then gives the software package to Z, X must destroy the backup copy. The right to have an archival copy belongs to Z, who is now the rightful owner of the TurboTax copy. X's possession of the backup copy, in effect, has ceased to be rightful.[37] Again, the rule codifies common sense. Only one copy of TurboTax was purchased, so only one person should be able to have it on her computer.[38]

The Act does not distinguish between types of software. Nor does it distinguish between stand-alone software and that which accompanies a purchased book or audiovisual work. Thus, under the law, a library also could make an archival copy of software that comes with another work which the library purchases.

In one of the few judicial interpretations of section 117, a federal court of appeals favored the copy owner.[39] Vault was a software protection company that marketed PROLOK, a program that enables a software company to protect its programs from unauthorized copying. PROLOK worked by encrypting a piece of software so that a computer could read only a purchased copy of the program; bit-by-bit copies of the purchased copy were rendered useless. Quaid was in the business of software decryption and in this connection developed RAMKEY, a program that would defeat the encryption of a program by PROLOK. Vault sued Quaid on the ground that Quaid's copy of PROLOK[40] was not authorized under section 117.[41] The court disagreed, holding that the language of section 117(1) did not confine the use of a program copy to that intended by the copyright owner.[42]

Another court of appeals, however, limited the application of section 117.[43] Sega manufactured video game cartridges for use in a game console which it manufactured.[44] Accolade wanted to market its own cartridges for use in Sega's console, but it needed to know exactly how the console executed the games' interface specifications. In this connection, Accolade bought three Sega cartridges to see what software segments in them were identical.[45] These "areas of commonalty" would alert Accolade to what software instructions would be needed if its own cartridges were to run successfully in Sega's console.[46] Accolade "wired a decompiler into the console circuitry, and generated printouts of the resulting source code" of Sega's game cartridges themselves.[47] On its own computers, Accolade analyzed and experimented with Sega's source code to discover the game-console interface requirements.[48] With respect to allowable copies, the court held that "it is clear that Accolade's use went far beyond that ... authorized by section 117."[49] The court continued,

> [s]ection 117 does not support to protect a user who disassembles object code, coverts it from assembly to source code, and makes printouts and photocopies of the related source code version.[50]

The holding thus appears narrower than that in *Vault v. Quaid* at least with respect to section 117.

In the same way that section 117 allows a user to make copies of a program, it allows the making of an adaptation of computer programs: it must be either "an essential step" or "for archival purposes only."[51] This right, like others in sections 107 to 118, operate to limit the rights of the copyright holder. For the copy owner, then, "[t]he conversion of a program from one higher-level language to another to facilitate use would fall within this right, as would the right to add features to the program that were not present at the time of rightful acquisitions."[52] It is important to emphasize, however, that the right is personal rather than commercial. CONTU saw the copy holder's right of adaptation as analogous to "marginal note-taking in a book." Thus, copyright law would not permit the note taker to market the book and the notes therein as one work.[53]

Two cases nicely illustrate this distinction. In *Foresight Resources Corp. v. Pfortmiller*,[54] the defendant purchased a commercial program and added five files to it to tailor use of the program to its business needs. The defendant used the adapted program only in-house. The court held that the plaintiff did not show a likelihood of success on the merits in its claim of copyright infringement. In a second case, *Allen-Myland v. Int'l Business Mach. Corp.*,[55] however, the defendant was unsuccessful. The plaintiff produced complex computer systems complete with configured software, and the defendant serviced these systems for third-party owners. For several customers who wanted reconfigurations of their systems, the defendant effectively spliced together different versions of the systems' software. The court read section 117 strictly and found that the adapted software was necessary neither for use nor for archival purposes. In support of the court's finding was language in the CONTU Report: "The adaptor could not vend the adapted program."[56]

D. License Agreements

License agreements typically govern what the owner of a copy of software may do with that program. Typical license terms include who is licensed to use the programs, for what machines, what use may be made of the program and for how long. Issues such as payment terms, delivery of the software, warranties and upgrades also are included in the typical license agreement. Increasingly, downloading provisions are included in typical license agreements.[57] The most common type of software license is an agreement for use on a single machine. The typical limitation on use in such a license is for the company and includes all of its facilities. Further, license agreements usually permit the user to modify the software for use within the company, but the copyright owner's warranty does not extend to such modifications.[58]

Software licenses provide for upgrades to the program and normally dictate that the company or individual cannot continue to use both a current version and an earlier version of the software for one purchase price. If the library wants to use both an upgrade and an earlier version, it should contact the software publisher for permission. Another common request from libraries that can be covered in a license is whether staff members may take library software home in order to work at home. Absent a provision in the license to permit home use of library-purchased software, the library should contact the publisher to request this use.

1. Site Licenses

Site license agreements normally state that a large user of a program (such as a company or a university) can make unlimited copies of software for use at a specific site or offer additional copies at low rates. Agreements also may limit the licensor to making a specified number of copies of the software, or they may limit copying to a certain class of individuals such as employees. Interestingly, there is no generally accepted definition of the term "site license" and license terms can vary substantially. For example, a site license might permit use of software on any machines at a particular location with no restrictions on the number of machines. The limitation might allow business use by employees of the licensee on an unlimited number of computers at any site, including their homes. A site license might restrict the number of computers on which the software may be used at one or more geographic locations. Additionally, some site licenses also permit the reproduction of manuals and other documentation by the licensee.[59] One of the reasons companies seek site licenses is to avoid copyright liability for any unauthorized copying of programs by employees;[60] another reason likely is economics.

2. Network Licenses

The growth of computer networks has had significant impact on the way computer software is licensed. Networks may be (a) local area networks (LANs) that normally are confined to a relatively small area such as a building or a group of buildings, (b) wide area networks (WANs) that span large areas such as a country or a continent or (c) metropolitan area networks or (MANs) that include an entire city or town.[61] From the first experimental networks in the 1970s, license agreements have been involved. Networks originally required both workstation software and network operating systems, but today there also is network application software.[62] Existing license practices may be inadequate for many networked environments because neither licensing by shrinkwrapped licenses nor site licensing is satisfactory, and vendors should have separate network licenses available. It is understandable that software companies fear loss of revenue when a network administrator purchases a copy of single-use software and loads it onto a network and permits simultaneous use by several users. Network administrators complain about the lack of uniformity in network software license agreements, which are a headache for the administrator to control.[63]

For networks, software can be licensed to individuals or to a particular server, but the most common is concurrent use licensing which requires the licensee to pay for a maximum number of users of the program on the network.[64] Concurrent use licenses have come to be the preferred type of license for networks because they provide the greatest use flexibility for the end user, and they simplify distribution for the supplier. It also frequently is a less expensive type of license since not all employees would ever simultaneously use the program. The only disadvantage for the user is that is requires some type of monitoring system such as license manager software that operates to ensure that no more than the specified number of users have simultaneous access to the program.[65]

3. Shrinkwrapped Licensing

Shrinkwrapped licensing has been particularly controversial. Sometimes referred to as "tear open" or "boxtop" licenses, the license is included on the package and is covered in plastic wrapping. The license should be readable through the wrapper, but this is not always the case. The license states that if one opens the box, she is bound by the terms of the agreement.[66] The package often is opened by pulling a tab or a piece of red tape.

There are two significant criticisms of shrinkwrap licenses. First, licensors frequently get carried away and attempt to impose burdensome, unreasonable and occasionally unlawful requirements on users.[67] Often these licenses purport to forbid the very kind of reproduction of programs that is expressly permitted under section 117 of the Copyright Act. The second problem is that a contract should be signed by two parties, each of whom has the authority to commit his organization to the terms of the agreement. With shrinkwrapped software, anyone can open the package. This includes library or company employees in the mail room or elsewhere in the company who have absolutely no authority to commit the company. Thus, many scholars question whether shrinkwrap licenses are valid contract.

In a shrinkwrap license case, a federal court of appeals held that a state statute that allowed software producers to restrict severely the uses of its software was preempted by federal copyright law.[68] The Louisiana statute[69] in question allowed producers to prohibit (a) any copying of the program for any purposes and (b) modifying and/or adapting the program in any way. This included adaptation by reverse engineering, decompilation or disassembly.[70] The matter has not ben litigated in other jurisdictions, so this case may well be an oddity.

Even in shrinkwrap licensed software, the copyright owner normally asks the user to sign and return the license to the licensor. To encourage compliance, the agreement often states that the licensee will not receive corrections, upgrades or other enhancements to the program unless she signs and returns the agreement.[71]

4. Executing License Agreements

Often librarians feel that the only choices they have when faced with a software license agreement are two: (a) sign the agreement and adhere to the terms or (b) reject the agreement and return the software to the licensor. Certainly, these are options. There is another one, however: the owner of a copy of a program can negotiate with the copyright holder. This begins with proposing new or altered terms in the license agreement.

Upon receiving software with an unacceptable license agreement, the librarian can alter the form by marking out terms that are unduly restrictive or unworkable and inserting new ones. It is also useful to initial each change (much as one does in a rental car agreement). Some attorneys recommend that the agreement be signed and a statement included to the effect "subject to our terms"[72] or "subject to the terms noted above." If the software publisher accepts the new terms, alteration of the agreement was well worth the effort as it converted unusable software because of restrictive license terms to usable software. If the publisher rejects the terms proposed by the librarian, the library can then sign the standard agreement. Thus, it is no worse off than if it had just agreed to the terms in the first place.

Reports indicate that reactions by software publishers vary considerably. Some licensees continue to receive upgrades just as if they had executed the standard license agreement form without marking out objectionable terms. On the other hand, others report always hearing back from the licensor which sometimes accepts new terms and sometimes rejects them.[73]

Licenses for library systems software such as NOTIS, DRA and Innovative Interfaces present particular challenges for librarians. Fewer standard licensing agreements are available, probably because the market is much smaller that for office application software. Library systems licenses require extensive negotiations to cover issues ranging from searching the library's own records to searching those of other libraries that use the same system. Often individual libraries do not have complete freedom in such negotiations as they are a part of a larger campus, library consortium, etc. It is also more difficult to learn about contractual terms offered to other libraries as vendors frequently impose a confidentiality clause that prohibits the sharing of copies of the contract with other libraries.

IV. LOANING COMPUTER SOFTWARE

A. General

With certain exceptions, section 109 prohibits commercial leasing and lending of phonorecords and software. Anyone who has worked with either knows how easy — and tempting — it is to copy either a record or a program. The purpose of this section of the Act is to prevent free-riding rental companies and other commercial concerns from taking sales away from manufacturers in each industry. In each case, congressional inquiry revealed what commonsense already knew: home users with stereos and computers did not rent records and software without making permanent copies for themselves which relieved them from the need to purchase legitimate copies.

Section 109 is important for two reasons. First, sections 107 to 118 of the Act limit the rights of copyright owners and operate in favor of the copy holders. Section 109 is an exception to this generalization. Second, section 109 is an exception to the first sale doctrine,[74] which derives from the broader common law rule forbidding restraints on alienation. The exception is significant. The first sale doctrine is a stalwart of American copyright law, so that altering it is "not a modest proposal."[75] Under the first sale doctrine, the purchaser of a copyrighted work is permitted to sell or lease his copy to others; the author has no authority over these practices after title to the physical copy is first transferred.[76] Thus, the purchaser of a novel is free to lend or even rent her physical copy to others.[77]

With the rise of information technology, however, such freedom ceased to be entirely benign. With respect to copying, the difference between lending a book and lending a software package is stark. To copy a book, one needs to photocopy hundred of pages. The cost — both financial and in terms of hours — is significant. One rarely hears of such efforts. To copy software, however, one needs only a few diskettes and a few minutes. The cost is small, and accounts of this practice are common. Compare also the cost of obtaining a legitimate copy. A hardbound biography may cost $30, while spreadsheet software may cost $500 or more. It is clear that the incentive to copy software, as opposed to purchasing it from the manufacturer, is great indeed.

The software rental industry met this demand in grand fashion. Rental organizations simply purchased original packages of software, then turned around and repeatedly rented the same copies multiple times to patrons for one or two nights in an attempt to recover their initial investment.[78] To the software publishers, then, the first sale doctrine was a giant "loophole in the U.S. copyright law that allowed or perhaps even encouraged pirates to structure themselves as rental outfits."[79] For less than $20, a renter could try a program at home for a night and return it the next day.[80] With the prices of certain commercial software packages approaching $1,000,[81] and with the cost of copying so low, the obvious motive to rent was expropriation. Software packages simply could not be mastered in a day.[82] In 1989 alone, estimates of losses to the computer software industry in the form of diverted sales ranged from $1.6 billion to $4.1 billion.[83] Congress responded to these excesses by passing the Computer Software Rental Amendments Act of 1990,[84] which proscribes the "rental, lease or lending" of computer software.[85]

As noted, before the Rental Amendments Act of 1990, software producers were faced with the naked first sale doctrine. Under the umbrella of this doctrine, the purchaser of a software package was free to dispose of it by any means such as sale, lease, rental, etc. For commercial purposes, this meant renting: once the original purchaser had a legitimate copy of a program, he could rent the program to someone for a fraction of the cost of the purchase price. This practice grew into an entire industry, which free rode on the efforts of software manufacturers with impunity. It bears emphasizing that a software manufacturer had no cause of action against the lessor, who had infringed none of the exclusive rights under section 106. Actual infringement, presumably, occurred in the privacy of the rentor's home. With the protection of the first sale doctrine, the lessors did not *need* to infringe the copyright to make a profit. The practice was thus legally risk free.

For the software manufacturers, the only cause of action was against the renters, the single patrons of the rental companies. This presented several obvious difficulties. First, the identities of those who had rented the software were not easily obtained since there was no recordkeeping requirement for software stores. Second, proving that a renter had actually copied the producer's software would entail going into his house and rifling computer files for illegal copies. Even if an illegal copy could have been found, it represents an isolated instance of infringement out of myriad others induced by the rental store. Third, the small-time renter is simply not a good legal target from a financial standpoint.

Clearly what the producers needed were legal grounds on which to sue the rental companies, who were much bigger, richer and slower targets than the isolated hacker. As shown, this was not possible under the copyright laws. The software license was thus born.

Aware that a license could not preempt federal law, the manufacturers exploited the language of the first sale doctrine itself, noting that it gave the right to dispose of copies only to the *owners* of those copies. The terms of the software licenses thus cast the purchaser of a software package (the rental store) as a licensee. As a mere licensee, the rental stores could not use the first sale doctrine as a shield from liability, since lawful possession of the program copy rested on the lease, not on outright ownership. Even after the transaction effecting the transfer of software from manufacturer to rental store, the beneficiary of the first sale doctrine was now the manufacturer. Or so it was argued. Purportedly having demonstrated that the licensee did not benefit from the copyright law, the manufacturers simply stipulated in their licenses that the licensee agreed not to rent or lend the software to others. Any violation provided the manufacturer with a cause of action in the common law of contract against the rental concerns. Software producers began shipping the licenses with the software, encasing both within a transparent wrapper. The plainly visible license indicated that the purchaser could accept its terms simply by tearing open the wrapper. Those terms, of course, included the stipulation that the purchaser was not the owner and agreed not to lend or lease the software.

Thus, the debate began. The software industry analogized itself to the motion picture industry, which had been using licenses for years to avoid the first sale doctrine. Commentators argued that the first transaction was a thinly veiled sale. They pointed to the absence of any time limit on possession by the licensee, which indicated that absolute ownership of the copy had in fact been conferred. Another issue was whether federal copyright law preempted state contract law.

Congress ended the debate with the 1990 Amendment, at least with respect to commercial renting and library lending. The former is expressly forbidden, the latter expressly allowed. At least one court has said as much, although not in a holding.[86]

Software is highly functional and because it is such an integral part of business and industry, its defects can be serious. A warranty is an implied promise that the delivered item will possess a certain minimum level of quality. So, where they once used licenses as a sword, software producers now use them as a shield: licenses attempt to distribute software "as is," with very limited warranties or guarantees, if any at all.[87]

B. *Nonprofit Library Exemption*

Congress modeled the Computer Software Rental Amendments Act of 1990 on the Record Rental Amendments Act of 1984,[88] which proscribes the commercial renting of phonorecords. Foreshadowing the events in the market for software, record stores in the 1970s had begun to lend records and albums, which a renter would promptly record onto cassette. Each time this occurred, the record industry lost a sale.[89]

Congress was keenly aware of the difference between a software package and a record album, however. The former could be copied in seconds,[90] while the latter required roughly an hour to reproduce. Augmenting this incentive to copy software was the fact that certain packages of software cost on the order of 40 times more than a compact disc.[91] The plain fact that "the differences between software and records make software a much likelier candidate for illegal duplication"[92] initially threatened libraries: while libraries and nonprofit education institutions are free to lend records,[93] as originally proposed, the Computer Software Rental Amendments Act made no such allowance.[94]

Two considerations appear to have so inclined Congress. First was the same rationale underlying the prohibition of commercial renting: software cannot be mastered in the short period of a loan, so that the clear motive behind such a loan was illegal duplication.[95] Second, and less obvious, was that libraries enjoy the privilege of lending records in part because "they [are] renting records which [are] generally not being recorded,"[96] i.e., are not commercially popular. This distinction is reflected in the stances of various software publishers on the issue of exemptions for *educational institutions.* The larger software publishing companies such as Microsoft, with sales of $800 million per year, opposed the exemption.[97] The smaller Software Publishers Association, on the other hand, supported the educational exemption.[98] More than half of SPA's members have sales of less than $1 million per year.[99]

The vast majority of those who testified, however, including the Register of Copyrights, were in favor of granting an exception for libraries.[100] SLA, along with other library associations, naturally favored the exemption.[101]

The exception, codified in section 109, seems to delineate clearly a library's rights and duties. First, a nonprofit library is expressly allowed to lend software to patrons for nonprofit purposes so long as the diskettes bear a warning of copyright law as promulgated by the Register of Copyrights.[102] Second, nonprofit educational institutions are free to transfer copies of computer programs to other schools or to their own faculty, staff and students. In fact, such a transfer is not even treated as a rental, loan or lease for direct or indirect commercial purposes.[103] In both instances a copy must have been "lawfully made."

Libraries in the for-profit sector thus may not loan software to users without either a license agreement that so permits or express permission from the holder of the copyright in the program. Most companies prefer site licenses for computer programs utilized anywhere within the company, so, in reality, this restriction may have little impact in the corporate world for libraries.

C. Software Warnings

Under section 109, a nonprofit library may loan software only if the library places on circulating software packages the warning promulgated by the Register of Copyrights:

SOFTWARE WARNING
NOTICE:
Warning of Copyright Restrictions

The copyright law of the United States (Title 17, United States Code) governs the reproduction, distribution, adaptation, public performance and public display of copyrighted material.

Under certain conditions of the law, nonprofit libraries are authorized to lend, lease, or rent copies of computer programs to patrons on a nonprofit basis for nonprofit purposes. Any person who makes an unauthorized copy or adaptation of the computer program, or redistributes the loan copy, or publicly performs or displays the computer program, except as permitted by Title 17 of the United States Code, may be liable for copyright infringement.

This institution reserves the right to refuse to fulfill a loan request, if in its judgment, fulfillment of the request would lead to violation of the copyright law.[104]

This warning is very similar to the warning libraries are required to post at the place where orders for photocopies are placed and on the order form itself under section 108(d).[105] The photocopy warning attacks the problem of copyright violations at the point of infringement: the desk where orders are taken. Further, section 108(f)(1) also requires some sort of notice on unsupervised reproduction equipment that the making of a copy may be subject to the copyright law. If the software warning requirement flowed directly from the policy underlying that of the photocopy warning, the copyright warning for software would appear not only on the software itself, but also either on or at the library's computers. To put a warning only on the software itself attacks only one side of the infringement problem. When libraries lend software to patrons to take home, the warning on the software must suffice. In the privacy of their own homes, users are then left to their own devices. The problem is that patrons may also try the software in the library itself. For the purposes of infringement, a computer is a much more efficient copier than a photocopier. Using a computer to pirate software is faster, cheaper (virtually free), and cleaner (the copies of a software program are perfect) than a photocopier.

For libraries that permit patrons to use software that the library has on a network, in order to comply with the spirit of the law, the warning should appear either on or near the computer or on the screen itself. Most public computers (networks or isolated workstations) offer a menu of choices to users. The warning could appear either on the menu itself, or, better yet, on the screen that immediately follows the menu, inserted in memory such that the sequence of screens is Menu - Warning - Application Program. By putting notices on the computers themselves, libraries would be attacking the problem of software piracy on both flanks. Also, this is more similar to the notice that libraries are required to put on unsupervised photocopy machines.[106]

The length of the warning has amused librarians. Just as personal computers moved from 5 1/4-inch to 3 1/2-inch disks, this lengthy warning was promulgated for attachment to the packages. In order to place the warning physically on the package, the wording must be inordinately small. It is also interesting to speculate why a software warning promulgated as late as 1991 did not envision computer networks in which libraries would not loan software packages to users but instead would load it onto a network in the library or in a computer laboratory where it is available to patrons or students.

At present, the Register of Copyrights is conducting hearings to determine how well the exemption for libraries is working.[107] Library associations have submitted testimony to the effect that the loan of software coupled with the required warnings is working as it should and that there is no evidence that patrons who check out software or use it on their network are copying it.[108]

V. DATABASES

A. General

Computer databases may consist of several types of material. A database consists of previously created copyrighted works now stored digitally. In that case, the copyright in the underlying work is not changed by the fact that the format in which the work is stored has changed from paper to electronic media. On the other hand, a database can include both copyrighted and uncopyrighted works; there are a variety of databases that even incorporate government-created public domain materials along with previously copyrighted works. Additionally, these databases may include some original works. A database could consist totally of originally created material; there are but few that do, however. In any event, a database is a compilation which the Copyright Act defines as "a work formed by the collection and assembling of preexisting material or data that are selected, coordinated or arranged in such a way that the resulting work as a whole constitutes an original work of authorship."[109]

One of the unique aspects of databases is how they are funded. Originally, databases were funded on a per search basis, and many still maintain that funding mechanism. Others, however, have begun to offer flat annual rates or even subscriptions on a fixed monthly rate. Many of the issues concerning misuse of database information relate to the manner in which they were originally funded. In other words, any attempt to avoid paying the search charges constitutes an unauthorized use of the database.

Because of user demand and the ability to predict actual use more precisely, many database vendors have moved beyond the per search basis. Obviously, if a library pays a flat fee for database access, then the answers to questions about whether certain uses or practices are infringement change. A per-user funded database creates the expectation on the part of the owner that the library will conduct a separate search for each individual user. If the database is funded on a flat rate, however, it makes no difference whether the library searches the database one or ten times for various users. It pays the same rate.

In the quarter-century or more since online databases first appeared, much has happened to facilitate the use of the valuable information contained in the database. Technological developments made it possible to search national databases from a local personal computer with a modem. With the advent of powerful personal computers and increasingly sophisticated communications software, the ease with which data can be moved from a central computer into a personal computer or onto a diskette has greatly increased.[110]

Librarians have led the way in developing databases, improving the documentation that accompanies a database and creating sophisticated teaching modules that utilize the database. Despite these developments, much remains problematic in how libraries access and use electronic databases. License agreements and access fees vary considerably, and it seems that use restrictions are proliferating.

B. *Copyrightability of Databases*

Unlike computer software, which has utilitarian aspects, databases are strictly factual. Facts are not copyrightable. Databases are copyrighted as compilations, much like other directories and almanacs. Despite the mental association most people have between computer software and databases, the applicable law in the two areas is distinct.[111]

Professor Arthur R. Miller[112] suggests two sources of copyright law that may govern databases.[113] The first is *Feist Communications Inc., v. Rural Telephone Co., Inc.*,[114] in which the Supreme Court held that the white pages of a telephone book did not meet the threshold of creativity required for legal protection. The second source is the merger doctrine, which holds that where an idea allows only one or very few forms of expression, the mode of expression is not copyrightable. The merger doctrine serves the policy of preventing the monopolization of ideas, which can be accomplished only through patent law. Both standards are applicable to databases.

Feist published telephone directories and wished to publish one that covered eleven telephone districts. One of these areas was served by Rural Telephone, which refused to license its white pages to Feist. Feist published the directory anyway and simply took the data it needed directly from Rural's directory. Rural sued for copyright infringement. The Supreme Court held that Rural's directory lacked any originality, a constitutional requirement for copyright protection. The Court stated that the two requirements for originality were independent creation by an author and a minimum degree of creativity.[115]

Turning to factual compilations, the Court recited the proposition that facts are not copyrightable and then examined the level of creativity necessary for a copyright. In a compilation of uncopyrightable material (facts), the Court noted that it was the selection and arrangement of the material that give it the requisite creativity. Rural's directory failed even this minimal copyright standard because its selection and arrangement of data were "so mechanical and routine as to require no creativity whatsoever."[116] What makes Rural's work significant, the Court noted, was the effort that went into amassing the data. That effort — the "sweat of the brow" — does not, however, invest in the work any creativity. Thus, the Court explicitly rejected the old sweat of the brow doctrine.[117] It noted that the requisite level of creativity is low, but definitely is above zero.[118]

The *Feist* decision surprised many commentators because it held that a particular compilation was not copyrightable. Previously, the copyrightability of compilations seemed beyond question simply by virtue of having been authored. With its focus on creativity, the decision generated a welter of commentary, including law review articles and symposia. The lower federal courts, on the other hand, have remained staid, treating Feist as a direct application of copyright law, unique only in its facts.[119] Probably the clearest example of the business-as-usual approach of the lower courts is *Key Publications*.[120] Key Publications assembled the "Key Directory," a phone directory that listed Chinese businesses and members of the community. The defendants published a competing directory that contained approximately three-quarters of the Key Directory material. The court held that despite *Feist* the Key Directory's copyright was valid and that the defendants had infringed. It reasoned that because of the judgment necessary in selecting which of a multitude of businesses would go into the work, the directory warranted copyright protection. The Key Directory had more than 9000 listings arranged into 260 different categories.[121]

The second source of copyright law that may govern databases is the merger doctrine. Merger describes the situation where the idea can be expressed in a very limited number of ways. When this is the case, the expression may merge into the idea, and ideas are not protectable subject matter under section 102(b) of the Act. The classic case of the merger doctrine is *Baker v. Selden*.[122] Selden developed a system for double-entry bookkeeping that he reduced to blank forms for ease of entry and on which he obtained a copyright. Baker used Selden's system in his business without obtaining permission or paying royalties. Selden sued for copyright infringement. The Supreme Court held that where a particular chart is necessary to implement an uncopyrightable system, the charts and diagrams are part of the public domain and thus are not copyrightable. The modern formulation of the merger doctrine is:

> When the "idea" and its "expression" are thus inseparable, copying the "expression" will not be barred, since protecting the "expression" in such circumstances would confer a monopoly of the "idea" upon the copyright owner free of the conditions and limitations imposed by the patent law.[123]

Whether the merger doctrine applies depends in some cases on how the court defines what constitutes the idea. In *Kregos v. Associated Press*,[124] for example, the plaintiff had assembled a baseball pitching form, i.e., a brief statistical summary of opposing pitchers' performances for use in predicting the game winner. A newspaper used the form, which was comprised of only nine statistics, without permission, and Kregos sued. Associated Press defended on the ground that the idea that "the nine statistics he has selected are the most significant ones to consider when attempting to predict the outcome of a baseball game"[125] merged with the expression of a nine-statistic pitching form. The court, however, agreed with Kregos that the idea was simply "an outcome predictive pitching form" that permitted a variety of expression. Thus, Associated Press infringed the copyright.[126]

In certain cases the sheer number of facts assembled seems to influence whether the merger doctrine applies. In *Eckes v. Card Prices Update*,[127] the selection of 5,000 of 18,000 baseball cards for the classification of premium was held to have permitted enough forms of expression to bar application of merger;[128] in *Financial Information, Inc. v. Moody's Investors Service*,[129] the selection of five items of commercial bond information was held insufficient for a copyright.[130]

There are two ways that the merger doctrine could undermine the copyrightability of a database. First, the arrangement of data in the database could be so obvious as to permit only a very few means of expression. Second, the purpose of the database could completely dictate its content such that only very few results are possible. In these two situations, a court could conclude that idea and expression have merged.[131] It is possible that bibliographic databases are of this last type.

Whenever one uses computers — either in assembling a database or in writing a program — a significant goal is efficiency, accomplishing the most in the fewest bytes or instructions. Because efficiency steers authors away from individual creativity and toward universal utility, the merger doctrine is likely to emerge more and more in discussions of the copyrightability of computer codes or databases:

> In the computer context, this means that when specific instructions, even though previously copyrighted, are the only and essential means of accomplishing a given task, their later use by another will not amount to infringement.[132]

It is conceivable that libraries which utilize online databases that contain either only bibliographic data or such data accompanied with abstracts to create customized bibliographies may be dealing with the merger doctrine. If the bibliography is given only to one user, either in print or in disk format or is sent electronically to a single user, the database has been used as intended unless the license agreement contains some restriction to prohibit such activity. The merger doctrine is not relevant in this situation. On the other hand, should the library wish to distribute widely the bibliography a librarian created using a database and software, it may be that under the merger doctrine, the idea of the bibliography and the alphabetical arrangement of bibliographic data is so merged that there can be no infringement by the library even for wide distribution. Abstracts that appear in the database, however, are copyrighted as derivative works and should be removed from any widely distributed bibliography unless the database owner has given permission for such distribution. The merger doctrine in no way affects the copyrightability of abstracts and cannot insulate the library from liability for infringement for using abstracts without permission of the copyright owner.[133] Fortunately, many database owners will license libraries to distribute abstracts widely although there may be an additional charge imposed.

C. Downloading Restrictions

Most commercial databases, whether available online electronically or on CD-ROM, include a copyright notice on the terminal screen and on printouts and downloaded files. Further, license agreements include a notice of copyright and usually refer to the fair use provisions of the copyright law. Virtually all of the basic license agreements forbid resale of data retrieved from online searching or from any kind of commercial use without permission from the online vendor.[134]

Such restrictions are included in the license agreement that accompanies subscription to the database. Many such restrictions harken back to the way in which the database is funded. Librarians should read license agreements carefully and work to negotiate terms most favorable to them and their users.

Librarians frequently ask about whether they are permitted to perform a search for one user and then reuse the results for subsequent users. The answer depends on how the database is funded. If the funding mechanism is based on a per use charge, clearly, the database owner anticipates a separate search and charge for each user. If, however, the database is funded on a flat-rate basis, it makes no difference economically to the vendor whether one or ten separate searches are conducted. Even for a flat-rate subscription database, librarians should be cautious about reusing search results since databases are updated so frequently, and in a matter of a few days one could be distributing old and outdated information. Conversely, it makes good sense to retain search strategies for complicated searches. There, the librarian is not retaining results but, instead, his own work in modeling the search.

Downloading is one of the most common types of restrictions. Downloading can be a reproduction of huge portions of a database onto a local computer so that the library can search the downloaded data without having to pay for access to that data. Or it can be the equivalent of making a printout to give to the user to report the results of the search. Most online databases permit a disk copy to be made for a user, although some are beginning to restrict how many lines or entries can be downloaded. Under the statute, if it is permissible to print information from the database for the user, it is permissible to give the user a disk containing the information unless the license agreement specifically prohibits downloading.[135] There generally seems to be some understanding on the part of vendors that users can download "insubstantial portions" of the database, but there is little information as to what particular database vendors believe constitutes an insubstantial portion of the database. Virtually all vendors say that the data is for personal use only and may not be transmitted or sold.[136]

On the other hand, some corporate libraries have received permission from a database that consist primarily of bibliographic entries with abstracts to download references of interest and load them on the company electronic mail system for mass distribution within the company. In other words, the library is creating an electronic newsletter that updates researchers on new materials published in fields of interest for their work. Some of the permissions have permitted the reproduction of abstracts while others have prohibited their inclusion. It is important that the library seek permission to download data from the database and to use it in the manner it needs. Again, it never hurts to ask. Many database vendors have now responded to demands for downloading substantial portions of their databases by making it possible to mount the databases locally. Fees for local mounting are likely to be high, but, depending on the use the library or company wants to make of the database, such fees actually could be a good investment for the library. Any library that wants to consider local mounting should contact the vendor directly.

University libraries have been experimenting with agreements to make certain databases are available through a campus network. Many vendors have been very willing to negotiate such agreements but almost always restrict access to students and faculty enrolled in the university. Such libraries generally have an affirmative duty to police access so that other users of the catalog via a network will not have access to licensed software or database information. This differs from the availability of collections in these libraries which often are available to the public at large, so there has been some concern on the part of the library community about restrictive access. To the extent that license agreements have facilitated the definition of user population, they have been useful. Several libraries have been successful in negotiating licenses with database proprietors to permit the loading of abstracts of journal articles into their online catalogs. Some corporate libraries have negotiated agreements to permit the loading of newsletters on the company network; such agreements generally permit libraries to use the newsletter in electronic format for a specified length of time after which they must be destroyed.

License agreements appear to be the standard for database access. To the extent that they have been used to extend a publisher's rights beyond those provided by the copyright law, they are detrimental.[137] Libraries should recognize that to the extent contract terms conflict with the copyright law, such agreements supersede rights under the Copyright Act. It is critical that librarians know what rights they are relinquishing in signing a license agreement.

VI. ELECTRONIC PUBLISHING AND OTHER ISSUES

A. *Electronic Publishing*

Within the past two years, the first two true electronic journals have become available. After 20 years of discussions about the coming revolution in publishing, the electronic era has finally arrived, although some journals continue to be available in both print and electronic format. The first two peer-reviewed journals, the *Journal of Clinical Trials and Postmodern Culture*, are very different, not only in content but also in how they are made available. Currently, *Postmodern Culture* is available free over the Internet; the *Journal of Clinical Trials* is available for a $110 annual subscription rate.

Many of the issues that a library is likely to raise concerning electronic journals have not yet been answered by either journal. For example, can one library borrow an article from a subscribing library through interlibrary loan? Obviously, the cost-free nature of *Postmodern Culture* means that the answer to the question has no economic consequence since there is no subscription rate. For the *Journal of Clinical Trials*, however, the matter is an important issue which the present license agreement does not answer. Once would expect that an electronic journal for which there is a subscription rate would limit access to employees of the subscribing library's corporate organization. Another question is whether access should be restricted through access codes monitored by the library.

It is likely that publishers will develop payment mechanisms on a per use basis. The nature of the electronic environment makes it possible to count uses, which never was possible before, and this raises concerns about a requirement to pay for each use, which is not required in the print environment. The assumption has been that a library which subscribes to a journal may circulate that journal to multiple users to read the article. In fact, the article may even be reproduced within the bounds of section 108 of the Copyright Act and fair use. Electronic information may not be so available in the future, but instead, publishers may resort to a "pay as you read system."[138]

Some predict that the electronic environment will forever change the contours of copyright law because of intrinsic factors such as the ease with which works can be replicated, transmitted and modified.[139] Others posit that copyright will continue to exist because of the need to regulate proprietary information in the marketplace and the public interest in knowledge.[140] Clearly, this is an area in which librarians must watch future developments.

B. *Internet and Copyright*

The increasing availability of the Internet to special librarians invariably raises copyright concerns. The Internet can be defined as "as a vast network of networks that interconnect thousands of computing sites in government, industry and academia."[141] It has moved beyond just an electronic mail system to become an infrastructure for broader information services, many of which we can only imagine today.[142] For copyright purposes, the Internet should be thought of as the equivalent of a telephone line; and, obviously, transmitting a copyrighted work over the telephone does not change the underlying copyright in the work.

If a librarian responds to an electronic mail request from another librarian or posts some original message to a listserv, that librarian holds the copyright in the message. A scrupulous person will seek permission to reuse the message and will cite it with credit to the original author. However, it is ludicrous to assume that the message will never be reused if there is a value in the content. Thus, if one is concerned about protecting words and messages, she should refrain from making them generally available on the Internet.

Of much greater concern, however, is whether one who finds a copyrighted work loaded onto the network may assume that it is there with permission. Certainly, an author can post his original article on the Internet and thereby give everyone else on the Internet access to it. Such a posting is with permission of the copyright holder. It is also technologically possible for someone to post an author's article without the author's permission, and then the work has received wide dissemination which would constitute copyright infringement. This means that librarians should determine the copyright status of articles located on the Internet before reproducing and using them. Just because something is found on the Internet does not mean that it was put there by permission of the copyright holder.

A good practice would be to indicate whether works posted are original with the poster or if they are posted with permission of another. The librarian who locates an article with no notice of copyright may not assume that the article is there with permission absent some indication to the contrary. The profession should police itself on this matter.

What effect the proposed National Information Infrastructure will have on these issues is not clear at present. Librarians should note that there are vocal advocates who seek a "pay for use" system similar to television "pay for view" schemes. Library associations should be posed to present the users' point of view at all relevant times including congressional hearings.

VII. CONCLUSION

The computer arena is the copyright area in which the most developments are likely to occur over the next 10 to 20 years. Already the section 101 definition of computer programs is dated, and new types of works such as CD-ROMS and multimedia works contain software along with other works. Librarians and their organization should watch for further congressional action in this area as well as consult corporate counsel as litigation concerning computer software and databases occurs.

Electronic publishing is increasing at a fantastic rate, as is Internet use. Key issues discussed in this chapter will continue to expand as librarians become even more dependent on the electronic environment.

ENDNOTES - CHAPTER 6

1   William M. Landes & Richard A. Posner, An Economic Analysis of Copyright Law, 18 J. Legal Stud. 325 (1989) [hereinafter Landes].

2   U.S. Const., art I, § 8, cl. 8.

3   Landes, *supra* note 1, at 326-27.

4   Making electronic copies of existing copyrighted works is discussed in chapter 7. *See also* text accompanying notes 134-42.

5   17 U.S.C § 101 (1988).

6   Brad Wright, Note, *Changing the Standard for Computer Software Copyright Infringement: Computer Associates Int'l v. Altai*, 13 Geo. Mason U. L. Rev. 663, 664 (1992).

7   In 1984 Congress passed the Semiconductor Chip Act to protect semiconductor chip products and mask works for a fixed period of ten years. 17 U.S.C §§ 901-14 (1988). A semiconductor chip is distinguished from computer software in the way that it is

        (I) ... the final or intermediate form of any product -

        (A) having two or more layers of metallic, insulating or semiconductor material deposited or otherwise placed on, or etched away or otherwise removed from a piece of semiconductor material in accordance with a predetermined pattern; and

        (B) intended to perform electronic circuitry functions. *Id.* at § 901(a).

8   Wright, *supra* note 6, at 665.

9   Steven L. Mandell, *Computers and Data Processing* 161 (2d ed. 1982) [hereinafter Mandell].

10   Wright, *supra* note 6, at 666.

11   Alan Freedman, *The Computer Glossary* 545-46 (5th ed. 1991).

12   Mandell, *supra* note 9, at 193.

13   *See* Alan Cohen, *Structure, Logic and Program Design* 13 (1983). Flow charts are a type of notation that depict the "flow" of sequencing of logic in a program, showing the various paths that can be taken. Id. An example of pseudocode is Program Design Language (PDL), a term used for informal, language-independent expression of program logic. It allows the programmer to express his thoughts and informally describe the operations to be performed without the rigid details of syntactically correct programming language code. Id. at 265. Structure diagrams are a notation for showing the pattern exhibited by the order in which events will occur in a program. *Id.* at 2-8.

14   There are literally hundreds of different languages, but a handful of them, COBOL, FORTRAN, C, Pascal, Ada and BASIC, account for most of the software produced. Language choice often depends on the problem to be solved as programming features and suitability vary among the languages. *See generally* Mandell, supra note 9, at 244-67.

15   *Id.* at 217.

16   *Id.*

17   *See* Apple Computer, Inc. v. Franklin Computer Corp., 714 F.2d 1240 (3d Cir. 1983).

18   Wright, *supra* note 6, at 665-66.

19   *Id.* at 665.

20   *See Software Engineer's Reference Book* II/3 (John A. McDermid ed., 1991).

21   *Id.* at 15/8.

22   *See generally* Wright, *supra* note 6, at 666-70. Judge Learned Hand developed his famous "abstraction test" approach for defining the limit of what is copyrightable in Nichols v. Universal Pictures, 45 F.2d 119 (2d Cir. 1930), cert. denied, 282 U.S. 902 (1930). "Upon any work ... a great number of patterns of increasing generality will fit equally well, as more and more of the incident is left out ... [B]ut there is a point in this series of abstractions where they are no longer protected." *Id.* at 121.

23   17 U.S.C § 101 (1988).

24   H.R. Rep. No. 1476, 94th Cong., 2d Sess. (1976) *reprinted in* 17 *Omnibus Copyright Revision Legislative History* 54 (1977).

25   17 U.S.C § 102(a) (1988).

26   CONTU completed both tasks and then passed out of existence. The interlibrary loan guidelines were published in the Conference Report that accompanied the Act. See H.R. Rep. No. 1733, 94th Cong., 2d Sess. (1976). The CONTU report was published as *The National Commission on New Technological Uses of Copyrighted Works, Final Report* (1979) [hereinafter cited as CONTU Report].

27   Act of Dec. 12, 1980, Pub. L. No. 96-517, 94 Stat. 3015, 3028 (codified at 17 U.S.C. § 117 (1988)).

28   17 U.S.C § 117(1) (1988).

29   *Id.*

30   "Memory" here refers to random access memory (RAM). This is not to be confused with two other forms of memory storage — computer diskettes and hard drives. RAM is electronic memory, while diskettes and hard drives are magnetic. Electronic memory takes up less space and can be accessed far more quickly than can magnetic memory. In addition, for a computer to execute a program, the program must be loaded in a precise location in RAM.

31   CONTU Report, *supra* note 26, at 24.

32   17 U.S.C § 101 (1988).

33   *Id.* § 106(1). Absent section 117, it is the right of reproduction that a user would violate each time she executed a program.

34   *Id.* § 117(2).

35   Atari, Inc. v. JS & A Group, Inc., 597 F. Supp. 5, 9 (N.D. Ill. 1983) (quoting CONTU Report at 31).

36   An example of such error is formatting one's hard drive which would delete all items stored on that drive.

37   17 U.S.C § 117(2) (1988).

38   For a discussion of lending copies of programs, *see* text accompanying notes 88-108.

39   Vault Corp. v. Quaid Software, Ltd., 847 F.2d 255 (5th Cir. 1988).

40   To develop RAMKEY, Vault alleged, Quaid must have executed a copy of PROLOK so that Quaid's computer programmers could study it. As noted, executing a program entails copying the program into computer memory. Quaid acknowledged as much but argued that it had purchased copies of PROLOK and that copying these was withing its rights under section 117. *Id.* at 258.

41   *Id.* at 259.

42   *Id.* at 261.

43   Sega Enters., Ltd. v. Accolade, Inc., 977 F.2d 1510 (9th Cir. 1992).

44   To run a game on one of these home entertainment systems, one needs the game itself (software embodied in the form of a read-only plastic cartridge), a console that runs the game and a television. The most popular of such systems is Nintendo.

45   Sega, 977 F.2d at 1514-15.

46   *Id.* at 1515.

47   *Id.*

48   *Id.*

49   *Id.* at 1520.

50   *Id.*

51   17 U.S.C § 117 (1988).

52   CONTU Report, supra note 26, at 13. This does not change the fact that the copyright owner retains the right to prepare a derivative work such as a new version of a program for commercial distribution.

53   *Id.* at 13-14.

54   719 F. Supp. 1006 (D. Kan. 1989).

55   796 F. Supp. 520 (E.D. Pa. 1990). Both this court and the *Foresight* court looked to the CONTU Report for guidance.

56   CONTU Report, *supra* note 26, at 13. Cited by the court at 746 F. Supp. 520, 536-37 n.17.

57   *User's Guide to Software Licenses*, 301-05 (W.A. Hancock ed., 1988) [hereinafter User's Guide].

58   *Id.* at 901-03.

59   Diane W. Savage, *Law of the LAN*, 9 Santa Clara Computer & High Tech. L.J. 193, 217 (1993) [hereinafter Savage].

60   User's Guide, *supra* note 57, at 1503.

61   Savage, *supra* note 59, at 194.

62   *Id.*

63   *Id.* at 212-13.

64   *Id.* at 213.

65   *Id.* at 217-19.

66   User's Guide, *supra* note 57, at 701.

67   *Id.* at 702.

68   *Vault*, 847 F.2d at 269-70.

69   Louisiana Software Enforcement Act, La. Rev. Stat. Ann. § 51:1961 (West 1987).

70   *Vault*, 847 F.2d at 268-69.

71   User's Guide, *supra* note 57.

72   *Id.* at 703.

73   *Id.* at 703-04.

74   17 U.S.C § 109(a)(1)(A) (Supp. III 1991).

75   *Software Rental Amendments of 1990: Hearings Before the Subcomm. on Courts, Intellectual Property and the Administration of Justice of the House Comm. on the Judiciary*, 101st Cong., 2d Sess. 15 (1990) [hereinafter 1990 Hearings].

76   17 U.S.C § 109(a) (1988).

77   1990 Hearings, *supra* note 75, at 8.

78   H.R. Rep. No. 265, 101st Cong., 2d Sess. 3-8 (1990) [hereinafter House Report 265].

79   *Computer Software Rental Amendments Act of 1988: Hearings on S. 2727 Before the Subcomm. on Patents, Copyrights and Trademarks of the Senate Comm. on the Judiciary*, 100th Cong., 2d Sess. 22 (1988) [hereinafter 1988 Hearings].

80   *Computer Software Rental Amendments Act of 1988: Hearings on S. 198 Before the Subcomm. on Patents, Copyrights and Trademarks of the Senate Comm. on the Judiciary*, 101st Cong., 1st Sess. 48 (1989) (testimony of Luanne James, Executive Director of ADAPSO) [hereinafter 1989 Hearings].

81   1988 Hearings, *supra* note 79, at 17.

82   House Report 265, *supra* note 78, at 3.

83   *Id.*; 1990 Hearing, *supra* note 75, at 16. These estimates reflect the amount lost due to inadequate legal protection. The U.S. International Trade Commission estimated the *absolute* amount lost at $25 billion per year.

84   Pub. L. No. 101-650, 104 Stat. 5134 (codified at 17 U.S.C §109(b)(2)(B) (Supp. III 1991)).

85   17 U.S.C § 109(B)(1)(A) (Supp. III 1991).

86   Step-Saver Data Systems v. Wise Technology, 939 F.2d 91, 96 n.7 (3d Cir. 1991).

87   The debate has shifted to whether software qualifies as goods under the Uniform Commercial Code (UCC). If so, the UCC places a limit on the contractual warranties. This debate is likely of very limited interest to libraries.

88   H.R. Rep. No. 735, 101st Cong., 1st Sess. 8 (1990).

89   *Id.*

90   1988 Hearings, *supra* note 79, at 10.

91   *Id.*

92   1989 Hearings, *supra* note 80, at 5.

93   17 U.S.C § 109(b)(1)(A) (Supp. III 1991). See Chapter 5 for a discussion of loaning recordings in libraries.

94   1989 Hearings, *supra* note 80, at 5.

95   1988 Hearings, *supra* note 79, at 16.

96   House Report 265, *supra* note 78, at 5.

97   1989 Hearings, *supra* note 80, at 40; *id.* at 56-57 (testimony of Jon Shirley, President of Microsoft).

98   *Id.* at 36 (testimony of Heidi Roizen, president of SPA).

99   *Id.*

100   *See, e.g., id.* at 13-72.

101   *See id.* at 60.

102   *See* text accompanying note 86; 17 U.S.C § 109(b)(2)(A) (Supp. III 1991). For the required text of the warning, *see* text accompanying note 104.

103   17 U.S.C § 109(b)(1)(A) (Supp. III 1991).

104   37 C.F.R. § 201.24 (1993).

105   *Id.* § 201.14.

106   17 U.S.C § 108(f)(1) (1988).

107   *See* 58 Fed. Reg. 37,757 (1993); *Comments Sought on Lending of Software by Non-Profit Libraries*, 46 Pat. Trademark & Copyright J. (BNA) 228, 234 (1993).

108   American Association of Law Libraries, the Association of Research Libraries and the Special Libraries Association. *Joint Statement to the Copyright Office As to Whether the Computer Software Rental Amendments Act of 1990 Has Achieved Its Intended Purpose with Respect to Lending by Nonprofit Libraries* (Oct. 7, 1993) (on file with authors and named associations).

109   17 U.S.C § 101 (1988).

110   Alan R. Greengrass, *Databases and Their Off-Spring*, Bookmark, Winter, 1992, at 147 [hereinafter Greengrass].

111   There is some overlap, however, especially with respect to the merger doctrine. *See* text accompanying notes 122-33.

112   Miller is a professor at Harvard Law School and was a member of CONTU. He is a prolific author of titles such as *Intellectual Property: Patents, Trademark, and Copyrights in a Nutshell* (2d ed. 1990).

113   Authur R. Miller, *Copyright Protection for Computer Programs, Databases, and Computer-Generated Works: Is Anything New Since CONTU?*, 106 Harv. L. Rev. 977 (1993) [hereinafter Miller].

114   499 U.S. 340, 111 S.Ct. 1282 (1991).

115   *Id.* at 1296.

116   *Id.*

117   *Id.* at 1291.

118   *Id.* at 1296-97.

119   For a recent sampling of federal appeals court decisions citing Feist, see Computer Assocs. Int'l, Inc. v. Altai, Inc., 982 F.2d 693 (2d Cir. 1992) (to support rejection of the sweat of the brow doctrine); Atari Games Corp. v. Oman, 979 F.2d 242 (D.C. Cir. 1992) (requisite creativity is low and the video game "Breakout" is copyrightable); Sega Enters., Ltd. v. Accolade, Inc., 977 F.2d 1510 (9th Cir. 1992) (copyright law encourages dissemination of creative works for public benefit); Atari Games Corp. v. Nintendo of Am., Inc., 975 F.2d 832 (Fed. Cir. 1992) (support of claim that copyright in a work does not entail a copyright in its constituent elements); Los Angeles News Serv. v. Tullo, 973 F.2d 791 (9th Cir. 1992) (noting that although Fiest held that the standard for creativity was extremely low, it did not undermine the rule that photographs are copyrightable as original works of creative expression); Arica Inst., Inc. v. Palmer, 970 F.2d 1067 (2d Cir. 1992) (citing *Feist's* two requirements for infringement: ownership of copyright and copying of constituent elements); Federal Election Comm'n v. International Funding Inst., 969 F.2d 1110 (D.C. Cir. 1992) (originality is a contitutional element); and Schiller & Schmidt, Inc. v. Nordisco Corp., 969 F.2d 410 (7th Cir. 1992) (a catalog of office supplies is an original compilation, but no infringement found).

120   Key Publications, Inc. v. Chinatown Today Publishing Enters., Inc., 945 F.2d 509 (2d Cir. 1992).

121 *Id*. at 515.

122 101 U.S. 99 (1879).

123 Herbert Rosenthal Jewelry Corp. v. Kalpakian, 336 F.2d 738, 742 (1971). The case found that a jeweled bee pin was more of an idea than an expression and thus did not infringe another jeweled bee pin.

124 937 F.2d 700 (2d Cir. 1991).

125 *Id*. at 706.

126 *Id*.

127 736 F.2d 859 (2d Cir. 1984).

128 *Id*. at 863.

129 808 F.2d 204 (2d Cir. 1986).

130 *Id*. at 208.

131 Miller, *supra* note 113, at 1041-42.

132 CONTU, *supra* note 26, at 40.

133 For a good general discussion of the merger doctrine in the context of computers, *see* Computer Assocs. Int'l. Inc. v. Altai, Inc., 982 F.2d 693, 708 (1992).

134 Greengrass, *supra* note 110.

135 DIALOG's standard agreement states that under no circumstances may customers copy or transmit data received from DIALOG in machine-readable form except as expressly authorized in advance. The standard NEXIS agreement permits subscribers to download and store in machine-readable for insubstantial protions of materials from an individual file unless specifically prohibited in the Supplemental terms for Subscribers or if it is contrary to fair use under the Copyright Act. *Id*. at 147-48.

136 *Id*. at 148.

137 Scott Bennett, *Copyright and Innovation in Electronic Publishing: A Commentary*, 19 J. Acad. Libr. 87, 89 (1993) [hereinafter Bennett].

138 *See* Jane C. Ginsburg, *Copyright Without Walls?: Speculations on Literary Property in the Library of the Future*, Representations, Spring, 1993 for a discussion of a myriad of issues that must be examined as information moves from print into electronic format.

139 Pamela Samuelson & Robert J. Glushko, *Intellectual Property Rights for Digital Library and Hypertext Publishing Systems*, 6 Harv. J. L. & Tech. 237, 239-40 (1993) [hereinafter Samuelson].

140 Bennett, *supra* note 137, at 87.

141 Samuelson, *supra* note 139, at 243.

142 *Id*. at 243-44.

CHAPTER 7

SPECIAL PROBLEMS: UNPUBLISHED WORKS, EDUCATIONAL COPYING AND LIBRARY RESERVES

I. FAIR USE OF UNPUBLISHED WORKS

A. *Introduction*

Unpublished works such as letters and journals are protected by the provisions of the Copyright Act of 1976. As is the case with published works, facts contained in unpublished works are not shielded from use, even if the copyright owner has refused permission; only expression is so protected. Similarly, since unpublished works are covered by the Copyright Act, the doctrine of fair use applies to them as clarified by a 1992 amendment to the Act. As interpreted by the courts, however, fair use is far narrower with respect to unpublished works.

B. *The Harper & Row Case*

The only Supreme Court case to date that addresses the fair use of unpublished material is *Harper & Row Publishers, Inc. v. Nation Enterprises*.[1] Even before President Gerald Ford left office he contracted with Harper and Row to publish his yet unwritten memoirs; by agreement the memoirs were to contain significant material not previously published concerning the Watergate affair, the pardon of President Richard Nixon and Ford's reflections on the history, events and people involved. In 1979, as the manuscript for *A Time to Heal* was nearing completion, the publisher contracted with *Time* magazine. *Time* agreed in 1979 to publish a serialized version of a small portion of Ford's memoirs concerning his pardon of Nixon. Before the *Time* article was published, however, *The Nation* magazine, which had obtained access to a stolen copy of the manuscript, published an article drawing on it, including verbatim quotes of 300 to 400 words and scooped the *Time* publication.[2] Harper & Row, which had a nonexclusive license to publish the material, sued *The Nation* to recover the $25,000 fee *Time* refused to pay for the 7,500-word excerpt after *The Nation* had printed its purloined account. *The Nation* raised the defense of fair use.[3]

The Supreme Court rejected this defense and found *The Nation* liable for infringement of copyright. In doing so, the Court announced a presumption against fair use of unpublished works: "under ordinary circumstances, the author's right to control the first public appearance of his undisseminated expression will outweigh a claim of fair use."[4]

The first section 107 factor is the *purpose and character of the use*.[5] The Court cautioned that if alleged infringement fits within one of the categories listed in section 107 as the type of activity courts tend to find as fair use, it does not necessarily follow that the use is fair. Rather, falling within one of the enumerated activities of section 107 merely creates a presumption in favor of the defendant for this factor.[6] *The Nation* also claimed that its use should be excused because of First Amendment values and the fact that its publication was of high public concern. In other words, because the information on why Ford pardoned Nixon was of such public concern that the use of the quotes should be excused as fair use even though scooping the first authorized serialization in this fashion ordinarily would not be fair use. The Court noted that if this line of reasoning were adopted, fair use would effectively destroy any expectation of copyright in any writings by a public figure.[7] "By establishing a marketable right to use one's expression, copyright supplies the economic incentives to create and disseminate ideas."[8] The Court found no reason to add to the fair use doctrine a public figure exception. The Court also addressed *The Nation*'s bad faith in accepting stolen documents to facilitate a "scoop" under this factor. If a defendant's conduct lacks good faith, then this factor may be tallied against the defendant even if the conduct fits within one of Congress's enumerated productive uses.[9]

The second factor, the *nature of the copyrighted work*, weighs heavily against use of unpublished material. As the Court phrased it, "(t)he fact that a work is unpublished is a critical element of its `nature,'"[10] since it means that the copyright owner has not yet exercised one of the section 106 exclusive rights, the right of first publication. The law recognizes a greater need to disseminate factual works, however, than works intended for entertainment.[11]

The third factor focuses on the *amount and substantiality used* in comparison to the work as a whole. This analysis is essentially the same whether the work is published or not; in both instances, the factor will be weighed against the defendant if the portion taken was the heart of the work, even though only a fraction of the plaintiff's work was used. This highlights the fact that the third fair use factor is both a qualitative test and a quantative test. Although the actual number of words taken from the larger memoirs was quite small, the part taken on the Nixon pardon was the central part of the work, the part in which the general public had the most interest.[12]

The Court held the last factor, the *effect on the market*, to be "undoubtedly the single most important element of fair use."[13] The right that was at stake was the right of first publication, and the Court found that "a fair use doctrine that permits extensive prepublication quotations from an unreleased manuscript without the copyright owner's consent poses substantial potential for damage to the marketability of first serialization rights in general."[14] Moreover, this analysis must look at not only the immediate impact of the market for plaintiff's work, but also the potential market for it, in addition to the demand for any derivative works.[15]

C. Subsequent Cases

Of the cases which have dealt with the issue of fair use of unpublished works since *Harper & Row*, *Salinger v. Random House, Inc.*[16] is the most important. *Salinger* involved the use of a number of unpublished letters in an unauthorized biography of the famous but reclusive author J. D. Salinger. The biographer found the letters deposited by recipients of the letters in several university libraries. He paraphrased many of them after Salinger contested the initial use of direct quotations in the biography. The U.S. Court of Appeals for the Second Circuit reversed the District Court's finding of fair use, holding the biographer liable for infringement of Salinger's copyright in his unpublished letters.[17] This case interpreted the fair use privilege with regard to unpublished works even more narrowly than the Supreme Court did in *Harper & Row*, especially in light of the absence of any bad faith in *Salinger* remotely similar to the defendant's in *Harper & Row*. Further, former President Ford was about to publish the memoirs, so the infringement took away income he had already contracted to earn. By contrast, Salinger had little or no interest in publishing the letters. Instead, his purpose in challenging their use in the biography was to see that they were not published at all. The nuances added to the section 107 factors by *Salinger* and other post-*Harper & Row* cases follow.

1. *The purpose and character of the use.* *Salinger* stresses that, even if a particular use falls within one of the categories listed by Congress in section 107 such as scholarship or research, this does not entitle the user to any special consideration,[18] although the user normally will have this factor weighed in her favor in the absence of any bad faith. Regarding the character of the use, a failure to request the right to use the copyrighted work does not qualify as bad faith, since "the lack of permission is beside the point as long as [the] use meets the standards of fair use."[19]

2. *The nature of the work.* The unpublished nature of the work seemed to insulate it from fair use claims. For unpublished material, the second fair use factor had never been applied in favor of an infringer.[20] This statement is no longer true today as there are cases which find fair use of unpublished works.[21]

3. *The amount and substantiality of the portion used.* *Salinger* emphasizes that a user cannot escape possible liability for infringement by paraphrasing expression. Moreover, "(t)he 'ordinary' phrase may enjoy no protection as such, but its use in a sequence of expressive words does not cause the entire passage to lose protection."[22] The ordinary phrase can be quoted, since it cannot be copyrighted, but any expressive words around it are protected and can cause the user to become an infringer.

4. *The effect on the market.* A use that hurts the market for an original work because of its criticism of the original, or its author leads people not to purchase it, is not infringement. It is infringement only if the market for the original is harmed because too much of the original is used.[23] *Salinger* was particularly disturbing since his objection to use of the quotations had the effect of censoring the work.

D. Recent Developments

Alarmed by the near disappearance of the fair use doctrine in these cases, librarians and other concerned groups such as biographers, literary critics and historians pushed for congressional action to secure unequivocally the privilege with respect to unpublished material.[24] Legislation was introduced in 1990 in both the House [25] and the Senate [26] to accomplish this goal. Passage of some version of this legislation appeared likely until the computer industry launched an effort to block adoption of any such bill, charging that expanded fair use in this area would jeopardize rights in unpublished source codes.[27] As a result of this last-minute lobbying, the House bill died in the subcommittee and the Senate bill saw no action.[28]

Again in 1991, identical bills were introduced in both houses of Congress to amend the Copyright Act. In September 1991, the Senate bill[29] was adopted and sent it to the House Judiciary Committee for its consideration.[30] The House, however, dropped the fair use provision from its bill[31] in October of that year.[32] In March 1992, Representative William J. Hughes (D-N.J.) introduced H.R. 4412, which added these words to the close of section 107: "The fact that a work is unpublished shall not itself bar a finding of fair use if such finding is made upon consideration of all the factors set forth in paragraphs (1) through (4)."[33] The resolution was slightly amended in subcommittee[34] before being sent to the House Judiciary Committee, which favorably reported the bill to the full House.[35] The bill passed both the Senate and the House and on October 24, 1992, President George Bush signed the bill into law.[36]

Thus, the fair use section of the Act now contains a provision that ensures unpublished works may be used in the same manner as published materials. Courts will apply the four fair use factors to any claimed infringement of an unpublished work. To date there have been no cases construing the 1992 amendment; however, courts are likely to look at *Harper & Row*, *Salinger* and their progeny to determine when use of an unpublished work is a fair use. Certainly, a prime consideration will be whether the author of the unpublished work seeks to protect works she later plans to publish or whether she wants to suppress writings that place her in an unfavorable light. If the author's motivation is the latter consideration, courts probably are more likely to find that a use by a biographer, critic or historian is a fair use.

II. EDUCATIONAL PHOTOCOPYING

A. *Introduction*

One of the most perplexing issues concerning reproduction rights arises in the area of multiple copying for education. Many people labor under the false impression that if the copying is for educational purposes it is per se a fair use. This is not true. In fact, even educational copying tends to be limited to copying within nonprofit educational institutions. Even within such institutions it is important to determine the proper scope of multiple copying on the part of schools and libraries. The extent to which educators can reproduce and distribute to their students copies of copyrighted works was thoroughly discussed during the revision of the Act. Ultimately, Congress dealt with the issue of copying for teaching under the fair use provisions of section 107. A number of exemptions for nonprofit educational uses of copyrighted works such as performances and displays within the course of instruction are sprinkled throughout the Act.[37] Congress dealt specifically with the issue of library photocopying setting out detailed exemptions to the rights of copyright holders in section 108.[38]

B. *Classroom Guidelines*

The House Report that accompanied the Act includes educational copying guidelines which define the reach of fair use with respect to photocopying by teachers in nonprofit educational institutions.[39] These limits establish the *minimum* range of protected activity; for that reason, the guidelines are not a figurative line in the sand beyond which fair use is exceeded. While the guidelines were drafted by advocates from both the author/publisher and educational communities, the American Association of University Professors and the Association of American Law Schools declined to endorse the result, claiming that the limits were "too restrictive with respect to classroom situations at the university and graduate level."[40] Even so, the guidelines are helpful, since a teacher is assured of complying with the law so long as he is within the guidelines. Further, although the guidelines are not law, they do have some stamp of approval by Congress. The House Judiciary Subcommittee had urged publishers and authors to get together with members of the educational community to try to reach a meeting of the minds on permissible uses of copyrighted works for education.[41] Congressional endorsement is further evidenced by incorporation of the negotiated guidelines into an official publication of the House of Representatives.

1. Single Copying For Teachers

Most single photocopying for personal use generally falls within fair use. Under the Guidelines teachers specifically are permitted to make one copy of the following items for scholarly research, preparation for class or use in actual teaching: (a) a chapter from a book; (b) an article from a journal, periodical or newspaper; (c) a short story, short essay or short poem, whether from a collective work or not; and (d) a chart, diagram, graph, drawing, cartoon or other picture from a book, periodical or newspaper.[42]

2. Multiple Copying

Multiple copies of a copyrighted material may be made for classroom use provided that the copies made do not exceed one copy per student, and that: (a) the copying meets the tests of brevity and spontaneity, which are defined in the guidelines; (b) the copying satisfies the cumulative effect test, also defined in the guidelines and (c) a notice of copyright is included on each copy distributed to students.[43]

While the third requirement is self-explanatory, the other two are vague without reference to the definitions contained in the guidelines. The definition of brevity differs depending on whether the work being copied is poetry, prose, an illustration or a so-called "special" work. With poetry, the guidelines permit reproduction of the complete work if it consists of less than 250 words and is printed on less than two pages; copying of longer works of poetry is limited to 250 words. Prose may be copied in its entirety if the article, story or essay is under 2,500 words, or one may copy an excerpt from a prose work of not more than 1,000 words or 10 percent of the work, whichever is less, a minimum of 500 words.[44]

Copies of illustrations are limited to one chart, graph, diagram, drawing, cartoon or picture per book or per periodical issue.[45] "Special" works such as children's books, those writings that combine words and illustrations may not be reproduced in their entirety. An excerpt comprising not more than two of the published pages and containing not more than 10 percent of the words found in the text, may be reproduced, however.[46]

Spontaneity means that the copying is done at the instigation of the individual teacher. In other words, the chair of the English Department cannot determine that all students in sophomore composition classes will receive a copy of a particular essay. Instead, the decision to reproduce the material for the class must be made by the individual teacher. Further, "(t)he inspiration and decision to use the work and the moment of its use for maximum teaching effectiveness are so close in time that it would be unreasonable to expect a timely reply to a request for permission."[47] Although this seems to run contrary to good teaching practices that stress lesson planning, it does make it clear that works reproduced for distribution to classes should be materials that fill in gaps in existing textbooks or are works that represent recent developments. The guidelines are not intended to provide a way for teachers to avoid having students purchase textbooks but are a way to provide needed supplementary materials without seeking prior permission of the copyright holder.

Cumulative effect requires that the copying be done only for one course in the school. If an instructor teaches multiple sections of the same course, she could reproduce copies of the same material for each student in those sections; however, she could not use the same item for different courses although the courses might be related. Moreover, only one short poem, article, story, essay or two excerpts may be reproduced from any one author or more than three from the same collective work or periodical issue during one class term. Finally, there can be no more than nine instances of such copying per course per class term.[48] Thus, if a teacher has permission from 20 copyright holders to reproduce and distribute their works to his students, the teacher may copy 29 items — 20 with permission and nine that satisfy the requirements of the Guidelines.

3. Prohibitions

The Guidelines contain several prohibitions that further clarify their application. It is impermissible to copy to create or replace an anthology. The restriction of one article, poem, etc., or two excerpts from an author further clarifies this point. It is also not permissible to copy any "consumable works." Consumable works are defined as standardized tests, workbooks, answer sheets and exercises. Also, copying shall not (a) substitute for the purchase of any published work, (b) be supervised by a superior, or (c) occur repeatedly for the same work from term to term.[49] The reason for the final prohibition is that a teacher who copies an article for distribution one term has ample time to seek permission for its use in subsequent terms. Thus, use the second time cannot satisfy the spontaneity requirement. Finally, students may not be charged more than the actual cost of making the photocopy.[50] The price of making the copy can include the cost of paper, toner and personnel costs associated with the copying activity. It is clear, however, that a school may not make a profit from the sale of copies to students.

4. Acceptance of the Guidelines

These "safe harbor" guidelines have enjoyed wide acceptance and they have influenced the development of copyright policies in schools and colleges. They certainly have shaped fair use standards at all levels of education including universities. Some university policies studied by Kenneth Crews in his survey are more lenient and stress that the guidelines express the minimum copying that may be done.[51] The guidelines were first invoked in the settlement agreement reached in 1983 between publisher members of the Association of American Publishers (AAP) and New York University (NYU) when NYU agreed to follow the guidelines and to adopt them as a part of an overall university copyright policy. The terms of the settlement agreement were widely publicized. Another much lesser known suit was filed against the University of Texas by publishers on similar grounds and the terms of its settlement were similar to those in NYU.[52] Both the NYU and Texas policies agreed not to exceed the guidelines unless the author first contacted the copyright holder to get permission from the university's legal counsel before exceeding the minimum guidelines. Interestingly, these universities appear to have accepted the minimums as maximum guidelines.[53]

At the same time publishers sued some commercial copy shops;[54] together with the university settlement this was a public relations coup for publishers. The suits had been rumored for some time before filed; then after settlement the AAP contacted hundreds of colleges and universities encouraging adoption of copyright policies that followed the guidelines so that they did not meet NYU's fate.[55]

The fact that the guidelines still are not well understood is evidenced by the litigation against the commercial copy shop, *Kinko's*.[56] It is surprising that after the NYU settlement and the related cases against the commercial copy shops *Kinko's* was even necessary.

B. Anthologies and Course Packs

For users of educational copyrighted materials the *Kinko's*[57] case has important ramifications. Professors, copy shops and campus copy centers generally considered most copying of materials for the classroom to be reproduction that fell within fair use which allows multiple copying of limited amounts of materials without permission under certain circumstances such as for teaching in a nonprofit educational institution. After this case educators are less sure. The national chain of photocopying stores was found to be infringing a group of publishers' copyrights by its course packet service.[58] The copy stores sold course anthologies copied from books to students without obtaining permission or paying royalties. Kinko's argued that its copying was fair use because it was done for educational purposes. The court found the chain's practices failed to meet three of the four fair use factors.[59] Kinko's is neither a nonprofit entity nor an educational institution.[60] The court recognized that the material was for educational use once it was in the hands of the students, but Kinko's purpose was primarily commercial and weighed against it on the first factor. Because the nature of the materials copied was primarily factual the court found in favor of Kinko's on the second factor, since factual works generally have greater fair use than do works intended primarily for entertainment.[61] Since Kinko's copied entire chapters, the judge determined that the copy store had crossed the quantitative line of the third factor. Sometimes as much as 25 percent of a work had been copied.[62] Finally, the court found that Kinko's anthologies hurt the market for both the sale of publishers' books and the royalties which users would have paid to copy the works.[63] The court awarded the publisher plaintiffs $510,000 in damages plus attorneys fees[64] of $1.36 million.[65] The damages and attorneys fees award along with the accompanying publicity about the case convinced many faculty members that copyright holders were serious about enforcing their rights and that they would no longer overlook all kinds of educational copying.

To develop customized publishing, Kinko's announced that it would enter into blanket agreements with publishers to pay agreed-upon royalties.[66] As a result of the suit, production of anthologies was delayed and costs increased while publishers and faculty attempted to deal with the overwhelming number of requests for permissions.[67]

Emboldened by the Kinko's settlement, the AAP announced plans to go after college anthology producers and has begun the enforcement phase of its efforts. On behalf of Princeton University Press, Macmillan's Free Press Imprint and St. Martin's Press the AAP filed suit against the Michigan Document Services in Detroit.[68]

Subsequently, Kinko's announced that it was discontinuing its course packet service.[69] In December 1993, Kinko's ceased fulfilling requests for course anthologies, announcing the company's intention to focus on electronic information services emphasizing photo and document transfer, video teleconferencing and electronic publishing.[70]

There remain some good alternatives to Kinko's for producing course packs. See Chapter 4 for a discussion of licensing mechanisms.

III. LIBRARY RESERVES

A. *American Library Association Reserve Guidelines*

Although there are no congressional guidelines on library reserves, the American Library Association (ALA) has promulgated suggestions for libraries regarding photocopying for library reserve as a part of a model policy for colleges and universities.[71] Since the reserve area is an extension of the classroom, the ALA views copying for reserve as permissible under conditions similar to the classroom guidelines, a position supported by the Association of American Law Schools but rejected by the Register of Copyrights.[72] In particular, the Register maintains that since the guidelines require spontaneity, libraries may not place material on reserve for consecutive terms.[73] Nonetheless, the reserve guidelines enjoy wide acceptance among libraries and presumably among publishers since they have not been litigated.

Single copies may be made for reserve use, the ALA believes, so long as the standards of the classroom guidelines are observed. When multiple copies are requested for reserve by a faculty member, the ALA makes the following recommendations.

1. The amount of material should be reasonable in relation to the total amount of material assigned for one term of a course. Matters such as the nature of the course, its subject matter and level [74] should be taken into account. This statement makes it clear that library reserves are not to take the place of a purchased textbook or course pack [75] on which royalties have been paid to the copyright holder. Materials photocopied for reserve generally are intended to supplement the other materials assigned for the course and not to serve in lieu of any other materials.

2. The number of copies should be reasonable in light of the number of students enrolled, the difficulty and timing of assignments, and the number of other courses which may assign the same materials.[76] This likely means that the library rather than the faculty member should determine what number of copies is reasonable.

3. The material should contain a notice of copyright.[77] If the article contains the notice of copyright on the first page of the article, the library need do nothing more than ensure that the notice is legible. If the notice is not printed on the article, then the library must write or stamp the notice.[78]

4. The effect of photocopying the material should not be detrimental to the market for the work.[79]

5. In general, the library should own at least one copy of the work.[80] This does not mean that occasionally a library could not place on reserve a photocopy that belongs to a faculty member or one the library obtained through interlibrary loan. The library should not make a general practice of this, however, if it is to comply with the ALA model policy.

Some writers have said that a reasonable number of copies would be six, although other factors may permit more copies to be made, including the difficulty of the assignment, the number of students in the class and the length of time the students have to complete the assignment.[81] If there is too little time for the professor to request permission from the copyright holder to make the copies, more copies may be placed on reserve than in the normal situation. A faculty member who is uncertain about placing copies on reserve should defer to the library's policy or obtain the copyright holder's permission.

A more puzzling part of the model policy is found in the materials that precede the requirements discussed above. As an introductory matter, the policy states that, in general, photocopying for reserve collections should follow the classroom guidelines. Then four specific requirements are listed: (a) the distribution of the same material does not occur every semester, (b) only one copy is distributed to each student, (c) the material contains a notice of copyright on the first page of the portion copied, and (d) students are not assessed any charge beyond the actual charge of making the copy.[82] When one examines these four requirements, it is difficult to see their applicability to reserves.

Regarding requirement (b), for library reserve situations, the library would practically never reproduce a copy for each student in the class. In fact, the very reason for placing materials on reserve is that a few copies are made and not one per student. The last requirement concerning a charge to students for the copies is totally irrelevant to reserve practices. No library charges a fee for library reserve use! Thus, the cost per student limitation is unnecessary. The requirement concerning the inclusion of a notice of copyright on the photocopied material is important; in fact, it is so important that the requirement is repeated in the specific portion of the guidelines that relate to reserve copying.[83]

Thus, the only requirement of the four that either has any relevance or is not repeated specifically in the reserve guidelines themselves is the first one which states that the same material should not be distributed every semester. Libraries have struggled with the meaning of this as it applies to reserve copying. Some libraries apply the requirement as if it were mandated by the law itself. Other libraries take a more liberal view and believe that when the model policy says that in general the classroom guidelines should be followed, that is what it means. Therefore, some libraries refuse to allow photocopies of copyrighted materials to remain on reserve more than one semester without written permission from the copyright owner. Libraries that follow this strict interpretation vary in whether they handle permission requests or whether they require faculty members who want the items placed on reserve to contact the copyright holder for permission. Some libraries go so far as to refuse a request to put an item on reserve the second semester without written permission from the owner which the faculty member must submit along with the request.

Among the libraries that apply a more liberal view of the policy, some encourage faculty to obtain permission to use photocopied material on reserve the second semester but stop short of an absolute requirement. Others take the tact of removing from reserve all materials each semester and returning them the faculty member in the hopes that over time he will pare down the amount of material placed on reserve.

The course chosen by the library concerning this requirement may depend more on space management issues than on a literal reading of the guidelines. In other words, if the library is concerned about finite reserve space, it may decide to interpret strictly the meaning of the first general requirement. This relates not so much to copyright as to an important administrative issue for the library. Unfortunately, there is no general guidance on this issue. No library has been sued over its reserve collections nor have there been other interpretations from the ALA, the Copyright Office or others.[84]

B. *Electronic Reserves*

Some libraries have substituted electronic copies for traditional photocopy reserve collections and many others are considering doing so. Electronic reserves could solve space and staffing problems currently associated with reserve collections of photocopies of copyrighted articles, chapters, etc. Is it possible to comply with the ALA Reserve Guidelines and still develop and maintain an electronic reserve system? Perhaps, but the publisher community is very concerned about retention and repeated use of electronically stored copies. The *ALA Reserve Guidelines* state that the amount of material a faculty member requests be placed on reserve as well as the number of copies should be reasonable. In the photocopy reserve situation, the library determines what number of copies is reasonable based on the number of students in the class, level of the class, length of the assignment, length of time before the assigned material must be read and the like. For example, based on these factors, a college library might decide that eight copies on reserve are sufficient for a class of 25 students.

When one thinks of an electronic reserve collection, the usual situation envisioned is that materials would be scanned and stored on a central library computer which students could access from terminals in the library or even from remote locations. There are other ways this could be done. For example, the library could scan the items and put them on a floppy disk that is circulated just as hard copies are circulated. Another method is to put the scanned copies onto the central computer and then make a "copy" for each user by putting the copies into the electronic mailboxes for each student.[85] For purposes of this chapter, however, assume that scanned copies are stored on a central computer in the library and that users access the material through terminals located within the library and from remote locations.

Where libraries have initiated electronic reserves, there is no uniformity in what types of materials are available and whether the library considers the activity to be fair use or one on which royalties should be paid. Both Rice and Duke universities have experimental electronic reserve collections of copyrighted materials. Rice is paying royalties on every copy made (i.e., for every use) of material in the electronic reserve collection; Duke believes an electronic reserve collection is fair use and is not paying royalties. The University of Pittsburgh has created an electronic reserve collection that consists of heavily used but uncopyrighted works created by faculty such as case studies, old examinations and the like. These and other libraries that experiment with electronic reserve collections will be able to help answer some of the unanswered questions.

Electronic reserve collections present several copyright concerns. First, what is the number of copies that are made and does the number raise fair use concerns? Instead of any reserve collection photocopy that might be read by several students, one electronic copy per student is made since the accepted definition of when an electronic copy is made includes whenever a copy is displayed on the screen in addition to when a copy is printed from the screen or downloaded to a disk.[86] Second, will the library erase the scanned copy at the end of the class term? Third, is it necessary for the library to restrict access to the electronic copies to students enrolled in particular classes? If so, will it be done through access codes or some other mechanism? Fourth, must the library require the professor to obtain permission to place the item on reserve for subsequent terms? For the present, these and other considerations have meant that few libraries actually have created electronic reserve collections, although the numbers seem to be increasing. It is possible that because of these concerns electronic reserve collections are more closely akin to course packets than to traditional reserve collections. If they are analogous to course packets, then the *Guidelines on Multiple Copying for*

Classroom Use must be met and royalties paid for copying in excess of fair use. The question then for the calculation of royalties becomes how many copies were made. For the library, an important question follows: who pays the royalty, the library or the student? Since the Copyright Clearance Center has not been authorized to collect royalties for electronic copies, if royalties are due, they must be paid directly to the publisher or copyright holder. This alone may discourage some libraries from converting traditional reserve collections to electronic format.

Another important concern arises under section 108(g) which states that the exemptions for library copying extend to isolated and unrelated reproduction of a single copy. This applies to the reproduction of the same material on separate occasions but does not extend where the library engages in related or concreted reproduction and distribution of multiple copies of the same material, on one occasion, or over a period of time. Section 108(g) applies whether the multiple copying is by aggregate use by one or more individuals or for separate use by the individual members of a group. The publisher community apparently does not object to reserve collections that carefully adhere to the ALA guidelines for photocopies as evidenced by the lack of complaints, articles challenging the guidelines or litigation. Why the same acquiescence is not present for electronic reserve collections appears somewhat inconsistent to many in the library community.

III. CONCLUSION

Many issues with respect to electronic copying and distribution have not yet been addressed. The advent of electronic publishing is likely to raise more issues such as the extent to which the *Interlibrary Loan Guidelines* are applicable to electronically published journals. The proposed National Information Infrastructure is likely to raise as many issues as it resolves. There may be other special problems which arise from time to time concerning the application of the copyright law, especially as it relates to new technologies and particular uses for education. Librarians must be diligent in their efforts to respond to future developments.

ENDNOTES - CHAPTER 7

1   471 U.S. 539 (1985). See Chapter 2 for a full explanation of the doctrine of fair use.

2   The Supreme Court opinion includes an appendix that contains the *Time* excerpt with the verbatim quotes taken by *The Nation* in bold face.

3   *Harper & Row*, 471 U.S. at 542, 544.

4   *Id.* at 555.

5   17 U.S.C. § 107(1) (1988).

6   *Harper & Row*, 471 U.S. at 560.

7   *Id.* at 555-57.

8   *Id.* at 558.

9   *Id.* at 562-63.

10   *Id.* at 564.

11   *Id.* at 563.

12   *Id.* at 564-66.

13   *Id.* at 566.

14   *Id.* at 569.

15   *Id.* at 568.

16   811 F.2d 90 (2d Cir. 1987).

17   *Id.* at 100.

18   *Id.* at 97.

19   Wright v. Warner Books, Inc., 953 F.2d 731 (2d Cir. 1991).

20   New Era Publications v. Henry Holt & Co., Inc., 873 F.2d 576, 583 (2d Cir. 1989).

21   *See e.g.*, Wright, 953 F.2d 731.

22   *Salinger*, 811 F.2d at 98.

23   New Era Publications Int'l, ApS v. Carol Publishing Group, 904 F.2d 152, 160 (2d Cir. 1990).

24   Written Statement of the American Association of Law Libraries Before the Subcommittee on Patents, Copyrights, and Trademarks of the Senate Committee on the Judiciary (July 11, 1990).

25   H.R. 4263, 101st Cong., 2d Sess. (1990).

26   S. 2370, 101st Cong., 2d Sess. (1990).

27   *Bill Would Apply Fair Use To Published And Unpublished Works*, 39 Pat. Trademark & Copyright J. (BNA) 405 (1990).

28   *Work's Unpublished Nature Is Important Under New Fair Use Bill*, 42 Pat. Trademark & Copyright J. (BNA) 44 (1991) [hereinafter *Work's*].

29   S. 1035, 102d Cong., 1st Sess. (1991).

30   *Work's*, *supra* note 28.

31   H.R. 2372, 102d Cong., 1st Sess. (1991).

32   *Senate Passes Legislation on Fair Use of Unpublished Works*, 42 Pat. Trademark & Copyright J. (BNA) 520, 523 (1991).

33   H.R. 4412, 102d Cong., 2d Sess. (1992).

34   *Fair Use Bill Is Introduced and Reported Out of House Panel*, 43 Pat. Trademark & Copyright J. (BNA) 407, 408 (1992). The words "all the factors set forth in paragraphs (1) through (4)were eliminated in favor of "all the above factors." *Id.*

35   *Briefs: Legislation, Copyright*, 44 Pat. Trademark & Copyright J. (BNA) 14 (1992).

36   *President Signs Bills on Fair Use and Late Payment of Patent Maintenance Fees*, 44 Pat. Trademark & Copyright J. (BNA) 686 (1992); correction, 45 Pat. Trademark & Copyright J. (BNA) 14 (1992).

37   17 U.S.C. § 110 (1988). See Chapter 5 for a discussion of using audiovisual works.

38   17 U.S.C. § 108 (1988). See Chapter 3 for further discussion of library photocopying.

39 H.R. Rep. No. 1476, 94th Cong., 2d Sess. (1976) *reprinted in* 17 *Omnibus Copyright Revision Legislative History* 68-71 (1977) [hereinafter House Report]. See Appendix C for text of Agreement on *Guidelines for Classroom Copying in Not-For-Profit Educational Institutions.*

40 *Id.* at 72.

41 *Id.* at 67.

42 *Id.* at 68.

43 *Id.*

44 *Id.* at 68-69.

45 *Id.* at 69.

46 *Id.*

47 *Id.*

48 *Id.*

49 *Id.*

50 *Id.* at 70.

51 Kenneth D. Crews, *Copyright, Fair Use and the Challenge for Universities: Promoting the Progress of Higher Education* 72-74 (1993) [hereinafter Crews].

52 *See id.* at 45-49, 72-74 for a discussion of the NYU settlement.

53 *Id.* at 73.

54 Harper & Row Publishers, Inc. v. Tyco Copy Service, Inc., 10 Copyright L. Dec. (CCH) ¶ 25,230 (D.Conn. 1981); Basic Books, Inc. v. The Gnomon Corp., 10 Copyright L. Dec. (CCH) ¶ 25,145 (D.Conn. 1981).

55 Crews, *supra* note 51, at 47.

56 Basic Books, Inc. v. Kinko's Graphics Corp., 758 F. Supp. 1522 (S.D.N.Y. 1991).

57 For additional discussion of Kinko's see Chapter 4 on licensing. *See also* David J. Bianchi, Grant H. Brenna & James P. Shannon, Comment, *Basic Books, Inc. v. Kinko's Graphic Corp.: Potential Liability for Classroom Anthologies*, 18 J. of C. & U. L. 596, 596-97 (1992).

58 Kinko's, 758 F. Supp. at 1529.

59 *Id.*

60 *Id.* at 1531-32.

61 *Id.* at 1533.

62 *Id.* at 1534.

63 *Id.*

64 *Id.* at 1545-47.

65 *See* Basic Books, Inc. v. Kinko's Graphics Corp., 21 U.S.P.Q. 2d (BNA) 1639 (S.D.N.Y. 1991).

66 Connie Goddard, *Textbook Authors Hear Kinko's View on Customizing*, Publ. Wkly., July 19, 1991 at 12.

67 Debra E. Blum, *Use of Photocopied Anthologies for Courses Snarled by Delays and Costs of Copyright-Permission Process*, Chronicle of Higher Educ., Sept. 11, 1991, at A19.

68 *AAP Files New Copyright Suit*, Pub. Wkly., Mar. 9, 1992, at 7.

69 *Kinko's Drops Course Packets to Focus on Electronic Services*, Pub. Wkly., June 14, 1993, at 21.

70 *Id.*

71 American Library Association, *Model Policy Concerning College and University Photocopying for Classroom Research and Library Reserve Use* (1982) [hereinafter Reserve Policy], *reprinted in* 4 Coll. & Res. Lib. News 127-31 (1982).

72 James S. Heller & Sarah K. Wiant, *Copyright Handbook* 28-29 (1984) [hereinafter Heller & Wiant].

73 *Id.*

74 *See* text accompanying notes 38-56.

75 Reserve Policy, *supra* note 71, at 129.

76 *See* text accompanying notes 57-70 for a discusson of course packs.

77 Reserve Policy, *supra* note 71, at 129.
78 *Id.*
79 See Chapter 3 for a discussion of library compliance with the notice requirement.
80 Reserve Policy, *supra* note 71, at 129.
81 *Id.*
82 Heller & Wiant, *supra* note 72, at 28.
83 Reserve Policy, *supra* note 71 at 129.
84 *Id.*
85 Kenneth Crews, in his survey of university copyright policies found that the amount of copying allowed for reserve was the widest variable among universities. *See* Crews, *supra* note 51, at 86-92. He points out that

> [L]ibraries in particular tend to implement the strictest legal interpretations, despite the potential constraints on library service, and despite their professed concerns about the law's inhibitng force. *Id.* at 89.

CHAPTER 8

INTERNATIONAL COPYRIGHT

I. INTRODUCTION

A. *General Matters*

Research is becoming increasingly interdisciplinary and multi-jurisdictional. Scientific and technical research and development has long been international in character, and other types of research have become dependent on publications and reports from other countries. This movement toward the globalization of research means that librarians must know something about international copyright law since so many valuable research publications are produced abroad. British and Western European countries publish much of this material, but increasingly it is published in Asia and other parts of the world.

Earlier chapters discussed types of works and particular uses of these copyrighted works by special libraries. The situation is complicated when the work is published abroad, however. Licensing differs although the Copyright Clearance Center (CCC) has agreements with a variety of foreign collectives to collect royalties far in excess of fair use copying and to distribute those royalties to the foreign collective.[1]

Developed nations view their intellectual property products as valuable property that should be protected by law both domestically and abroad. The U.S. law and that of Canada, the United Kingdom and other members of the European Union (formerly the European Economic Community) reflect this belief that copyrighted works should be protected. There are four arguments commonly advanced to justify copyright. First is the principle of natural justice which entitles an author to the fruits of his labor. To some extent, this is the equivalent of an author receiving wages for her endeavors. Second, an economic argument may be made. It takes a considerable investment to produce a copyrighted work, especially works like motion pictures, videotapes, sound recordings and computer programs. No company is likely to invest in producing such works if there is no economic reward or at least a reasonable expectation of recouping costs and making a profit. Third, creative works can be viewed as cultural assets by the country in which they are produced. Some countries that recognize public lending rights view those rights as a companion to copyright in protecting cultural assets. So, it is in the public interest to reward this creativity as a contribution to national culture. Finally, a social argument favors wide dissemination of works because they forge links between social classes, various races, ages, genders, etc. The sharing of the ideas contained in the works contributes to the advancement of society.[2] A nation's copyright law is based on these arguments to varying degrees. Although the scope of the rights provided by a country's copyright laws may vary, the copyright system must balance two public interests: the rights of the copyright owner and the "reasonable demands of an organized society."[3]

Through various copyright treaties discussed below there has been an attempt to harmonize the basics of copyright law, but there are differences in both owners' rights and users' rights abroad. As a general matter, the United States treats works published in this country by foreign authors as if they were produced by an American author.[4] Works by foreign authors but published abroad in a country which is a member of the Universal Copyright Convention (UCC) or the Berne Convention receive the same protection in the United States as works published domestically by U.S. authors.[5]

International copyright law developed from both public and private international law. Public international law is based on customary law, which has a superstructure of international treaties. It encompasses international economic law that governs the distribution of goods and services through trade agreements, international financial agreements, and the like. Private international law is not really international law but rather is part of national law. "Conflict of laws" is another name for private international law; it is a discipline within each country's legal system. International copyright law is a hybrid. The main source of international copyright law is found in treaties between nations, and as such, is a part of the public international law. The principles applied in these treaties or conventions come from public international law, but the situations to which they are applied are principles of private international law. The other major source of international copyright law is national legislation interpreted by judicial decisions.[6]

B. Matters Generally Covered by a Country's Copyright Law

Foreign copyright law exhibits the same basic framework as does U.S. copyright law. Certainly, there are differences, but U.S. copyright law provides an excellent focal point from which to shade the differences in foreign law. In their work analyzing and comparing the copyright law of the twenty-two nations,[7] including the United States, Nimmer and Geller[8] selected countries that produce a significant number of copyrighted works each year. The countries represent all areas of the world; only Poland and what was Czechoslovakia are included from the countries of the former Soviet bloc. Despite a rapidly changing international political climate, the copyright laws from these nations present a good general overview. Matters compared in this section include originality, creativity, term of copyright, moral rights, fair use, registration, categories of works, and adaptation to new technology, i.e., software protection.

1. Fixation, Originality and Creativity

In the United States, a work must be "fixed in any tangible medium of expression" to merit copyright protection.[9] With the exception of Belgium and Poland, European countries do not require the fixation of a work as a prerequisite to its legal protection. This division may reflect the continental emphasis on an author's moral rights, which is distinct from the Anglo-American emphasis on broad economic rights, including the rights of the public. A work in France, for example, has earned protection "by the mere fact of the author's conception being implemented, even incompletely."[10] Neither China nor Japan require fixation.[11]

All countries require a work to be original as a preliminary condition of copyright protection. Originality, however, does not have the same meaning in every country. Two conceptions dominate. First, originality may imply that the work originated with the author. That is, the author did not copy the work from someone else. In this sense, originality goes toward individuality. The starkest example of this is found in countries that subscribe to the theory of statistical uniqueness, which holds that where the independent creation of the same work by another author is statistically improbable, the work is original and deserves protection.[12] The concern is clearly that the work be the author's own. Second, originality may imply a minimum degree of creativity. Continental countries generally frame this aspect of originality as bearing the imprint of the author's personality. In the United States, the creativity threshold appears to be no higher than not "devoid of any creative thought, obvious."[13] The requirement of creativity, wherever applicable, is always in addition to the requirement that the author be the originator of the work. The originality requirement is universal; creativity is not. Despite these differences, nations that impose the creativity criterion agree that the requisite level of creativity is very low. No country's law requires a subjective evaluation of a work's artistic merit.

Moreover, all countries appreciate the same distinction between copyright and patent law and are careful to keep patent law's more stringent requirement of novelty from encroaching on the domain of copyright law. This separation of the two regimes for intellectual property protection works in reverse, as well: no country allows an author to copyright an idea or procedure. In every country it is the *expression* itself that is protectable under copyright. Where creativity is required, the notion of novelty is avoided.[14]

The differences among countries over the meaning of originality do not reveal themselves in the law governing compilations. All countries afford some form of protection to compilations. The bases of protection vary however. The copyright statutes of many countries enumerate compilations specifically as protectable subject matter.[15] In other countries the courts have included compilations within the meaning of traditional copyright subject matters, usually adaptations or collections; a few countries simply extend protection to compilations as literary works as does the United States. Alone on this matter is Sweden, which employs a "catalogue rule" for compilations, a *sui generis* form of protection available for ten years to works containing "large blocks of data." In all countries a compilation still must satisfy the originality requirement.[16] Usually originality is satisfied by the judgment necessary for the selection and arrangement of the data. Some countries[17] specify that sufficient judgment or labor will suffice for protection of a compilation. The U.S. Supreme Court has explicitly rejected the labor theory (called the sweat of the brow theory),[18] as has Greece, which bars the protection of obvious arrangements such as chronological or alphabetical lists. There is further worldwide agreement that the protection of a compilation does not extend to the preexisting material, i.e., the individual data or components that make up the compilation.[19]

2. Term of Copyright

As a general rule, the duration of a copyright on a literary work extends 50 years following the death of the author. The original idea was that two generations of the author's heirs ought to enjoy the benefits of copyright on the work following the author's death: Poland, 25 years; Brazil and Spain, 60 years; Germany and Israel, 70 years.[20] At the present time, the European Union and the United States are considering expanding the term of copyright to life plus 70 years.[21] Possible justifications for expanding the term are that the longer life expectancy together with the increasing age of couples when they begin to have children has changed the number of years of what constitutes two generations. Of course, a change to life of the author plus 70 years also would harmonize the laws of European Union (EU) countries to be consistent with the longest term offered by a member country (Germany).[22] Universal life has the advantage of equalizing the duration of protection granted the works of a single author. For example, author Isaac Asimov has written or edited over 250 books, and the copyright on each title will expire on the same date in any given country.

3. Moral Rights

Moral rights traditionally include paternity and integrity. Paternity gives the author the right to claim copyright in any work he created; an author may also deny authorship. Integrity is the right of the author to prevent any distortion of a work that might damage her reputation in any way. The continental tradition of recognizing the moral rights may be fairly said to have taken hold worldwide with three notable exceptions: Australia, the United Kingdom and the United States Among these, only Australia flatly rejects the notion of "moral rights,"[23] although it does recognize rights what could be characterized as such, e.g., the right of attribution. Britain and the United States recognize them to a limited degree only, and the latter has done so only recently (since adherence to the Berne Convention in 1989).[24] The resistance of the Anglo-American tradition to the recognition of moral rights reflects the primary emphasis of copyright law in these countries: authors' rights are secondary to the rights of the public.[25]

4. Fair Use

This distinction is plainly visible in the rights of the public with respect to protected uses of copyrighted works. The same three countries that resist moral rights protection explicitly allow the public in broadly drawn circumstances to use an author's work without permission. In the United States this is called fair use and in Britain and Canada it is called fair dealing. Canada is the only nation that explicitly recognizes both moral rights and the fair use doctrine. With respect to moral rights, "doctrine" is the key word. In the other 18 countries, for a use to qualify as fair it must fit an often narrowly drawn statutory description, e.g., classroom use. Two countries, Belgium and Argentina, do not recognize fair use at all.[26]

5. Registration

On the whole, the registration of a work as a prerequisite to copyright protection has vanished. This is a direct result of the Berne Convention. Where it plays a part at all, registration seems to influence the availability of certain remedies[27] or certain economic rights.[28] It does not affect whether the work is copyrighted.

6. Subject Matter

In terms of what subject matter is protected by copyright law, the distinction between the French and German traditions,[29] on the one hand, and the Anglo-American tradition,[30] on the other, has been noted widely. The former countries employ the broad notion of art and literature to bring new types of work within the sweep of copyright law, while the latter continue to enumerate statutory categories of subject matter worthy of protection. U.S. copyright law protects literary works (including computer software); musical works (including lyrics); dramatic works (including music); pantomime and choreographic works; pictorial, graphic, and sculptural works (with a host of enumerated subclassifications such as prints and labels); motion pictures and audiovisual works; sound recordings; architectural works; and "other works."[31] It is fair to say that nations worldwide protect the vast majority of these works, as well. What differs is the means and extent of protection.

To see how the laws of various countries adapt to protect newly arising types of works, computer programs serve as a good illustration. Presently, eight of the 22 countries do not cover software by statute.[32] Of these, only Brazil does not have a significant body of case law that wrestles with the copyrightability of software, however. In fact, in most of the countries without statues specifically designating software as protectable, the courts have found protection in the general copyright law anyway. These courts have done this by classifying software as a literary work,[33] a work of cinematography,[34] or within the catchall clause of the copyright statute.[35] One court simply held that computer software was copyrightable but that is also had to meet the requirements of originality and fixation.[36]

The remaining countries have provisions in their copyright statutes specifically to protect software. Within these countries, some include software in a more traditional subject matter classification, usually defining it as a literary work, while others have established software as a uniquely protectable subject matter.[37]

The broad uniformity among worldwide nations in the protection of software by copyright seems to stem from two considerations. First, courts and the legislatures seemed to believe that software was worthy of some type of legal protection. Second, governments also considered patent law to be an inappropriate form of protection. Thus, copyright law was the natural choice. Two countries,[38] however, are seriously considering *sui generis* protection of software because of difficulties in applying the notion of originality to software information (object code) that is intelligible only to a machine.

The software situation reveals the highly adaptable nature of copyright law. As technologies continue to emerge, such adaptability will be critical to extend legal protection to new works that likely are to be of considerable value.

C. Rights Afforded to Foreign Works Generally

In most countries, a copyright entitles its holder to a bundle of rights. France does not grant this bundle; instead, it prefers to grant a general right of economic exploitation, which includes control over reproduction and performance.[39] So, too, the former Czechoslovakia, which quit trying to enumerate specific rights that attend a copyright.[40] When a country is dealing with a question of the rights of foreigners authors, the following questions should be asked in this order:

1. Can the foreigner claim protection under one of the international conventions to which country X is a party?
2. If not can he claim protection under a bilateral agreement to which country X is a party?
3. If not, can he claim protection under the national law of country X relating to foreigners?[41]

Generally, a country will apply its own law to an infringement claim by a foreigner. The tort principle of *lex loci delicti* is operative: the law of the state in which the wrong occurs governs. Once the applicable body of law is known, three other criteria must be satisfied before an action will lie. The first criterion requires that the work be copyrightable under national law. A computer program in Brazil, for example, is not necessarily protectable. Second, the work must not be within the public domain. The avenues by which a work may reach the public domain vary worldwide and must be examined. Third, a "connecting factor" must be present. The connecting factor will be found if either (a) the work was published first in the country where the alleged infringement took place, or more likely, (b) the author is a resident of a country with which the host country has an applicable treaty or with which the host country is a co-member in a convention such as Berne.[42] First publication is an important concept, but it may be amended by a provision that provides for simultaneous publication in a treaty country and thereby may enlarge the protection a work may receive especially if first publication is in a country that would grant little or no protection.[43]

Where a treaty or convention applies, the effects of its provisions depend on the law of the country in which the wrong occurred. In France, for example, a claim can be brought directly under the treaty provision, while in the United States it is necessary to bring the claim under domestic law that implements the treaty.[44]

II. THE BERNE CONVENTION

A. *Introduction*

An author's rights under copyright would be greatly reduced if they were not recognized internationally. The author would receive no royalties for the use of her work outside of her own country, and this would greatly reduce the value of the copyright. In fact, absent international recognition of copyright, an author's work could be exploited commercially in another country where the author would receive no royalty payments and reimported back into his own country at a cheaper rate.[45] In some Pacific Rim countries this practice is still a concern and the United States continues to encourage countries without a copyright law to enact legislation or, if such legislation is in effect, to enforce it. Prior to the nineteenth century, copyright law was a matter of domestic concern for most countries. As creative and scientific works began to be exchanged beyond national borders with increasing frequency, the desirability for providing transnational protection for an author's work was recognized. The result was a large number of bilateral treaties between countries to provide for protection of copyrighted works from across national borders.[46]

After considerable discussion among European countries, a copyright treaty to replace the existing bilateral treaties was the choice of a number of countries. The first effort to create a European copyright treaty was in 1858 when the Congress of Authors and Artists met in Brussels.[47] Touted as the solution to the problems caused by lack of uniformity, in 1886 the Berne Convention was established with ten member countries.[48] Among international treaties, the Berne Convention is considered a real success as it has remained in effect for over a century. In fact, the areas to which the Convention applies have changed very little, although the level of protection has increased. The major difference is that many of the developing countries were members only as colonies of European countries. Today they are members in their own rights.[49] Initially, the United States was not a signatory to Berne. Until the United States amended its law with regard to term of protection, notice and registration, and, to a lesser extent, its position on moral rights, it could not become a member of the Berne Union. As discussed below, it was years before the United States adhered to the Berne Convention.

Diplomatic conferences attended by member nations of Berne (known as the Berne Union) convene to consider revisions to the Convention. Over the years there have been five revisions and two significant additions to the Convention[50] that enhance the level of protection for copyrighted works in member states.[51] Because of the number of revisions, countries that join the Berne

Convention at different times belong to different versions.[52] This complicates any discussion of Berne and examination of whether an individual country adheres to a particular provision of the treaty.

When the United States joined the Berne Convention on March 1, 1989, discussion of Berne ceased to be solely academic exercises. Two significant changes were wrought by joining Berne: (a) the United States acquired copyright relations with 24 additional countries and (b) the level of copyright protection in this country increased.[53] Congress has amended the U.S. law to conform to the provisions of Berne;[54] and further amendments are likely as the Berne Convention continues to evolve through the years.

B. *Major Provisions and U.S. Adherence*

The two main principles of the Berne Convention are national treatment and minimum rights. The national treatment provision requires each Berne Union member to extend its national copyright laws to protect works originating from other member nations. For example, Greece, which is a Berne country, must treat works from Germany (also a Berne member) as if the work had originated in Greece and was subject to Greek copyright law.[55]

The other dominant Berne principle is that of minimum rights. While the Berne Convention is based on the idea of national treatment, there are some rights that all Berne Union members must guarantee to authors of other member nations. This floor of rights is necessary to prevent economic free-riding, i.e., enjoying the strict protection under the copyright laws of other countries while allowing a culture of piracy to thrive under lax laws in the home nation. So, minimum rights are guaranteed under Berne regardless of national law.[56]

Berne does not allow formalities to stand in the way of protection of a work. Minimum procedural rights guarantee that an author will enjoy copyright protection regardless of compliance with any formalities. Thus, an author need not include a notice of copyright on her work or register the work to receive protection.[57] This provision of Berne traditionally was one source of U.S. resistance to joining the Union since prior to 1989 notice and registration were required for enjoyment of copyright protection in the United States American law required notice on a work before copyright would even issue, and it required registration as a prerequisite to filing an infringement claim in federal court.[58] With the American version of copyright law rooted in the notion of public welfare rather than an author's integrity, resistance to eliminating these requirements appears natural. Thus, the Berne Implementation Act (BIA),[59] although it dropped the required formalities, retains shades of the old regime. While registration and notice are no longer prerequisites of copyright protection for an infringement action, the thrust of U.S. law strongly favors both. For example, notice on a work totally defeats the defense of innocent infringement. Conversely, the absence of notice allows courts to reduce statutory damages.[60]

Aspects of registration exhibit an even stronger imprint of American policy. First, registration remains a prerequisite to an infringement claims for both U.S. citizens and citizens of non-Berne Union countries. Only Berne Union authors outside the United States need not register their works here.[61] Second, registration of a work entitles its author to tactical legal advantages. As a matter of evidence, registration is presumptive proof of both the author's identity and the copyright's validity. As to remedies, registration within three months of publication is a prerequisite to the recovery of both statutory damages and attorneys' fees.[62]

At the present time, Congress is considering amending the statute to bring U.S. law even closer to the Berne Convention provisions. If implemented, registration would no longer be necessary even for eligibility for certain damages and the deposit requirement would be eliminated.[63]

The BIA eliminated entirely the requirement that transfers of copyright or exclusive rights under copyright be recorded with the Copyright Office. As with registration, an author could not file an infringement action without compliance. Following the BIA, compliance with this formality is no longer necessary for any authors, whether citizens of a Berne country or not.[64]

Also, as part of its minimum rights scheme, Berne requires the protection of certain subject matter. Every Berne member must protect books, pamphlets, addresses, choreographic works, sculpture and architecture. The requirement that architecture be protected generated controversy in the United States, where domestic law protected architectural blueprints and models, but not the buildings themselves. Nevertheless, some commentators believed that U.S. law was in compliance with Berne on this point, citing the absence of specific criteria for the copyrightability of architectural works in the laws of many long standing Berne Union members.[65] Congress amended the statute in 1990 to provide for the copyrightability of buildings.[66]

Other features of the minimum rights scheme relate to the protection of both economic and moral rights of an author. Whatever protection a country grants to an author through its copyright laws must extend for the life of the author and 50 years thereafter. The United States for many years extended protection for 28 years, with renewal option for an additional 28 years.[67] Even prior to the BIA Congress lengthened statutory protection and adopted the Berne standard of life plus 50 years when it passed the 1976 Act.[68]

Out of the bundle of rights guaranteed under U.S. copyright law, only the right of adaptation is even arguably mandated by the Berne Convention; all nations must grant the right of translation to authors. Most nations, however, grant the same bundle granted by the United States That the world's oldest international copyright convention requires the protection of the right of translation should surprise no one.

Finally, Berne protects two major moral rights, paternity and integrity. Paternity is the right to claim or deny authorship, and integrity is the right to prevent distortion or mutilation that tends to harm an author's reputation.[69] Two important aspects of moral rights warrant discussion. First, Berne considers moral rights and economic rights to be distinct. That is, despite transferring his economic rights[70] to someone else, an author does not cede the rights of paternity or integrity. Second, while the moral rights remain with an author even upon the transfer of economic rights, moral rights are alienable via an express agreement.[71] When the United States joined the Berne Convention, there was considerable debate over whether the United States granted adequate protection to an author's moral rights. Despite the fact that Article 6*bis* of the Berne Act specified that national legislation would determine the means of protecting moral rights, the BIA did not explicitly do so. Many argued that existing U.S. legislation was elastic enough to cover moral rights, but today the point is moot. Congress passed the Visual Artists Rights Act of 1990, which guarantees protection of moral rights (but only for works of visual art).[72] Whether the rights of paternity and integrity will be extended to other subject matter in the United States is a matter of continuing speculation.

Any discussion of minimum rights requires a notation that the Berne Convention entails more than countries agreeing to enforce a body of copyright law, although that is part of it. The minimum rights are simply the *threshold* at which membership in Berne is predicated. Beyond the minimum rights, the Berne Convention is more of a system than a treaty listing rights and obligations.

The general rule under Berne is that a member nation must extend the protection of its copyright laws to authors whose works originate in other Berne Union nations. Copyright laws in the United States, for example, protect the object code of a computer program. French law does not. Nevertheless, a French computer programmer may bring an infringement action against an American infringer in federal district court. As Berne Union members, France and the United States implicitly agree to provide national protection to each other's authors.[73] Clearly, the protection is not reciprocal, or federal courts would not recognize the claim.

One requirement of Berne membership is that a country must adopt measures necessary to ensure application of the treaty. Any state that does not do so will first be the subject of diplomatic negotiations between the nations involved. If diplomatic negotiations fail, the matter may be referred to arbitration. If the arbitration is not successful, the dispute may be submitted to the International Court. Before the Court decides the issue, the country that brings the dispute must inform the World Intellectual Property Organization (WIPO); WIPO will then notify all Berne member countries to give them an opportunity to intervene in the dispute. The International Court has limited power only. It can issue opinions that interpret the Convention, but the opinion has no executory force, which leaves settlement of the dispute to diplomatic means based on the opinion or to the country found to be in violation of the Convention to amend its national legislation to conform with the Convention.[74]

C. New Technologies Under Berne

The Berne Union has been in existence for over one hundred years and convenes about once a decade to devise methods for dealing with new developments; often such new developments are technological in nature. Each meeting results in another Berne Act, to which not all Berne members necessarily adhere. A country may be a Berne member without subscribing to each act. Romania, for example, last agreed to the Brussels Act in 1928, yet it is still a Berne member and extends protection to authors from other Berne Union nations.[75] When each of two nations has agreed to the most recent Berne Act, the governing law is that Berne Act to which both have agreed. When two nations have signed no common Berne Act, each is bound by the last act it individually signed.[76]

When the United States joined the Berne Union in 1989, it subscribed to the 1971 Paris Act. The reasons the United States joined the Union after one hundred years of resistance are various, but certainly necessity was one of the more prominent motivating factors. The U.S. needed to agree with other developed nations, which are net exporters of copyrighted works. Prior to joining Berne, protection of U.S. works in many foreign countries was limited. To obtain protection in Berne nations, U.S. authors relied on "back door" protection available under Berne Act language and through its membership in the UCC discussed below. Under Berne, however, Union members protect works originating in other Berne countries. A U.S. author thus published a work simultaneously in the United States and in a Berne Union country, usually Canada (for proximity's sake). Because the work was technically "first published" in a country belonging to Berne, other Berne countries had to accord it national treatment. This proposition was costly, however. IBM, for example, estimated it would save ten million dollars annually by avoiding duplicate publication in Canada.[77] Moreover, the method achieved only uncertain protection since national treatment resulted in differing interpretations of the words "publications" and "simultaneous."

The future of the Berne Convention is likely to be positive. Any treaty that still is effective more than a century after enactment has proved that it is elastic enough to embrace new technologies.

Some amendments and expansion likely are needed in particular areas such as (a) satellite and cable distribution, (b) compulsory licensing and levies, and (c) collective administration of rights.[78]

III. THE UNIVERSAL COPYRIGHT CONVENTION

A. *Introduction*

When the United States joined the Berne Union in 1989, it had already been a member of the Universal Copyright Convention (UCC) for 34 years. Although not as venerable as the Berne Convention, the UCC had membership requirements that were attractive to the United States. The weakness of the Berne Convention was noted immediately after World War II. The primary weakness, as seen by the United States, was that Berne lacked universality; quality of rights was stressed over quantity of membership. The two superpowers after the war, the United States and the Soviet Union, were not members, but neither were many other states that were members of the United Nations, especially many Asian and African countries. Another difficulty with Berne was that its high level of protection excluded many countries from membership. The real impetus for a universal convention came from the newly founded cultural arm of the United Nations — UNESCO. The aim was to attract countries from all five continents and to include states at different levels of economic development.[79] This country was not alone in its resistance to the requirements of Berne; it was, however, the singular influence of the United States that led to a copyright convention that strongly favored U.S. law.[80] Other nations joined because either, like the United States, their public policy for promoting copyright was fundamentally different from that underlying Berne, or because their national laws fell short of the Berne requirement (such as copyright duration of at least 50 years plus life of author.[81] In either case, it is no accident that the UCC complements the Berne Convention.[82] It should be noted initially that when two countries belong to both the UCC and the Berne Convention, the latter governs the legal relationship between the two. This convention is part of the UCC agreement itself.[83]

The United States joined 14 other nations in the formation of the UCC in 1952. The treaty underwent a major revision in 1971. Over the next several decades, the Berne and UCC Conventions competed with each other for new members.[84] With 84 member states today, the UCC owes some of its success to its lower membership requirements. Many developing countries could not meet the Berne standards, yet desperately wished to trade in copyrighted materials, particularly those for education. The solution in many cases was to join the UCC.[85] The result was not always good, however. The Soviet Union, for example, joined the UCC in 1973 as a bargaining chip in trade negotiations,[86] but its compliance has been minimal.[87]

B. *UCC Requirements*

The basic framework of the Universal Copyright Convention resembles that of the Berne Convention. Eligibility for copyright protection extends to authors who are citizens of a member nation or who first publish in a member nation.[88] Further, the protection is national.[89] Thus, if Tom Wolfe learns that *Bonfire of the Vanities* has been infringed in Algeria and brings suit there, Algerian law applies.

The primary difference between the Berne Convention and the UCC is with respect to minimum rights. Under Berne, a country must extend to Union authors an extensive body of minimum rights regardless of its national laws. Under the UCC, these rights, added at the 1971 Paris Convention, include only the rights of reproduction, public performance and the right of broadcasting.[90]

Prior to 1971, the UCC requirements were, if not vague, quite minimal. The "requirements" of the UCC were actually guidelines for signatory nations to use when drafting domestic law.[91] A signatory country had to assure "adequate and effective protection."[92] The UCC also contained a provision to deal with formalities. Without totally dismissing formalities as a prerequisite to copyright protection (as did Berne), the UCC stipulated that a work satisfied the formalities required by any member nation when it was marked with a UCC notice which consists of the symbol Ⓒ, the name of the copyright holder and the year of first publication.[93] Thus a musician from Senegal who attaches UCC notices to phonorecords shipped to the United States has met the formal requirements of notice and registration. Nothing more need be done to earn protection.

The UCC also specified that the minimum duration of a copyright would be the life of the author plus 25 years.[94] There is a slight wrinkle in the UCC requirement for copyright duration, however. Treatment under the UCC is national *except* for the duration requirement. Thus, it would seem that if Country A established the copyright duration at life plus 30 years and Country B set it at life plus 50 years, Country B would have to protect works originating in Country A for the life of the author plus 50 years. This is not the case, however, due to the "rule of the shorter term."[95] Under this rule, treatment is reciprocal rather that national. Country B would thus protect works from Country A for life plus 30 years, the shorter term under the UCC. Reciprocal treatment applies only to copyright duration, however.

The 1952 Geneva Convention specified that the right of translation must be guaranteed.[96] Thus, another requirement was added for adherents to the UCC. There are a few other features of the UCC that distinguish it from Berne. As might be expected from the genesis of the UCC, the Convention does not mention moral rights.[97] The 1971 revision of the Convention granted minimum rights of an exclusively economic nature.[98]

The UCC also allows countries to claim copyright in works produced by their governments. Under the system of national treatment, this leads to the strange result that works by the U.S. government are protected in certain UCC countries while they are not protected domestically in the United States[99] Conversely, this means that the U.S. would have to protect government works from abroad although its own are classified as public domain materials.[100]

Lastly, although minimum rights under the UCC are not as extensive as those under Berne, only the UCC explicitly recognizes computer software as a protectable subject matter.[101] The UCC classifies computer programs as a scientific writing.[102]

C. Developing Nations

The UCC has special provisions that apply to developing nations before they can implement a compulsory licensing system for the translations and reproductions.[103] Whether a country is determined to be a developing nation is based on established practice of the General Assembly of the United Nations. The decisive factor appears to be whether the country files a notification at the time it certifies the UCC or thereafter declaring that it will avail itself of the benefits for translations or reproductions. Although this is a very flexible criterion, other member nations would reject the declaration if the country's economic and cultural situation were not one considered to be "developing."[104]

The compulsory license is relevant to libraries since the nonexclusive license can be granted only for educational purposes that are defined as "systematic instructional activity." The translation license is "for the purpose of teaching, scholarship or research." After three years, if the work has neither been published nor translated into a language in general use or after one year if no publication or translation into such a language in a developed country, the compulsory license comes into play. If the copyright owner can be found, he must be consulted since a compulsory license can be issued only for payment of an equitable remuneration. The developing country is not permitted to export copies made under a compulsory license; in fact, all copies made under the license must bear a notice which specifies that distribution is allowed only within the country.[105] Although research purposes are included in the translation license, libraries in corporations with foreign subsidiaries in a developing nation may not distribute translations of U.S. works under the compulsory license because industrial research institutes and industrial and commercial companies that undertake research for commercial purposes are excluded. If the research is in the educational sphere, however, the library may be able to take advantage of the compulsory license. [106] "The purpose of these licenses is to facilitate the translation of school books, text-books of all kinds, encyclopaedias, technical and scientific manuals etc."[107]

There also is available a compulsory license for translation for use in broadcasting intended exclusively for teaching or the noncommercial dissemination of research results to experts in a particular profession. The broadcast must be intended for person *within* the country. [108]

IV. EUROPEAN ECONOMIC COMMUNITY COPYRIGHT LAW

A. *Introduction*

The Treaty of Rome, signed in 1957, established the European Economic Community (now EU) as a legal entity. The EU had 12 member states in 1991: Belgium, Denmark, France, Germany, Greece, Ireland, Italy, Luxembourg, the Netherlands, Portugal, Spain, and the United Kingdom. With the dissolution of Soviet control of Eastern Europe, the number of EU members likely will increase. The population of EU countries in 1991 totals 320 million.[109] All members of the EU belong to both the Berne Union and the Universal Copyright Convention.

One of the primary goals of the EU is to create a uniform market conducive to the free flow of goods, services, capital and labor. A significant impediment to a smooth and unified market is differing national laws that relate to trade. Such differences were identified as trade barriers that increase the transaction costs of doing business.[110] The EU thus faced two problems: the conflict of national laws and the differences in national laws. The conflict of laws was resolved by instituting a system of national treatment. Just as under Berne and the UCC, the applicable law is the law of the nation where the wrong occurs. Resolving the conflict of laws, however, only eliminated the cost of uncertainty. A greater part of the high transaction costs among nations was the differing legal regimes themselves, not merely determining which law applies.[111] The EU thus undertook to harmonize its members' national laws in key areas. One area of concern was imports and exports. Articles 3 to 34 prohibit national restrictions on imports and exports, including measures that have a *tendency* to restrict trade.

One problem with a flat bar on national measures having a tendency to restrict trade is that all intellectual property schemes have such a tendency by their very nature.[112] A copyright and a patent both vest their owners with a limited monopoly of rights that affect trade.[113] Nevertheless, Article 36 excepts "industrial and commercial property" from the ban on measures restricting trade. The legal conflict between protecting national interests and fostering a smooth and uniform market cannot be understood without first understanding the legal framework of the EU.

B. The EU Generally

The EU is a fundamentally different legal entity than is either the Berne Union or the Universal Copyright Convention. Members of Berne and the UCC must make sure their national laws are not incompatible with a few mandatory provisions of those treaties.[114] That is the only requirement for membership. As noted above, each country's national law is sovereign under the conventions' systems of national treatment.

Membership in the EU is more difficult. First, when the lawmaking body of the EU speaks on a matter, member states must change their laws to conform to the pronouncement. A member state cannot sit idly and trust that its national law is not incompatible with EU law; it must be made compatible. Second, EU treaty law is supreme over national law.[115] Unlike a member of Berne or the UCC, then, a member of the EU is not the sovereign lawmaker where the international body has spoken.

The legislative branch of the EU is the Council. It can pass or reject laws that are proposed by the Commission, which is the administrative and executive branch of the EU. Members of the Commission come from each nation, for which they serve as national representatives.[116] Before passing a law, the Council must get a "reading" of the law from the European Parliament, an elected body that serves in an advisory capacity only. The Parliament, however, may propose amendments to the law.[117]

Either the Commission or the Council, most often the latter, may issue directives. A directive orders member states to implement laws to achieve certain goals.[118] The members must comply with the directive, but they are free to select the legal means of achieving the legislative goals. Unlike treaties and Court of Justice decisions, directives have no direct effect in member states except where they create rights for individuals. One of the EU's most influential directives has been the Software Directive.[119]

Once enacted, enforcement of the laws is left to the Commission. When a member state fails to adopt or amend law to conform to Community law, the Commission may bring suit against the state in the Court of Justice. In addition, the Commission represents the EU in international trade negotiations. With respect to intellectual property, it sets the standards for unfair competition and sets restrictions on the licensing of intellectual property rights. Despite the fact that it is the executive arm of the EU, the Commission may not penalize member states. It may, however, penalize individual legal entities within member states such as a corporation or a person.[120]

The Court of Justice is the ultimate interpreter of EU law. It has jurisdiction over three kinds of cases: those brought by member states, those brought against member states and those from national courts seeking an interlocutory ruling on EU acts or the treaties themselves. Together with EU treaties, decisions of the Court of Justice constitute the main body of law. This law is enforceable in national courts, regardless of national law, or in the Court of Justice. Just as American state courts may rule on the U. S. Constitution, so may national courts in the EU rule on Community law.[121]

C. Copyright in EU Member States

Although the term "copyright" does not appear in the Treaty of Rome, the phrase "industrial and commercial property" does appear and has been interpreted by the Court of Justice to embrace national intellectual property rights.[122] As mentioned earlier, the recognition of national intellectual property rights conflicts with the general scheme of eliminating trade impediments. Gray market goods often highlight the conflict. A gray market good is one that is produced in one country and imported into another without permission.[123] (Permission may be required because the producer is a licensee exporting the good into the licensor's country, or because the import is "parallel" to a similar patented good in the receiving country.) The producer in the importing country claims that it has the exclusive right of distribution, while the exporting producer claims that the exercise of the right inhibits free trade. Thus, two conditions must be met for Article 36, protection of "industrial and commercial property," to apply. First, protection of the good must be justified for reasons listed in the article such as a national interest. Second, such protection must not "constitute a means of arbitrary discrimination or a disguised restriction on trade between Member States."[124]

While the treaty allows individual nations to protect intellectual property rights, the Court of Justice has held that the exercise of certain rights is inconsistent with the free movement of goods.[125] With respect to copyrights, the Court has recognized that certain rights in the bundle of rights are compatible with the principle of free trade, while others are not. In attempting to balance a nation's interest in protecting its intellectual property with the Community's interest in eliminating trade barriers, the Court of Justice has recognized several copyright principles and doctrines.[126]

1. First Sale Doctrine

The first sale doctrine applies to a single copyrighted good. Once the good is sold, the doctrine holds, the copyright owner has relinquished the right to control its fate. The buyer of the good may then dispose of the good in any manner, including sale, rental or lease. No further royalty is paid to the copyright holder. The first sale doctrine derives from the common law forbidding restraints on alienation.[127] It applies only within the national territory, but it applies to patented goods as well. Thus, a patent holder may sell a patented good and subsequently prevent its import into a second country.[128]

2. Principle of Exhaustion

The principle of exhaustion, like the first sale doctrine, limits the monopoly of an intellectual property right holder. The doctrine holds that once the rights holder places a protected good into the stream of commerce, the right of exploitation has been exhausted. Once the good is in the stream of commerce, the right holder may not bar its import anywhere, either at home or abroad. Just as under the first sale doctrine, the right holder may utilize the monopoly of the product just once. Also, like the first sale doctrine, the principle of exhaustion applies to both patented and copyrighted goods.[129]

3. Principle of Territoriality

Despite the similarity of the first sale doctrine and the principle of exhaustion, applying these rules to imports produced two different results: the owner was allowed to bar imports in the first case but was not allowed to do so in the second. This highlights the principle of territoriality; intellectual property rights are confined to national territories. The principle of territoriality is thus consistent with the first sale doctrine, which holds that a right holder uses her rights only within national territory. By contrast, the principle of exhaustion says that a rights holder exhausts his rights throughout the entire EU.[130]

Today the principle of exhaustion governs importing goods protected by national intellectual property laws. The Court of Justice has held that for purposes of the first sale doctrine, the relevant market is the entire EU and not just the country in which intellectual property rights were obtained.[131] The distinction between the first sale doctrine and the principle of exhaustion is thus largely academic since the two terms are now used interchangeably.

4. Principle of Proportionality

The principle of proportionality is the means by which the Court of Justice most directly pits the recognition of national rights under Article 36 against the promotion of free trade under Articles 30 to 34. Under it, a national measure protecting an international property right will be held invalid if there is another means of achieving the underlying aim of the measure that is less inhibiting on Community-wide trade.[132]

D. *Software Directive*

Although treatment of copyrights in the EU is national, subject to exceptions for trade discussed above, as of January 1, 1993, member states must abide by the Computer Software Directive.[133] Consistent with Article 52 of the European Patent Convention, which forbids patent protection of computer software, the directive mandates that EU members protect software as a literary work under their copyright laws.[134] The Commission saw several advantages of copyright protection. Copyright protection is cheap and accrues immediately upon fixation of the work;[135] it saw quick protection as necessary to defend against piracy. The Commission had noted that development of a computer chip cost about $100 million, while reproduction of the same chip would cost just $100,000, representing one-tenth of one percent of the research and development costs.[136] The underlying reason for protecting software was the EU's desire "[t]o retain its place in the forefront of technical advance, and to maintain its competitiveness generally ... to ensure that it has a competitive, dynamic software industry."[137] Again, the policy of enhancing trade was foremost. Prior to the directive, the means of protecting software varied widely. This inconsistency was found to have a negative impact in the transnational software trade.[138]

The directive remains true to the Economic Community's policy of free trade by protecting a software producer's economic but not moral rights, which were seen as impediments to the development and trade of software. The moral rights prohibition is no small sacrifice since ten of the EU members recognize moral rights under their national laws.[139]

The directive also adopts the stance of the United Kingdom with respect to originality. To meet the originality requirement, software must be the author's own intellectual creation, i.e., it cannot be copied from a preexisting source. The work may not be judged by an aesthetic standard, however.[140] Because thresholds of originality varied widely throughout the EU, many countries were required to amend their laws on this matter. France and the Netherlands, for example, had to raise their requirements, while Germany had to lower its threshold of protection.[141]

As noted, software is protectable immediately upon fixation and, consistent with the Berne Convention, extends for the life of the author plus 50 years. Protection also extends retroactively: countries must protect software written prior to 1993 as if it had been written after the directive took effect. Further consistent with Berne, the directive implements a system of national treatment. Because the directive harmonizes the law governing computer software, the resulting system of law is both mutual and reciprocal throughout the EU. This differs from Berne, where copyright protection is reciprocal only. For example, Germany has to protect a novel written in the United Kingdom, but under its own copyright laws. Now a country must not only protect computer programs written abroad, it must protect them under harmonized law.[142]

Absent a contract, ownership is determined by two provisions in the directive. Under the normal circumstances, the programmer is the copyright owner, or ownership is joint if two or more people work together to write the program. For a program written by an employee in the ordinary course of a job, however, copyright belongs to the employer. Parties are free to modify these arrangements by contract.[143]

Under Article 4 of the directive, the author of a program enjoys three exclusive rights: reproduction, adaptation and distribution. Like copyright law in the United States, the directive gives the author the exclusive right to control rental of the program; unlike the United States, this right is included under the right of distribution.[144] Like the right to control distribution, the general right of distribution under the directive is limited by the doctrine of exhaustion. After the first sale of the software package within the EU, the copyright owner may not prevent third parties from importing the software. The directive, however, specifies that the copyright owner retains the rights to control rentals, an important exception to the exhaustion principle.[145]

There also is an important exception to the author's rights. As direct infringement of a program's copyright is fast, cheap and hard to detect, Article 7 of the directive penalizes secondary infringement. Strictly speaking, the prohibited acts are not infringing per se, but they nevertheless devalue an author's economic rights by facilitating direct infringement. The directive declares illegal the acts of putting infringing program copies into circulation and of possessing infringing copies for commercial advantage. Further, the copyright owner has the right to seize such copies. The same acts are illegal when the program is not an unauthorized copy but is a decryption device. Decryption programs may be used to circumvent copy protection schemes on copyrighted software. The penalty for possessing or putting into circulation such software is also seizure.[146] Decryption programs are illegal only if their only purpose is to infringe copyrighted programs. The limitation is important. Advertisements of Copy ii PC, for example, announce the program only as a means of backing up copy-protected software, a legitimate act. Economically, such software is more valuable for its ability to reproduce expensive copy-protected programs such as Lotus 1-2-3. Under the directive, however, the purpose is a legal one and thus the decryption program is legal.

The Computer Software Directive further qualifies the illegality of both acts with a knowledge requirement. That is, the secondary infringer must have known or should have known that the programs were infringing. Otherwise the acts outlined above are legal.[147]

The most dramatic provision of the directive, and the one that generated the most debate, allows third parties to decompile and translate the object code of a program "to achieve the interoperability of an independently created interoperable program."[148] Programmers write software in high-level languages such as BASIC. These languages are close to English.[149] Once a program is written, it is compiled into a stream of ones and zeros, which turn on and off a computer chip's transistors. This is how a computer chip runs the program. A commercial software package often has many programming innovations that its producers want to keep away from its competitors. If the program were marketed along with its source code (the high-level language), competitors could quickly write similar programs to compete with the original package. Instead, software producers often market only the compiled version of its product — the working stream of ones and zeros. They keep the source code a secret. On the software market, however, numerous decompiler programs are available. These programs are capable of converting streams of ones and zeros back into high-level languages. Decompilers are analogous to code breakers used to descramble coded messages sent by spies in that they covert the unintelligible into the intelligible. The process is the software equivalent of reverse engineering.

In the United States, decompiling can easily amount to direct infringement. While the copyright law says nothing about decompiling, it does bar unauthorized reproduction. Reproduction of a program is a necessary step for its analysis and eventual decompilation. Many courts have examined decompilation, but the majority have done so only by comparing the two finished programs, the original and the allegedly infringing one. Only one case directly addresses whether decompilation amounts the infringement.[150] The court there found that decompiling a program necessarily infringed the right of reproduction, but found for the defendant on the basis of the fair use defense.[151]

The likely effect on the United States of the Computer Software Directive is somewhat unclear at present. Certainly, having all EU countries applying a uniform law relative to software will make it easier to determine what European law applies. Most libraries in the United States likely rely primarily on domestically produced computer programs. Corporate libraries, whose parent organizations have European branches or subsidiaries probably will be confronted first with any problems created by the Software Directive.

V. CONCLUSION

Clearly, while the copyright arena is complicated by the interplay of copyright law from different countries, the existing copyright treaties help clarify the situation for librarians. The best advice for a librarian in the United States is to consider any work published abroad as if it were published in this country. If the library should pay royalties for reproducing the work as if it were a U.S. work, then royalties are owed to the foreign publisher. The Copyright Clearance Center has agreements with many of the foreign collectives and is licensed to receive royalty payments for publishers in those countries. If one is unsure about how to submit royalty payments, she should contact the CCC and ask whether it has an agreement with that country's reproductive rights collective. If it does not, then the librarian should contact the foreign publisher directly.

Although a work published abroad may not contain a notice of copyright, a librarian should presume that the work is copyrighted unless he can determine that it is in the public domain.

ENDNOTES - CHAPTER 8

1   Stanley M. Besen, Sheila N. Kirby & Steven C. Salop, *An Economic Analysis of Copyright Collectives*, 78 Va. L. Rev. 383, 387 (1992). See Chapter 4 for a discussion of licenses.

2   Stephen M. Stewart, *International Copyright and Neighboring Rights* 3-4 (2d ed. 1989) [hereinafter Stewart].

3   *Id*. at 5.

4   17 U.S.C. § 104(1) (1988). *See* text accompanying notes 45-78 for a discussion of Berne and text accompanying notes 79-108 on the UCC.

5   *Id*. § 104(2), (4).

6   Stewart, *supra* note 2, at 29-34.

7   The 22 countries are Argentina, Australia, Belgium, Brazil, Canada, China, Czechoslovakia, France, Germany, Greece, Hungary, India, Israel, Italy, Japan, Netherlands, Poland, Spain, Sweden, Switzerland, United Kingdom, and the United States. References to Czechoslovakia refer to the country prior to its gaining independence and dividing into two countries.

8   Melville B. Nimmer & Paul Edward Geller, *International Copyright Law and Practice* (4th Release 1992) [hereinafter Nimmer & Geller].

9   17 U.S.C. § 102 (1988).

10   *See* Nimmer & Geller, *supra* note 8, at France § 2[1][a].

11   *Id*. at China § 2[1][a] and Japan § 2[1][a].

12   Greek law holds this view; *id*. at Greece § 2[1][b].

13   Feist Publications, Inc. v. Rural Telephone Co., Inc., 499 U.S. 340, 111 S.Ct. 1282 (1991).

14   *See* Nimmer & Geller, *supra* note 8, at Introduction § 4[1][1][a]. There is some overlap between copyright and design patent, but a discussion of the similarities and differences in protection under copyright and patent is beyond the scope of this book.

15   Argentina, Australia, Canada, China, Czechoslovakia, India, Japan, Netherlands, Spain, Switzerland, and the United States. *See* Nimmer & Geller, *supra* note 8.

16   *See* Nimmer & Geller, *supra* note 8, at Introduction § 4[1][c][1].

17   E.g., India and Britain. *Id*. at India § 2[3][b] and United Kingdom § 2[3].

18   *Feist*, 499 U.S. 340.

19   Nimmer & Geller, *supra* note 8, at Introduction § 4[1][c][iii].

20   *Id*. at Introduction § 2[4].

21   *Hearings Held on Possible Extension of Copyright Term*, 46 Pat. Copyright & Trademark J. (BNA) 466, 466-67 (1993). The European Union is refered to as the Economic Community (EC) when it takes a technical action and as the European Union as a political entity.

22   *See* Nimmer & Geller, *supra* note 8, at Germany § 3[1][a].

23   *Id*. at Australia § 7.

24   *See* 17 U.S.C. § 106A, 113 (Supp. III 1991). The U.S. recognizes the rights of attribution and integrity only for works of visual art. For a brief discussion, see Chapter 2.

25   *See* Nimmer & Geller, *supra* note 8, at Australia, United Kingdom and United States.

26   *Id*. at Argentina § 8[2][a] and Belgium § 8[2].

27   United States. *See id*. at United States § 5[3][a] (citing 17 U.S.C. § 412 (1988)).

28   Argentina. *Id*. at Argentina § 5[3].

29   Included here also is the thrust of law in Scandinavian and socialist countries. *Id*. at Introduction § 2[3][b].

30   Italy and Spanish-speaking countries fall into this lot, as well. *Id*.

31   17 U.S.C. § 102 (1988). See Chapter 6 for a discussion of computer programs.

32   Argentina, Belgium, Brazil, Israel, Italy, Netherlands, Poland, and Switzerland. *See* Nimmer & Geller, *supra* note 8, *passim*.

33   Argentina and Israel. *Id.* at Argentina § 2[4][d]; Israel § 2[4][d].

34   Italy. Case law in Japan prior to its statutory protection of software held the same position. *Id.* at Japan § 2[4][d].

35   Netherlands. *Id.* at Netherlands § 2[4][d].

36   Poland. *Id.* at Poland § 2[4][d].

37   *See generally id.*

38   Czechoslovakia and France. *Id.* at Czechoslovakia § 2[4][d]; France § 2[4][d].

39   *Id.* at France § 8[1]. French law also grants broad moral rights to authors. *Id.*

40   *Id.* at Czechoslovakia § 8[1].

41   Stewart, *supra* note 2, at 35.

42   Nimmer & Geller, *supra* note 8, at Introduction § 3[3][b][i].

43   Stewart, *supra* note 2, at 46.

44   Nimmer & Geller, *supra* note 8, at Introduction § 3[1][a].

45   Stewart, *supra* note 2, at 98.

46   Carol Motyka, *U.S. Participation in the Berne Convention and High Technology*, 39 Copyright L. Symp. (ASCAP) 105, 110 (1992) [hereinafter Motyka].

47   Paul Goldstein, 3 *Copyright Principles, Law and Practice* § 16.7 (1984) [hereinafter Goldstein].

48   The original member countries are Belgium, France, Germany, Italy, Luxembourg, Monaco, Spain, Switzerland, Tunisia and the United Kindom.

49   Stewart, *supra* note 2, at 101.

50   Motyka, *supra* note 46, at 110-11, n.14.

51   Goldstein, *supra* note 47.

52   Motyka, *supra* note 46, at 110-11, n.14.

53   *Id.* at 115.

54   Relevant amendments necessitated by Berne adherence are discussed in various chapters of this book.

55   Nimmer & Geller, *supra* note 8, at Introduction § 3[3][b][i].

56   *Id.*

57   Berne Convention, art. 5(2) (1971).

58   17 U.S.C. §§ 401, 410-11 (1988).

59   Pub. L. No. 100-568, 102 Stat. 2853 (1988) (codified at various sections in 17 U.S.C. (1988)).

60   17 U.S.C. § 504(c)(2) (1988).

61   *Id.* § 411(a).

62   *Id.* § 412(2).

63   *See House Passes Copyright Bill Ending Registration as Condition for Suing*, 47 Pat. Copyright & Trademark J. (BNA) 78 (1993).

64   *Id.* at 79.

65   *See* Natalie Wargo, *Copyright Protection for Architecture and the Berne Convention*, 65 N.Y.U. L. Rev. 403, 440 (1990) (citing *Berne Convention Implementation Act of 1987: Hearings on H.R. 1623 Before the Subcomm. on Courts, Civil Liberties, and the Administration of Justice of the House Comm. on the Judiciary*, 100th Cong. 1st & 2d sess. 76 (1988) (testimony of Paul Goldstein)).

65   17 U.S.C. § 102(8) (Supp. III 1991).

67   17 U.S.C. § 24 (1970).

68   17 U.S.C. § 302 (1988).

69   Berne Convention, art. 6*bis* (1971).

70   In the scheme of national treatment, these will vary except for the right of translation.

71   Berne Convention, art. 6*bis* § 2 (1971).

72   Pub. L. No. 101-650, 104 Stat. 5128 (Dec. 1, 1990) (codified at various sections in 17 U.S.C. (1988)). See Chapter 2 for a discussion of these rights for works of visual art.

73   *See* Nimmer & Geller, *supra* note 8, at Introduction § 3[3][b][i].

74   Stewart, *supra* note 2, at 139-40.

75   *See* Nimmer & Geller, *supra* note 8, at Introduction § 3[3][b][i].

76   Stewart, *supra* note 2, at 104.

77   Motyka *supra* note 46, at 117.

78   Stewart, *supra* note 2, at 142.

79   *Id.* at 146.

80   Martin B. Levin, *Soviet International Copyright: Dream or Nightmare* 31 J. Copyright Soc'y 99 (1983) [hereinafter Levin].

81   Nora Maija Iocupa, *The Development of Special Provisions in International Copyright Law for the Benefit of Developing Countries*, 29 J. Copyright Soc'y 405 (1982) [hereinafter Iocupa].

82   *Id.* at 406.

83   UCC, art. XVII(b) (1971).

84   Iocupa, *supra* note 81, at 406.

85   *Id.* at 407.

86   Levin, *supra* note, at 80.

87   Natasha Roit, *Soviet and Chinese Copyright: Ideology Gives Way to Economic Necessity*, 6 Loy. Ent. L.J. 53, 63 (1986).

88   Berne has the identical provision, which led to U.S. efforts at "back door" protection. *See* the above discussion of the Berne Convention.

89   UCC, art. II (1971).

90   Maria Francoise Gilbert, *International Copyright Law Applied to Computer Programs in the United States and France*, 14 Loy. U. L.J. 105, 111 (1982) [hereinafter Gilbert].

91   *Id.* at 112.

92   UCC, art. I (1971).

93   Barbara Ringer & Louis I. Flacks, *Applicability of the Universal Copyright Convention to Certain Works in the Public Domain in Their Country of Origin*, 27 Bull. Copyright Soc'y 157, 169 (1980) [hereinafter Ringer].

94   UCC art. III (1971).

95   Ringer, *supra* note 93.

96   UCC, art. V (1971).

97   Frank P. Angel, *France, More Ded Arts ... at Ded Laid; Also for Foreign Works in France*, 30 J. Copyright Soc'y 335, 338 (1983).

98   Gilbert, *supra* note 90.

99   Ringer, *supra* note, 93 at 167.

100 17 U.S.C. § 105 (1988). Works produced by the federal government are not protected. The Act does not say that no works by any government may receive protection, however.

101 June M. Stover, *Copyright Protection for Computer Programs in the United Kingdom, West Germany, and Italy: A Comparative Overview*, 7 Loy. L.A. Int'l & Comp. L.J. 278, 284 (1984).

102 Gilbert, *supra* note 90, at 110.

103 UCC, art. V*bis* - V quarter (1971).

104 Stewart, *supra* note 2, at 171-72.

105 *Id.* at 172-73, 176.

106 *Id.* at 175.

107 *Id.*

108 *Id.* at 178.

109 Beryl R. Jones, *An Introduction to the European Economic Community and Intellectual Properties*, 18 Brook. J. Int'l L. 665, 666 (1992) [hereinafter Jones].

110 John E. Somorjai, *The Evolution of a Common Market: Limits Imposed on the Protection of National Intellectual Property Rights in the European Economic Community*, 9 Int'l Tax & Bus. Law 431, 433 (1992) [hereinafter Somorjai].

111   Victor Vandebeek, *Realizing the European Community Common Market by Unifying Intellectual Property Law: Deadline 1992*, 1990 B.Y.U. L. Rev. 1605, 1625.

112   *Id.* at 1676.

113   A copyright owner, for example, has the exclusive right of distribution, while a patent holder has the exclusive right to make, use or sell the patented device.

114   Leo J. Raskind,*Protecting Computer Software in the European Economic Community: The Innovative New Directive*, 18 Brook. J. Int'l L. 729, 730 (1992). [hereinafter Raskind].

115   Judgment of July 15, 1964, Case 6/64 (1964), E.C.R. 585.

116   Jones, *supra* note 109, at 669.

117   *See id.* at 673.

118   Treaty establishing the European Economic Community, as amended Mar. 25, 1957, art. 189.

119   *See* text accompanying notes 133-51.

120   *See generally* Jones, *supra* note 109, at 671-72.

121   *See* Nimmer & Geller, *supra* note 8, at European Community § 1[2][c].

122   Coditel v. Cine Vog Films S.A., Judgment of March 18, 1980, Case 62/79, (1980) E.C.R. 881.

123   Jones, *supra* note 109, at 682.

124   Art. 36. *See generally* Somorjai, *supra* note 110.

125   *See* Jones, *supra* note 109, at 681.

126   *See generally* Somorjai, *supra* note 110.

127   "Alienation" is defined as the transfer of property rights in real or personal property such as a copy of a book. *See* Black's Law Dictionary (6th ed. 1990).

128   *See* Somorjai, *supra* note 110, at 460.

129   *Id.*

130   *Id.*

131   *Id.* at 446 (citing Deutsche Grammophon Gmb. H. v. Firma Pap Import, 1971 C.M.L.R. 631).

132   *Id.* at 439.

133   32 O.J. Eur. Comm. (L 91) 4 (1989).

134   *Id.*

135   *See* Raskind, *supra* note 114, at 732.

136   *See* Somorjai, *supra* note 110, at 435. The low cost of copying software was also a concern of the U.S. Congress, which eventually passed the Computer Software Rental Amendments Act of 1990, which forbids a software buyer from renting to others. Here, however, the Council seems to be talking about protecting the chip itself, and not the software. The stance of the U.S. on this matter is clear: The Semiconductor Chip Protection Act (17 U.S.C. § 904 (1988)) does not allow the mask work, which is the physical embodiment of a program on a chip, to be protected.

137   Green Paper on Copyright and the Challenge of Technology, Com (88) 172 final (June 7, 1988), § 5.2.11. *See also* Jorg Reinbothe & Silke Von Lewinski, *The EC Directive on Rental and Lending Rights and Piracy* (1993).

138   *See* Christopher Voss, *The Legal Protetion of Computer Programs in the European Economic Community*, 11 Computer/L.J. 441, 442-44 (1992) [hereinafter Voss].

139   *Id.* at 442-43. The two exceptions are England and Ireland.

140   *Id.* at 447.

141   *See* Nimmer & Geller, *supra* note 8, at France § 2[1][b][ii]; the Netherlands § 2[1][b].

142   Voss, *supra* note 138, at 447.

143   *Id.* at 450.

144   *Compare* art. 4(c) *with* 17 U.S.C. § 109(b)(A) (Supp. III 1991).

145   *See* Voss, *supra* note 138, at 450-51.

146   Economically, these are no penalites at all. The entire rationale for penalizing was the low cost of copying. Thus, the lower the cost of copying, the weaker the penalty — seizure of program copies that are cheap to start with.

147 *See* Council Directive 91/250 on the Legal Protection of Computer Programs, 1991 O.J. (L 122) 45, at 7.1(a)-(c), 7.2-.3.
148 *Id*. at art. 5(a)(1).
149 The BASIC command to cause a program to execute at line 1,000, for example, is GOTO 1000.
150 Sega Enters., Ltd. v. Accolade, Inc., 977 F.2d 1510 (9th Cir. 1992).
151 *Id*. at 1527.

CHAPTER 9

CANADIAN AND BRITISH COPYRIGHT

I. CANADIAN COPYRIGHT LAW

A. *Introduction*

It should be easier to describe the copyright law of Canada than it is for any other country besides the United States. This is not the case, however, since the recent revision of the law in Canada is incomplete. The Constitution Act of 1867[1] gave exclusive jurisdiction to the Canadian Parliament for matters of copyright. Thus, as in the United States, copyright is a federal matter. The Copyright Act of 1924,[2] along with its various amendments, rules, schedules and annexes, remains in force today. Although the Act was passed long before most of the technological advances that have such impact on copyright law were developed, there is general agreement that it has been remarkably flexible to encompass new technology. Today, however, the Act is seriously dated and the government has begun to revise it.[3]

The revision process has been underway for some years; the situation has been much studied, commented upon and criticized. The revision process in Canada has been just as tortuous as it was in the United States, perhaps even more so for the library community. Studies of Canadian copyright resulted in a legislative plan to revise the 1924 Act in two phases. As planned, Phase I was enacted by Parliament as the Copyright Amendment Act of 1988[4] and has been implemented. Phase I covers issues such as what works are eligible for copyright protection, rights of copyright holders, moral rights, penalties for infringement, and the like.

Phase II has never been introduced in Parliament, and this portion of the planned legislative revision will deal with significant issues of primary interest to librarians and users of copyrighted works. These include what institutions will be defined as libraries under the statute, reproduction of materials for archival purposes and making single photocopies for private study, research and scholarship.[5] Librarians and library groups have lobbied to ensure that Phase II will strike a proper balance between the rights of copyright holders and the needs of users. "In the meantime, libraries work in a kind of vacuum, either possibly breaking the law or acting as the `Police of the Copyright Act' or trying to make the best of a bad situation."[6]

The statute establishes rights for private copyright holders to enjoy but it does not provide any government enforcement. Individual owners generally must enforce their own rights. There are two federal departments charged with responsibility for various aspects of copyright. The Department of Consumer and Corporate Affairs administers the Act and such matters as registration of works. The Department of Communications is responsible for developing policy for the revision of the act as part of its general responsibility for culture in Canada.[7]

B. Formalities and International Treaties

In Canada, a work is eligible for copyright immediately upon creation. Following the Berne tradition, Canadian copyright law does not require an author either to publish a work or to adhere to formalities for the work in order to hold a valid copyright. Likewise, as a member of both the Berne Union and the Universal Copyright Convention (UCC), Canada does not require authors residing in member nations to adhere to any formalities for the protection of their works in Canada. Certain formalities, however, do work in favor of authors.

Although registration is voluntary, it is evidence of both valid copyright and ownership.[8] The Act itself states that a registration certificate creates two presumptions: (a) that the copyright exists and (b) that the person registered is the owner of the copyright in the work.[9] Voluntary registration neither confers protection nor guarantees the existence of it, but it does confer some important advantages such as presumptions in favor of ownership. In litigation another presumption is that the alleged infringer knew the copyright existed. Registration also entitles the owner to certain remedies including monetary damages.[10]

Although not required for domestic protection, notice of copyright earns protection in UCC countries other than Canada. Placing a notice of copyright (referred to as "marking a copyrighted work" by some Canadian writers) is useful even though it is not mandatory. It reminds users of the work that the author claims copyright in the creation, and it can assist in locating copyright owners to obtain permission to use a work.[11]

There is no deposit requirement in order to perfect the copyright although there is one under the National Library Act.[12] The National Library Act has nothing to do with copyright but it requires that two copies of every book published in Canada and one copy of every sound recording produced with some Canadian content be sent to the National Library.[13]

Canadian adherence to the basic provisions of the Berne Convention comes despite the fact that the last Convention to which it subscribes is the Stockholm Revision of 1967, and there only in part (Articles 22 to 38). The last full Berne Revision that Canada recognizes is the Rome Act of 1928. Like the United States, Canada also is a member of the UCC. Unlike the United States, however, Canadian courts have shown reluctance to allow domestic law to have outright primacy over a treaty.[14]

C. Protected Works, Duration and Rights Afforded Owners

A work or writing must exhibit certain characteristics before copyright protection will attach. The requirements closely resemble those in American copyright law; for example, a work must be original, i.e., not copied from preexisting sources.[15] Cases from Canadian courts have interpreted what originality means in the copyright context to include that a work must (a) originate with the author, (b) not be copied from another work, (c) be more than a mechanical or automatic arrangement and (d) be the fruit of independent labor. A further requirement is (e) that in order to create the work, the author must use "skill, experience, labour, taste, discretion, selection, judgment, personal effect, knowledge, ability, reflection, imagination."[16] Neither great thought nor literary skill is required, but independent creation is essential. The threshold for originality is fairly low. Further, only expression is protectable, ideas are not.[17] Fixation is required for some types of works — musical, dramatic and computer programs, specifically.[18] The term "fixation" is not defined in the

Act but through case law has been defined as requiring a material form that is both capable of identification and which has a more or less permanent duration.[19] Thus, fixation means the same thing in Canadian law as it does in the United States.

Like American law, Canadian law recognizes the author as the prima facie owner of the work. Photographs are the exception where the owner of the negative owns the copyright in the resulting photograph.[20] The presumption is basically reversed in the employment situations, however. When an author creates a work in the course of employment, the employer owns the copyright. Only when an employee's work is not written "in the course of his employment" does the employee retain copyright in a work.[21]

Canadian law extends copyright protection to a wide variety of works detailed under four categories: (a) literary works, (b) dramatic works, (c) musical works and (d) artistic works. Further, these include

> ... [E]very original production in the literary, scientific or artistic domain, whatever may be the form or mode of its expression, such as books, pamphlets and other writings, lectures, dramatic or dramatico-musical works, musical works or compositions with or without words, illustrations, sketches and plastic works relative to geography, topography, architecture or science.[22]

The Act does not mention audiovisual works at all, and newer revisions surely will substitute the term "audiovisual works" for the old term "cinematographic work." Neither does the Act mention videotapes; however, it is likely that they are protected as cinematographic works, photographs or mechanical contrivances works. Although broadcasts are not specifically protected, the music, films, etc., contained in a broadcast are protected.[23]

Perhaps no aspect of American copyright law has had more influence in Canada than that which governs computer works. Canada's 1988 Revision defines a computer program as "a set of instructions or statements ... stored in any manner, that is to be used directly or indirectly in a computer in order to bring about a specific result."[24] The form of a program, application program, or operating system does not bear on its copyrightability. Nor does the medium in which it appears, it can be a diskette, a microchip or a tape. Also, like U.S. law, Canada has a separate section of its copyright statute devoted to mask works, which are protected for 10 years.[25] Databases are protected as compilations, and the fact that databases are electronic makes no difference. Compilations generally are protected so long as certain elements went into creating the compilation, such as knowledge of the compiler, experience, research, skill, judgment, time, labor, thought, and the like.[26]

The duration of a Canadian copyright follows the Berne standard (life of the author plus 50 years), with certain exceptions. The most notable of the exceptions is for sound recording which are protected for 50 years from the time the recording is first fixed.[27]

Canadian law recognizes the rights of reproduction, public performance, publication, translation and telecommunication.[28] Before 1988, the moral rights influence could be seen in the way the law dealt with unauthorized derivative works: such works often received no copyright at all.[29] Under American law, a copyright would issue and would protect original expression in the work, i.e., that expression not contained in the original work; but the United States leaves deterrence of such unauthorized works simply at the very real threat of an infringement suit.

Moral rights have been recognized in Canada for many years. Nevertheless, the 1988 Revision was significant in its allowance for monetary relief when moral rights are violated. The only requirement is that the author show that her reputation has been prejudiced. When the work the "author" creates is a painting, sculpture or engraving, no such prejudice need be shown.[30] Moral rights include the right of paternity which can be further divided into three separate rights: (a) the right to claim authorship, (b) the right to remain anonymous and (c) the right to use a pseudonym. The right of integrity includes both the traditional right to prevent changes to a work and the right to prevent the use of a work in association with a product, service, cause or institution.[31] Moral rights reside only in the author. Considering the underlying rationale for moral rights, the rule has two logical implications. First, moral rights may not be assigned to another even when the author does assign the copyright to a third party. Second, moral rights may be waived by the author (usually for sufficient compensation).[32]

D. Infringement and Remedies

Infringement analysis in Canada is similar to that in the United States According to the statute, "copyright in a work shall be deemed to be infringed by any person who, without the consent of the owner of the copyright, does anything that, by the Act, only the owner of the copyright has the right to do."[33] Copyright infringement is a question of fact and each violation must be examined on its own facts. An alleged infringement for reproducing a book, for example, is limited by the statute which defines reproduction rights of a copyright holder as applying of only to a *substantial part* of the original work.[34] Whether a substantial part of a work has been taken is a qualitative judgment, not a quantitative one, and it is decided by the hypothetical lay observer, who must recognize that the appropriation has indeed occurred.[35] Factors used to judge substantiality of reproduction include both quantity and quality of the portion used, the nature of the reproduction and the degree to which the reproduction competes with the original. Further, courts examine the similarities between two works and not the differences.[36]

Along with the United States, Canada recognizes several limitations on the rights of copyright holders that permit use of protected works without obtaining permission from or compensating copyright owners. Because Canada takes moral rights more seriously than the United States, the field of fair use is narrower.[37] As in Great Britain, fair use in Canada is known as "fair dealing," and it is intended to allow copying "for purposes of private study, research, criticism, review or newspaper summary."[38]

Fair dealing is not defined in the Act and, despite the number of cases that have considered fair dealing as a defense to copyright infringement, no case has established its precise limits.[39] In determining whether the provision applies, Canadian courts consider four factors: (a) the amount of the copyrighted work taken, (b) the amount taken in proportion to what the defendant has added, (c) whether the defendant's work competes with that of the first author and (d) whether the first work was published.[40] Fair dealing is the only provision that exempts a private use of a copyrighted work, and it applies primarily to the reproduction right. There are no provisions that permit home videotaping from television, or the like.[41]

Some uses are per se legal regardless of fair dealing. For example, similar to section 117 in U.S. law, computer users may make a backup copy of software they have purchased and may make a copy in order to adapt or modify a program. The Act does not permit copying of a program in order to evaluate it, however. While rental of software is allowed, the renter may not copy a rented or borrowed program.[42] Some reproduction of art works is permitted, such as taking a photograph of a painting displayed permanently in a public place.[43] Another exemption allows newspapers to publish public lectures so long as there was no written or printed notice affixed before and during the lecture to or near the main entrance to the building in which the lecture is delivered.[44]

Unlike the United States, Canada recognizes certain defenses separately from the fair dealing analysis: freedom of expression, public policy, implied consent and various equitable defenses which have very limited applicability, however.[45] Although not mentioned in the Act, there is a developing defense of "public interest" which, in some cases, may override the private right of copyright. Cases that have offered this defense typically have been concerned with "serious inequities."[46]

The basic premise of remedies under Canadian copyright law is to cure the violation through administrative, civil and criminal remedies. Administrative remedies primarily relate to stopping the importation and spread of infringing work. Civil remedies enable a copyright owner to take direct action against an infringer. If the owner succeeds in obtaining a judgment against an infringer, he may obtain an injunction or monetary damages. The Act does establish the range of damages; instead, a court determines the proper monetary award on a case-by-case basis.[47]

E. Libraries and Schools

Phase II of the Act, which has not yet been enacted, will address the major issues dealing with libraries, but there are some provisions of the current law that address legitimate uses of copyrighted materials by schools. These are fairly narrow, however, which points out the limited scope of fair dealing in Canada as compared to fair use in the United States.

1. Educational Institutions

Canadian law contains no umbrella provision concerning use of copyrighted works by schools. While fair dealing generally does not permit multiple copying for classroom use, schools are mentioned three times in the Act. First, "collections" of short passages or works for use in schools may contain copyrighted material, but only under certain conditions: (a) the collection must predominantly comprise noncopyrighted material; (b) its title must indicate that the work is intended for school use; (c) the source must be acknowledged; (d) the passage must be short; (e) passages must come from literary works that were not published for use in schools (f) and the collection publisher may use no more than two passages from the same author within five years.[48] Because these requirements are so stringent, licenses are available to schools from various copyright collectives to permit the reproduction of multiple copies of copyrighted works for use in schools.

The best known of these is the Canadian Reprography Collective, CANCOPY, for English language materials.[49] It licenses schools in the province of Ontario. The purpose of the CANCOPY license is "to permit the nonprofit reproduction of published works for any purpose within the mandate of an educational institution under the Ontario Education Act."[50] The license permits the reproduction of one copy for each student and two for each teacher for a wide variety of purposes at educational institutions such as instruction, research, administration, recreation or professional purposes. A school is permitted to copy up to 10 percent of a published work or an entire work from the following list:

1.   A short story, play, essay, poem or article or issue of a periodical,
2.   A newspaper or page,
3.   An entry from a dictionary, encyclopedia, annotated bibliography, etc.,
4.   A reproduction of an artistic work from a book that contains other works, or
5.   A chapter from a book.

The copying may not be done for profit and the photocopy must reference the author and source of the material. CANCOPY produces a regularly updated poster that contains a list of works excluded; the poster is entitled "Exclusions List" and is posted near photocopy machines. The license for schools specifically prohibits (a) systematic, cumulative copying, such as an entire book over a year, (b) permanent binding of reproduced materials into an anthology and (c) entering materials into an electronic database to which anyone not licensed may have access.[51]

The second mention of schools in the Act permits schools, along with churches, colleges, religious, charitable and fraternal organizations to perform a musical work publicly, without paying royalties, if the purpose of the performance is for educational, religious or charitable purposes.[52] The statute further requires a connection between the performance and the object or purpose. For example, music performed in a classroom for a music class is exempted, and playing recorded music, in this instance, also is permitted. Music performances for entertainment are not allowed, however, without seeking permission and paying royalties if requested by the copyright owner. Technically, this exemption simply means that no royalties have to be paid to the owner, but she still could obtain an injunction and stop the performances.[53] The third mention of schools permits teachers or students to import up to two copies of a work from a Berne country so long as the copies are for personal use.[54]

2. Libraries

Librarians find themselves in particularly vulnerable positions with respect to copyright compliance in Canada. They purchase a wide array of copyrighted works, make them available to their users and serve as repositories of Canadian culture. Librarians and library associations have lobbied hard for provisions to permit library reproductions. Through letter-writing campaigns, appearances before the government and participation in a country-wide survey of photocopying practices librarians have made their voices heard.[55] Despite all of their efforts, the official position remains that there should be no exception for library reproductions.[56] Nevertheless, the Department of Communications in 1988 specified certain library uses should not be subject to the Copyright Act:

1.   Copies made for archival purposes;
2.   Copies made to replace damaged material;
3.   Single copies of periodical articles for patron use;
4.   Works outside of collectives, and
5.   Patron use of unsupervised photocopy machines on the premises.[57]

To date these agreements have not produced Phase II of the needed legislation. Sometimes it has appeared that the Minister of Communications has favored exemptions for libraries but still there has been no action. Changing governments have further dimmed the prospects of Phase II.[58]

A recent report indicates that the new Minister of Canadian Heritage does not believe that the work done by the Conservative government that would extend greater rights to users of copyrighted works should be continued. One commentator points to the Minister's statements on remuneration for artists and fears that the current government's views on Phase II will be even more pro-creator than was Phase I.[59] Librarians much need a clear definition of what qualifies as a library as well as specific statutory authority to permit reproduction for preservation purposes and to provide the public with a "working" copy of such works. Additionally, Phase II should contain language to ensure that libraries can reproduce one copy of journal articles to be used for scholarship, research or private study. Further, if a library patron infringes copyright by utilizing an unsupervised photocopy machine in the library, that user should be liable for the infringement and not the library.[60]

Thus, currently, the law is of little assistance to Canadian libraries or librarians. There is a provision that permits the importation of works by any public library or educational institution before a work is printed in Canada.[61] Otherwise, there is little guidance other than seeking a CANCOPY license. The license available to libraries permits the reproduction of rare or fragile works or to replace missing pages from a work owned by the library or a damaged out-of-print work so long as an effort is first made to secure a permanent replacement. The license even specifies that the copy sought must be available "within a reasonable time and at a reasonable price." In this situation, however, the library must notify CANCOPY that a copy has been made.[62] In addition to the copies that can be reproduced for students and teachers under the license, a reasonable number of copies may be made for reference in or for loan by the library.[63] The license appears to be limited to libraries in educational institutions, however. With these severe restrictions, it is not surprising that copying a work in toto is almost never covered by the fair dealing provision.[64]

A survey of Canadian libraries conducted in 1987 revealed that many had implemented copyright policies and limited the quantity they would copy for users. Libraries varied widely in how they interpreted the "less than substantial portion" stricture; the range was 10 to 25 percent of a publication although most used a periodical article as the unit of measurement.[65] Some librarians criticize these policies as unduly narrow and lament the fact that librarians have become the copyright police. Canadian librarians and users seek certainty such as is found in many other countries around the world which have special provisions relating to libraries.[66] Until then, librarians in Canada are operating without much guidance on copyright.

II. BRITISH COPYRIGHT LAW

A. *Introduction*

Copyright law in the United Kingdom is governed by the 1988 Copyrights, Designs, and Patents Act (CDPA).[67] With respect to copyright, the CDPA modifies the 1956 Copyright Act[68] and moves Britain's law closer to the continental European law of copyright and away from that of the United States. In fact, the change reflects the increasing requirements of the Berne Convention.[69]

As discussed in the previous chapter on Berne and the Universal Copyright Conventions, foreign works are protected provided they meet the criteria of the CDPA. Any member of one of these organizations need not rely on bilateral agreements. In addition, copyrighted law protects the works of authors who are residents of countries that belong to the Rome Convention[70] and the

European Agreement on the Protection of Television Broadcasts.[71] An author who is a resident of a country that belongs to none of these organizations may nonetheless achieve copyright protection in the United Kingdom if she abides by the rule of first publication: publishing in the country within thirty days of first publication elsewhere.[72] Finally, Britain no longer observes the rule of the shorter term, that is, a foreign work that is protectable under the 1988 CDPA is protected for 50 years plus the life of the author.[73]

B. Protected Works, Duration and Rights Afforded Owner

The Act protects three classes of work: (a) literary, dramatic, musical and artistic works; (b) sound recordings, films, broadcasts and cable programs; and (c) typographical arrangements of published editions.[74] Many other types of work are also covered but not explicitly named. The traditional designations are interpreted flexibly, often to accommodate emerging forms of expression. "Literary works," for example, includes tables and compilations as well as computer programs and software. Computer software has long been protected by the courts, but it first earned statutory protection in the 1985 Copyright (Computer Software) Amendment Act.[75] The 1988 CDPA affirms the statutory protection afforded three years earlier allowing a copyright owner to "stor[e] the work in any medium by electronic means."[76] The Act also protects copies "which are transient or are identical to some other use of the work."[77] This provision indicates Parliament's intention to protect telefacsimiles[78] which are protected as a type of photograph (a subcategory of artistic works); microforms are likewise covered.[79] Electronic databases are protectable not as literary works (as they are in the United States), but as cable programs.[80] Like the 1956 Act, the 1988 CDPA does not extend protection to either titles or literary characters. An author who wishes to protect these must rely on the common law action of passing off.[81]

The Act retains the requirement of originality. That is, a work must originate in the author and not be copied from a preexisting source.[82] The requirement is similar to that found in U.S. copyright law, which adds a minimum level of creativity as a requirement.[83] The U.S. Supreme Court barred protection of an alphabetized telephone directory of white pages, which the Court found "devoid of even the slightest trace of creativity."[84] The same directory would most likely be protectable under U.K. copyright law.[85]

The 1988 CDPA grants authors five exclusive rights: reproduction, distribution, performance, broadcast and adaptation. For certain works, the right of distribution encompasses a new right — the right to control rentals. Under the Act, the author of a sound recording, film or computer program has the exclusive right to control the renting of the work.[86] This new right is Britain's response to technological innovations that make pirating these works cheap, fast, efficient and clean. Videocassette recorders (VCRs), tape players and computers (especially the latter two) have turned the practice of renting phonorecords, movies and computer programs into an invitation to pirate copyrighted works. This new right is broader than its equivalent in the U.S. While U.S. law forbids the commercial rental of phonorecords and computer programs without subsequent payment to the copyright holder,[87] it does not restrict the rental of films. U.S. law makes an exception for nonprofit libraries, which may lend records and software.[88] The rental ban in the United Kingdom is absolute, and libraries must get permission from the copyright holder to lend sound recordings, films and computer programs.

Normally, a British copyright endures for the life of the author plus 50 years. There are some important exceptions for this rule, however. From the time they are first marketed, industrial designs are protected for 25 years. Works generated by a computer with no real human author are protected for a period of 50 years. Government works, if they are commercially published, are protected for 50 years from the date of publication; otherwise, works produced by the government are protected for 125 years.[89]

In most cases, ownership of copyright rests in the author. Where a work has multiple authors, and their contributions are indistinct, ownership is joint. An exception to the general rule of ownership occurs when an employee authors a work in the ordinary course of employment. In that case, ownership rests in the employer, absent any agreements to the contrary.[90] The most significant effect of this provision is on journalists. Copyright in their articles now vests in the newspapers or magazines for which they write. Where quasi-employment blurs the dividing line, courts look to indicia of employment such as salary, whether the employer deducts income taxes and who contributes to the employee's pension plan.[91]

The 1988 CDPA changed earlier law concerning the ownership of photographs. Before 1988 the owner of a film became the owner of the copyright in the photograph. The new law simply switches ownership of the copyright to the photographer.[92]

In the United Kingdom, a copyright owner may transfer ownership of the copyright. A copyright is treated as personal property; as such, it may be assigned or transferred by will. If it is transferred, however, the transfer must be in writing, but it need not be publicly recorded.[93]

Despite its status in the Berne Union as a charter member, Britain has long resisted the recognition of moral rights. Instead, Britain has placed heavy emphasis on economic rights, which are considered to be hindered by the exercise of moral rights. Officially, Britain's position was that moral rights were adequately protected by a variety of common law actions, such as defamation, injurious falsehood and passing off. These actions, it was held, roughly protected the right against derogatory treatment of a work, the right to be identified and the right against false attribution. Moreover, it was argued that some moral rights were protected under certain areas of economic rights, which are protected by statute.[94] A significant criticism of this approach was that it protected only the commercial value of the author's name, and not his creative personality.[95]

The drafters of the 1988 CDPA tried to remedy the overall deficiency by specifically enumerating the moral rights. The Act protects the right to be identified as the author of a work or director of a film. Interestingly, this right does not extend to computer software or government works. The right of integrity is also protected. This right gives the author a cause of action against those who add to, destroy or mutilate his work.[96] Finally, the rights against false attribution is protected.[97] This right prevents third parties from attaching the author's name to a work that the author did not produce, akin to passing off. Further, these rights attach immediately upon the production of the work.[98]

The author can waive these rights in two general ways. First, the author may sign a contract waiving these rights. Conversely, she may bargain for additional rights in the work. Second, the author may consent to acts that infringe the moral rights. If the author does not assert the rights at the time of their violation, they are waived.[99] This latter provision has been criticized as contrary to Berne's stance against formalities as prerequisites to copyright protection.[100]

Despite British efforts, criticisms remain. The 1988 CDPA does not afford authors much more protection than does prior law. To this end, one commentator has recommended that authors continue to enter contracts for the protection of their moral rights.[101]

An issue of some importance in Great Britian is Crown Copyright. A work created by a Crown employee in the course of her duties or by Her Majesty belongs to the Crown and anyone who wants to copy material owned by the U.K. government should act as if the copyright were privately owned.[102] A separate Parliamentary copyright also exists. Although the copyright exists, the Crown does not always enforce its copyrights and allows such things as the Acts of Parliament to be copied. The official attitude toward copyright Crown materials is described in an HMSO letter.[103]

C. Infringement

A person infringes an author's copyright when he violates one of the author's exclusive rights. Violating the right of reproduction, for example, entails reproducing the work in a material form, but there is no requirement that the reproduction be of the same type as the original.[104] Thus, one may violate the copyright of a painting by taking a photograph of it, and vice versa. The right of distribution is the right to issue the work to the public; the right is violated when one puts the work into circulation without the author's permission. The right of adaptation prevents third parties from "translating" the work, that is, from changing its form.[105] One may not, for example, write a screenplay from a novel without permission or write an opera from a piece of sheet music.[106]

Usually, infringement actions center on the right of reproduction. Whether infringement has occurred is analyzed qualitatively rather quantitatively. How important the copied segment of a work is to the work as a whole is far more important than the percentage of the entire work that the copied segment constitutes. The legal test for infringement is whether a "substantial part" of the work has been taken. Unfortunately, no definition of "substantial part" exists.[107]

British law also has a cause of action for secondary infringement. One who commercially exploits a work with reason to believe that the copy is infringing may be held liable. Secondary infringement encompasses a wide variety of acts, since commercial exploitation includes many stages, including importing the work. Other acts of secondary infringement, following the course of exploitation, include "possessing in the course of business, selling, letting for hire, exposing for sale or hire, exhibition in public in the course of business and distribution."[108] Some parts of the law of secondary infringement closely resemble European Union law. Selling anti-copy protection software, for example, is an act of secondary infringement, as is publishing an article that shows computer users how to circumvent a copy-protect scheme. Secondary infringement, moreover, gives authors a cause of action against authoritative parties who facilitate infringement.[109] This is particularly important to librarians and to organizers of events where music will be played.

Given the seemingly antipodal position of U.S. law on the matter of secondary or contributory infringement, British and American courts are much more in accord on the matter than one would expect. In the United States, for example, where the copyright statute does not provide for contributory infringement, the Second Circuit Court of Appeals nevertheless has held a defendant liable for copyright infringement where he knew (or should have known) of a third party's infringing activity and actively facilitated it.[110] Conversely, in the United Kingdom, where secondary infringement is well defined, a court nevertheless found that a record store which lent albums to a customer and authorized copying them was held not liable for secondary infringement.[111]

With respect to videocassette recorders, devices that facilitate the infringement of copyrights in television shows and movies, U.K. law recognizes the validity of time-shifting the tape of a program so that it may be viewed later[112] just as U.S. law does.

D. Defenses

The United Kingdom recognized several defenses to copyright infringement, three of which are examined here. The distinction between idea and expression is recognized in the United Kingdom as in the United States In many cases, this is a defense involving databases or compilations as protectable forms of expression; it allows them to be copyrighted as literary works. Facts cannot be copyrighted because they are not original in any author. The standard of originality for compilations requires that an author exercise sufficient "skill, judgement, and labor" in assembling data, but the data need be vast. Sixteen different statistics compiled in a betting form for soccer matches, for example, has been held to be copyrightable.[113] The arrangement of facts was held distinct from the previously existing facts themselves; the originality was in selecting sixteen statistics that the public considered important for predicting the outcome of soccer games. Thus, while the facts were not copyrightable, their arrangement — the expression of facts — was.[114] Nevertheless, the dichotomy between idea and expression constitutes a defense simply because ideas cannot be copyrighted.

In the United States, the idea/expression dichotomy has produced different results. Where, for example, a statistician compiled a thirteen-statistic pitching form used to predict the outcome of baseball games, a federal court of appeals held that idea and expression had merged.[115] The court reasoned that given the space limitations in the newspaper (where the form was published), the number of ways to assemble statistics to predict a game's outcome was so restricted as to merge expression into idea.[116]

U.K. and U.S. law concerning compilations is further inharmonious. The U.S. policy that encourages the dissemination of information dominates over the value attached to raw intellectual effort. Thus the "sweat of the brow" theory is explicitly rejected in U.S. law,[117] so that a second author may use the data obtained by the first so long as the second expression of it is sufficiently different from the first. This is not true in the United Kingdom where another requirement is imposed for the valid duplication of data. The second author under U.K. law must not only express the facts in a different manner, she also must uncover the original facts as well.[118] Thus, where U.S. law considers such second efforts inefficient and unnecessary, U.K. law not only accepts the sweat of the brow theory but actually requires an author to sweat. The obvious aim of the law is to discourage intellectual free riding.[119] Where a defendant copied one-eighth of plaintiff's phone directory to use in a solicitation by mail, a U.K. court analyzed whether the copying "was for the purpose of overcoming difficulties and was of benefit to defendant."[120]

A second defense available under U.K. law is fair dealing. Like fair use in the United States, fair dealing acknowledges that some infringements are socially beneficial. There are far-reaching differences between the two, however. Although codified in U.S. law, fair use is an equitable defense that is applicable in a wide variety of settings.[121] Courts analyze whether a use is fair by weighing four factors enumerated in the statute.[122] In the United Kingdom, however, fair dealing applies in narrowly construed settings enumerated in the 1988 CDPA.

First, fair dealing applies to research, private study and educational uses.[123] Even within these contexts, however, the Act distinguishes between moderate copying and multiple photocopying, for which the royalties must be paid. Thus, not all copying for purposes of research is considered fair dealing. Moreover, fair dealing does not protect photocopying for profit.[124] Second, fair dealing permits infringement for the purpose of criticism, review and news reporting.[125] The only stipulation here is that the critic or reporter acknowledge the original author. Fair dealing as a defense in this context has its limit, which is measured by the public's need for the information and the qualitive value of the work taken. When criticism is involved, however, the fair dealing provision heavily favors the borrower: in some instances, even the whole work may be taken.[126]

Fair dealing also differs from fair use in the subject matter for which the defense may be exercised. Fair use is a generally applicable defense. If the four factors, when weighed, favor the borrower, the use is fair. In the United Kingdom, however, fair dealing is applicable only to literary, dramatic, musical and artistic works.[127] Thus, fair dealing simply does not apply in certain cases regardless of the socially beneficial character of the use. There is no fair dealing in nonprint media, with the exception of computer programs.[128] A borrower may not copy sound recordings, films, broadcasts or cable programs without obtaining permission from the copyright owner or paying for a license. A second exception to this rule is "time-shifting," which allows a user to record a television program for a later private viewing.[129]

E. Library Uses

The fair dealing provision has sections that specifically apply to libraries.[130] As an initial matter, if the library copying is not covered in these sections, the copying is not allowed.[131] Further, artistic works are not included by the provision for libraries.[132] Therefore, a library may not copy maps or photographs for users under the fair dealing doctrine.[133] Even for a work that is covered and a use that is fair, the library user must sign a declaration form stating that the copy is to be used for legitimate (noncommercial) purposes and must pay a fee to cover the library's cost of making the copies. If the library provides copy machines (but not copy service), it must post warnings nearby to alert users to the limits of fair dealing. Where copy service is provided, the declaration form shields the library from liability for copyright infringement. The library's reliance on a declaration form, however, must be reasonable; in other words, there must not be a reason that the library should have believed the declaration to be false.[134]

The 1988 CDPA distinguishes between prescribed and nonprescribed libraries (nonprofit and for-profit, respectively). All libraries may make copies if the fair dealing provisions apply, but only prescribed libraries may receive copies of materials from other prescribed libraries under the fair dealing provisions. For periodicals, the amount of copying allowed is one article per issue. For books, however, the library must make a "reasonable inquiry" of the author to obtain permission to make a copy. Only if the author cannot be found will the library's reproduction of a part of a book be fair.[135]

A further limitation on libraries applies to old or dilapidated works. Fair dealing *does not* sanction the copying of these works even for preservation. When a copy is worn, the author's permission must be obtained, or royalties paid, for it to be legally reproduced.[136]

By U.S. standards, copying for classroom uses also appears restrictive. The Act limits classroom copies to no more than one percent of the work every three months. If a license is available, however, more may be copied under the license depending on its terms. The critical difference, of course, is that the school will pay royalties under a license.[137] This provision reflects the fact that fair dealing is meant to apply only to individuals.[138]

A third defense relates to licensing, which plays a quirky role in the United Kingdom. Under the Berne Convention, a country may not have a scheme of compulsory licenses, which compel a copyright owner to allow others to copy the work or adapt it, albeit for payment.[139] Without explicitly compelling a copyright owner to license his work, Britain does offer some heavy incentives to license. For example, in some instances, the law states that by offering a license, the copyright owner cuts off third parties' rights to copy with impunity.[140] The choice for the author, then, is to license and receive money for copying or refuse to license and receive nothing. The absence of a license is a defense to infringement for an educational institution, however.[141]

The temptation for a copyright owner reluctant to license the protected work would be to offer a license with unreasonable terms (e.g., $100 per copied word). This is not possible, however, because all copyright licenses fall under the jurisdiction of the Copyright Tribunal, which examines the licenses in light of three factors: (a) availability of the work, (b) the amount of work being copied and (c) the nature of the licensee's use.[142]

Aside from fair dealing, under which a user may be allowed to copy with impunity and without cost, there are three uses under the 1988 CDPA that are per se lawful, so long as no license exists for the use of the work. First, abstracts of scientific articles, which have a copyright separate from that of the article itself, may be copied and circulated. Second, educational establishments are allowed to record broadcasts or cable programs for later playback in classes. Third, broadcasts and cable programs with subtitles for the deaf may be copied and circulated.[143]

III. CONCLUSION

Canadian and British law is similar to American law in many respects, but it also differs at some key points. U.S. law appears to grant more rights to the user than does either the United Kingdom or Canada. Perhaps this is because the law in these countries is more closely patterned after Berne. The United States has been amending its law, however, to align it more closely with the law in traditional Berne member countries. Thus, the laws of these three countries relating to copyright should be even closer in the future.

ENDNOTES - CHAPTER 9

1   *Reprinted in* R.S.C., App. II, No. 5 (1985) (Can.).
2   1 R.S.C., c. C-42 (1985) (Can.).
3   Lesley Ellen Harris, *Canadian Copyright Law* 10-11 (1992)[hereinafter Harris].
4   Act of June 8, 1988, ch. 15, 1988 S.C. 279 (Can.).
5   Susan Merry, *Canada Calling*, 15 SpeciaList, June, 1992, at 1, 7 [hereinafter Merry].
6   Lillian B. MacPherson, *The State of Copyright Legislation in Canada and Its Impact on Libraries* 1 Commonwealth L. Libr. 59 (1992).
7   Harris, *supra* note 3, at 12.
8   1 R.S.C., c. C-42, § 53(2) (1985) (Can.).
9   *Id.*
10 Harris, *supra* note 3, at 12.
11 *Id.* at 29-30.
12 1 R.S.C., c. N-12 (1985) (Can.).
13 *Id.*; *see* Harris, *supra* note 3, at 29-30.
14 Melville B. Nimmer & Paul Edward Geller, *International Copyright Law and Practice* (4th Release 1992) at Canada, § 1 [hereinafter Nimmer & Geller].
15 University of London Press, Ltd. v. University Tutorial Press, Ltd. [1916] 32 T.L.R. 698; 17 U.S.C. § 102(a) (1988).
16 Harris, *supra* note 3, at 18-19.
17 *Id.* at 18.
18 For definitions of these terms *see* 1 R.S.C., c. C-42, § 2 (1985) (Can.).
19 Canadian Admiral Corp., Ltd. v. Rediffusion Inc., et al [1954], Ex C.R. 382.
20 Moreau v. St. Vincent [1950] Ex. C.R. 198 (discussing the unavailability of copyright for the idea of a particular photograph).
21 1 R.S.C., c. C-42, § 13(3) (1985) (Can.).
22 Harris, *supra* note 3, at 59-63.
23 1 R.S.C., c. C-42, §§ 2-3 (1985) (Can.).
24 *Id.* § 1(5). The definition closely parallels that found in 17 U.S.C. § 101 (1988): "... a set of statements or instructions to be used directly or indirectly in a computer in order to bring about a certain result."
25 *Id.* § 64.2 (imported by the Intergrated Circuit Topography Act (1990)).
26 Harris, *supra* note 3, at 46, 64.
27 1 R.S.C., c. C-42, § 11 (1985) (Can.). The other important exceptions are photographs and government works each of which receives 50 years of protection.
28 *Id.* § 3(1).
29 T.J. Moore v. Accessories de Bureau de Quebec, Inc. [1973] 14 C.P.R. (2d) 113, 116 (Fed. T.D.).
30 1 R.S.C., c. C-42, § 28.2(2) (1985) (Can.).
31 Harris, *supra* note 3, at 102-05.
32 The Act does not explicitly allow for waiver, but case law has interpreted the statute as allowing waiver. *See* Nimmer & Geller, *supra* note 14 at Canada § 7[4].
33 1 R.S.C., c. C-42, § 27(1) (1985) (Can.).
34 Harris, *supra* note 3, at 148-49.
35 Nimmer & Geller, *supra* note 14, at Canada § 8[1][a].
36 Harris, *supra* note 3, at 149.
37 *Id.* at 109.
38 1 R.S.C., c. C-42, § 27(2) (1985) (Can.).
39 Harris, *supra* note 3, at 110-11.

40   Nimmer & Geller, *supra* note 14, at Canada § 8[2][a].

41   Harris, *supra* note 3, at 112-13. The author posits that is is likely that a court would find home videotaping to be infringement under the current law. *Id.*

42   *Id.* at 113-15.

43   1 R.S.C., c. C-42, § 27(2)(c) (1985) (Can.).

44   *Id.* § 27(2)(e).

45   Nimmer & Geller, *supra* note 14, at Canada, § 8[3].

46   The Queen v. James Lorimer & Co. Ltd. [1984] 1 F.C. 1065. In this case a public interest defense was recognized but failed.

47   Harris, *supra* note 3, at 153-55, 160.

48   1 R.S.C., c. C-42, § 27(2)(d) (1985) (Can.).

49   The collective in Quebec is l'Union des ecrivains quebecios (UNEQ).

50   Harris, *supra* note 3, at 117.

51   *Id.* at 119-22 *citing* Pamphlet, Ontario Ministry of Education, *Copyright in Ontario: What Teachers Should Know*, CANCOPY, 379 Adelaide Street West, Suite M1; Toronto, Ontario Canada, M5V 1S5. (416-366-4768).

52   1 R.S.C., c. C-42, § 27(3) (1985) (Can.).

53   Harris, *supra* note 3, at 116.

54   1 R.S.C., c. C-42, § 44(3)(c) (1985) (Can.).

55   MacPherson, *supra* note 6, at 61-62.

56   House of Commons, Standing Committee on Communications and Culture, Subcommittee on the Revisions of Copyright, *A Charter of rights for Creators*, Charter 22, Recommendation 25 (1985) *reprinted in* Judith McAnanama, *Copyright Law: Libraries and Their Uses Have Special Needs*, 6 Intel. Prop. J. 225, 232 (1991).

57   *Id.*

58   MacPherson, *supra* note 6, at 62.

59   Bernard Katz, Comment on CNI-Copyright Listserv, *quoting* Dec., 1993, Globe and Mail article.

60   Merry, *supra* note 5.

61   1 R.S.C., c. C-42, § 44(3)(c) (1985) (Can.).

62   Harris, *supra* note 3, *quoting* Pamphlet, Ontario Ministry of Education, *Copyright in Ontario: What Teacher Should Know.*

63   *Id.*

64   *See* David Vaver, *Copyright Phase 2: The New Horizon*, 6 Intel. Prop. J. 37, 41 (1991).

65   MacPherson, *supra* note 6, *see* n.11.

66   *Id.* at 64.

67   Copyrights, Designs, and Patents Act, 1988, ch. 48, § 17(2) (Eng.) [hereinafter CPDA].

68   *Id.*

69   Nimmer & Geller, *supra* note 14, at United Kingdom § 1[3][d].

70   Membership in Berne or the U.C.C. entitles a nation's authors to protection of all works except broadcasts and cable programs. CDPA, *supra* note 67, at § 206. Protection under the Rome Convention extends only to sound recordings, however.

71   Nimmer & Geller, *supra* note 14, at United Kingdom § 6[2].

72   *Id.*

73   *Id.*

74   *Id.* § 3[d].

75   Copyright (Computer Software) Act, 1985, ch. 41 (Eng.).

76   CPDA, *supra* note 67 § 17(2).

77   *Id.* § 17(6).

78   *See* Graham P. Cornish, *The New United Kingdom Copyright Act and Its Implications for Libraries and Archives*, 83 Law Libr. J. 51, 53 (1991) [hereinafter Cornish].

79   *See* Raymond A. Wall, *Copyright: The New Act of 1988*, 21 Law Libr. 18, 19 (1990) [hereinafter Wall].

80   *Id.*

81   Nimmer & Geller, *supra* note 14, at United Kingdom §1[1]. "Passing off" is defined as selling goods that the seller claims were produced by another manufacturer. It usually involves lower quality goods being passed off as ones of higher quality.

82   University of London Press, [1916] 32 T.L.R. 698 (construing Copyright Act's originality requirement).

83   Feist Publications, Inc. v. Rural Telephone Services Co. Inc., 499 U.S. 340, 111 S.Ct. 1282, 1294 (1991).

84   *Id.* at 1296.

85   See the discussion of the idea/expression distinction below.

86   CDPA, *supra* note 67, §18.

87   17 U.S.C. § 109(b)(1)(A) (Supp. III 1991).

88   See Chapters 5 and 6 for a discussion of U.S. law regarding the exemption for nonprofit libraries to loan records and software.

89   CDPA, *supra* note 67, § 163(3).

90   *Id.* § 11(2).

91   *See generally* Nimmer & Geller, *supra* note 14, at United Kingdom § 4(1)(b)(i), Cornish, *supra* note 78.

92   CDPA, *supra* note 67, § 48(1).

93   *See* Nimmer & Geller, *supra* note 14, at United Kingdom § 4[2][d].

94   *See* Shiela J. McCartney, *Moral Rights Under the United Kingdom's Copyright, Designs and Patents Act of 1988*, 15 Colum.-VLA J.L. & Arts 205, 210 (1991) [hereinafter McCartney].

95   J. M. Cavendish & Kate Pool, *Handbook of Copyright in British Publishing Practice* 13 (3d ed. 1993) [hereinafter Cavendish].

96   *Id.* at 104.

97   CDPA, *supra* note 65, § 84.

98   *See id.*

99   *See* McCartney, *supra* note 94, at 239.

100   *See id.* at 227-34.

101   *Id.* at 229-30.

102   Robin Jacob & Daniel Alexander, *A Guidebook to Intellectual Property; Patents, Trademarks, Copyright and Designs* 143 (4th ed. 1993).

103   Cavendish, *supra* note 95, at 69-73. *Citing* HMSO also publishes a letter on *Photocopying Crown and Parlimentary Copyright Publications*, Reference no. PU 15/108, Nov. 1989, available from HMSO, St. Crispins, Duke Street, Norwich NR3 1PD.

104   CDPA, *supra* note 67 § 17(2).

105   *Id.* § 21.

106   Nimmer & Geller, *supra* note 14, at United Kingdom § 8[1][a].

107   *Id.* § 8[1][b].

108   *Id.* § 8[1][a].

109   *See id.* and discussion of libraries at text accompanying notes 130-43.

110   Gershwin Publishing Corp. v. Columbia Artists Mgmt., Inc., 443 F.2d 1159 (2d Cir. 1971).

111   C.B.S. v. Ames Records, 2 W.L.R. 973 (1981).

112   *See* Sony Corp. of Am. v. Universal City Studios, Inc., 464 U.S. 417 (1984); CDPA *supra* note 67, § 70.

113   Ladbroke v. William Hill, 1 W.L.R. 273 (1964).

114   *Id.*

115   Kregos v. Associated lPress, 731 F. Supp. 113 (S.D.N.Y. 1990).

116   *Id.* at 119.

117   Feist, 499 U.S. 340.

118   *See* Thomas P. Arden, *The Conflicting Treatments of Compilations of Facts under the United States and British Copyright Laws*, 19 AIPLA Q.J. 267, 276 (1991). "Courts applying British law also have not recognized that facts cannot originate with an "author" ... [and] have not recognized the policy of promoting the dissemination and further use of information ... " *Id.*

119   *See generally* Richard A. Posner & William Landes, *An Economic Analysis of Copyright Law*, 18 J. Legal Stud. 325 (1989).

120   Waterlow Directories v. Reed Information Services Ltd, Chancery Division, Times, Oct. 11, 1990. *See generally* Thomas P. Arden, *The Conflicting Treatments of Compilations of Facts Under the United States and British Copyright Laws*, 19 Am. Intell. Prop. L. Ass'n Q.J. 267 (1991).

121   Harper & Row Publishers, Inc. v. Nation Enters., 471 U.S. 539 (1985).

122   17 U.S.C. § 107 (1988). See Chapter 2 for a general discussion of fair use.

123   CDPA, *supra* note 67, § 29(1).

124   The first factor of fair use — purpose and character of the use — bars this entirely in the U.S. 17 U.S.C. § 107(1) (1988).

125   CDPA, *supra* note 67, § 29(1).

126   *See generally* Nimmer & Geller, *supra* note 14, at United Kingdom § 8[2][c].

127   *See* Cornish, *supra* note 78.

128   *See* Wall, *supra* note 79, at 21.

129   CDPA, *supra* note 67, § 70.

130   *Id.* §§ 37-43. Fair dealing generally is covered by sections 29-30, while educational fair dealing is in sections 32-36. For a good succinct overview of the Act, *see* Wall, *supra* note 73, at 25.

131   *See* Cornish, *supra* note 78, at 54.

132   *Id.* (citing CDPA, *supra* note 67, § 41(1)(b)).

133   This is similar to U.S. law which generally excludes pictoral, graphic and sculptural works from the library section. *See* 17 U.S.C. § 108(h) (1988). Despite this exclusion, libraries may be able to justify such copying as fair use.

134   *See* Cornish, *supra* note 78, at 55 (citing CDPA, *supra* note 67, §§ 38-39. This is similar to the § 108-(d) requirements).

135   *Id.* at 56.

136   *Id.* at 57. This is contrasted with U.S. law; *see* 17 U.S.C. § 108(c) (1988).

137   *See* Cornish, *supra* note 78, at 54.

138   *See generally id.* at 51.

139   *See* the discussion of the Berne Union in Chapter 8.

140   *See generally* Wall, *supra* note 79, at 23.

141   CDPA, *supra* note 67, § 35.

142   *Id.* § 130.

143   *Id.* §§ 60, 35 & 74.

CHAPTER 10

A PUBLIC LENDING RIGHT FOR THE UNITED STATES

I. INTRODUCTION

Long recognized in the western world as a copyright or copyright-like principle, the Public Lending Right (PLR) gained impetus for recognition in the United States in the 1980's.[1] PLR is the term used for a claim or right to compensation on the part of an author or other creator of a work for loans of their books by libraries. Inherent in this rather straightforward definition are many questions; for example, is there a right to compensation for loans or is there another basis besides a "right" for payments to authors? If so, what libraries should be included, and which loans, what authors, and which books should be considered eligible to receive compensation under a PLR scheme? What administrative mechanism is necessary to accomplish the collection and disbursement of PLR funds? What are the likely effects on libraries and on authors?

Twelve countries have enacted PLR through legislative or administrative regulation: Australia (1974), Canada (1986), Denmark (1946), Finland (1961, implemented 1964), Federal Republic of Germany (1972), Great Britain (1979, implemented 1982), Iceland (1967), Israel (1987), the Netherlands (1971, implemented 1986), New Zealand (1974), Norway (1947), and Sweden (1954).[2] If legislatively created, PLR may exist as a part of the public library act, as copyright legislation or as a separate enactment. PLR's general purpose is to improve the economic position of authors who, as a group, earn inadequate sums to support themselves solely on sales of their writings. There is a widespread belief among American authors that their financial picture is bleak and getting worse and that their situation is similar to that experienced by authors in other countries. PLR is seen as a partial solution to authors' economic woes.

One common theme is prevalent among PLR countries: the government provides a fund for compensating authors and sometimes other creative persons for loans of their works.[3] Three methods have been utilized to determine how benefits to authors should be measured: library loans, library holdings and library acquisitions.[4] Generally, PLR is limited to circulation at public libraries but may include loans in other types of libraries as well. Another common theme is that library users do not pay direct fees;[5] thus, only publicly funded libraries have been included to date.[6]

The history of PLR represents the struggle for acceptance of the idea that the borrowing of a copyrighted book from a library constitutes a use for which an author has a right to be compensated. The principle began to emerge shortly before 1920 and several events combined to create a climate ripe for the adoption of PLR. The idea of free public libraries had become firmly entrenched in society, copyright law had been expanded, government support of cultural affairs had increased, and in some countries, an awareness of a need to protect and encourage the development of a national literature and language had grown. Additionally, in Europe, there was a movement toward collective activism on the part of writers.[7]

II. LEGISLATIVE ACTION

A. *General*

In November 1983, Senator Charles McC. Mathias of Maryland introduced S. 2192 to create a commission to study whether PLR could be enacted in the United States without adversely affecting public libraries and readers.8 The title of the act was National Commission on the Public Lending of Books Act of 1983; it was referred to the Senate Committee on Rules and Administration. "I believe the time is ripe to study the desirability and feasibility of compensating authors in this country."9 This bill died in committee. In 1985, Mathias introduced another bill, S. 658, which generated some interest in the idea of PLR.[10] In introducing the bill, Senator Mathias recognized that the concept was not new and that ten countries had systems for compensating authors for public loans of their works at that time. S. 658, entitled "A bill to establish a commission to study and make recommendations on the desirability and feasibility of amending the copyright laws to compensate authors for the not-for-profit lending of their works," was referred to the Committee on the Judiciary, where no action was taken. Senator Mathias proposed a commission to give the Congress the expert guidance it needed to evaluate the public lending right in the American context,[11] but no hearings were held on this bill. Since Senator Mathias retired no new champion of PLR has come forward.

Although Germany is the only country which enacted PLR as a part of its copyright legislation, the principle focuses on uses of copyrighted works which traditionally are considered copyright matters. Under American copyright law the first sale doctrine entitles the copyright holder to royalties only on the first sale of his work. After that sale, the copyright owner has no further control over the work.[12] In other words, the owner of a copy may sell it, give it away, destroy it, etc., without paying any further royalties to the copyright holder. For many years no exceptions to this doctrine existed; however, the Record Rental Amendment of 1984[13] amended section 109 of the Copyright Act by restricting the first sale doctrine with respect to the rental, lease or lending of sound recordings. The Act requires authorization by the copyright holder before a record or tape is rented so that royalties can be paid. Due to the enactment of this amendment, the copyright owner now may require additional royalties each time a record is rented or leased.[14] The enactment of the Computer Software Rental Amendments Act of 1990 further amended section 109.[15] An important exemption exists for private borrowing, noncommercial lending and renting or loaning of records in nonprofit libraries and nonprofit educational institutions.[16] These acts represent exemptions to the first sale doctrine and ultimately could open the door for PLR legislation.

B. *The PLR Study Commission*

The 1985 Mathias bill proposed that the commission consider whether compensating authors for public lending would promote authorship in the United States without adversely affecting the reading public.[17] Further, the bill directed the commission to make recommendations on procedures for determining the method of disbursements from funds to be appropriated by Congress or from a national trust established for that purpose. The commission was to examine and evaluate existing and proposed foreign PLR systems and compile data on the various types of loans for books and publications by libraries and other institutions. If after such examination the commission were to recommend a PLR system, it should (a) consider whether PLR should be created by legislative and/or administrative action; (b) determine what criteria should be used for setting compensation, (i.e., the number of copies sold to libraries or the number of loans); and (c) identify collection and disbursement procedures that would not impose burdensome administrative requirements on public libraries. Further, the commission would address the issue of author eligibility such as whether the copyright status of a book should be considered, whether citizenship or domicile of the author is important and the alienability or descendability of the right to compensation. Finally, the commission would determine whether PLR should be administered by an existing agency or by one newly created for that purpose.[18]

The commission would have had 11 voting members — the Librarian of Congress plus ten appointed by the president. Of the ten appointed members, two would be selected from authors, two from publishers, three from librarians and three from the public generally. The commission's findings would be reported to the president and to Congress.[19] According to Mathias, the bill's introduction in the previous Congress spurred a great deal of debate and interest both by the public and by the press. Also, several symposia on the subject were held, thus indicating the interest generated by the proposal.[20]

The American Library Association (ALA) adopted a resolution on January 11, 1984, concerning the proposed PLR Commission.[21] The resolution recognized the importance of the issue and vowed the support and involvement of the ALA in the study of the PLR issue if a commission was created. ALA resolved to communicate its interest to association members, members of Congress and other appropriate agencies.[22] The ALA Legislation Committee reported the following June that the Ad Hoc Copyright Subcommittee was continuing to review the bill and had referred the issue to state and division liaisons for further action.[23]

Some members of the library community have expressed doubt that library borrowing actually interferes with book sales. Studies in several countries regarding the effect of library circulation on sales have produced inconclusive results; surprisingly in fact, some studies indicate that book borrowers are heavy book purchasers as well. Still, most observers believe library lending hurts sales, although the seriousness of the harm is unclear.[24] Robert Wedgeworth, former executive director of ALA, announced that ALA had contracted with the National Center for Educational Statistics to design a more effective system for data compilation in order to generate some of the information needed to determine the effects of library circulation on book sales, but no results were ever announced.[25]

C. Authors' Position

Historically, the first U.S. PLR bill was introduced in 1973 with the backing of the Authors Guild.[26] That bill also proposed the establishment of a study commission; it died in the Committee on House Administration. In 1979, the guild revived its interest in pursuing PLR legislation with a series of programs and materials sent to its members. Lord Willis, a member of the British Parliament and a PLR advocate, addressed the guild in 1980. He stated the writers' theory, "... paying authors for library loans is not a charity, it is a right - a payment for the service of borrowing an author's work."[27] Authors generally assert that their income has decreased as the book trade has declined; they allege that the decline is a direct result of public library expansion. In other words, they claim that free libraries have negatively affected the free market.[28] In 1981, the Authors Guild commissioned a study to be conducted by the Columbia University Center for Social Studies. The results, published in the *Columbia Economic Survey of American Authors*, indicated that the average American author who had published one book earned less that $5,000 annually from income derived from book royalties, magazine and newspaper articles, as well as television and motion picture work.[29] In addition to sales royalties, other sources of income for writers such as subsidiary rights from movies, television, serialization and translations produce minimal income for most authors.[30]

Many writers feel that even if PLR produced only an additional $5,000 annually, it would be an important contribution to the well-being of American authors.[31] William Goodman, Editorial Director for David Godine, noted that the book world loses many authors because they are unable to earn a living from their writings.[32] Authors do not uniformly support PLR, however, Norman Spinrad, past president of the Science Fiction Writers of America, stated that much of the dissatisfaction with authors' incomes is due to "... all kinds of people writing things that nobody wants to read."[33]

Some of the concerns over the economic status of writers may be due to changes within the publishing industry itself. For example, growing corporate control is said to pose a threat to serious book publishing. Corporate control may eliminate quality books in favor of those with mass appeal. The modern conglomerates do not buy middle-range sellers as publishing houses did in the past. Also, because of vertical integration, separate bidding for paperback rights has evaporated as publishing houses have gained control of paperback publishers.[34]

In any event, many writers believe that PLR legislation is at least a partial answer for their economic woes. Librarians continue to question the validity of the claim that the income of writers is adversely affected by library loans of their books. In fact, librarians argue that libraries serve as a showcase for the book trade. Frequently, a borrower will become acquainted with a book in the library and later buy it.[35] As a result, the interests of authors and libraries are closely related: author Barbara Tuchman stated, "... authors and libraries are in the same game... without authors there would be no reading."[36]

III. PUBLIC LENDING RIGHT ISSUES

A. *Purpose of PLR*

The Mathias bill did not articulate the purpose of PLR. It recognized that a national and cultural identity is enriched and shaped by authors' works,[37] but no real statement of purpose was included. Before PLR legislation could be recommended, the commission surely would have to identify the purpose of such a scheme for this country. A look at other nations offers a variety of purposes. For example, PLR is viewed as a form of social security for writers in Scandinavia, as a right accruing under copyright law in Germany and as a natural right in some other English-speaking countries.[38] If it is a right, then arguments about the economic position of authors are irrelevant.[39] If the purpose is social welfare, however, then net earnings are crucial to a determination of who should receive payments.[40]

B. *Who Pays*

Under a "natural right" argument, the logical person to pay is the borrower; one generally assumes that the person receiving the benefit from reading the work should pay. Of course, this idea is diametrically opposed to the long-standing norm of the free use of public libraries. In the early days of interlibrary loan photocopying, the library usually even paid that charge for a borrower. The volume of interlibrary loan photocopying quickly made it impossible for public libraries to pay photocopy charges without reimbursement from the borrower, however. The idea that one should pay for information is more acceptable today than it was a decade ago. As library budgets have been eroded by inflation, few publicly funded institutions can afford to pay fees for searches of computer databases without charging it back to the user requesting the search. Thus, paying for use of materials may be more acceptable today than it was in the past.

On the other hand, there is something distinctly distasteful about the idea of charging users a fee to read books that were purchased with public funds. Many of the poorest readers would be disenfranchised and limited to reading older works, not a pretty prospect in a country that expends vast resources for mass education. During the PLR debate in Britain, editorials in *The Daily Telegraph* consistently advocated borrowing fees to provide authors with an adequate income and to pay the administrative costs of administering a PLR scheme; the editorials met with significant opposition to such fees.[41] Most authors do not want borrowers to pay.

An alternative is for libraries to pay the fees. This does not fit the above model where the person receiving the benefit is responsible for the fees. It might, however, be construed as a welfare model where publicly funded libraries pay for benefits accruing to users. The fees charged to libraries could be based either on loans or on acquisitions. Although this certainly is easier to administer than direct user fees, the effect on library budgets is likely to be devastating. If somehow library budgets would be increased in an amount equal to the levy, then this alternative would not be so objectionable to the library community.[42]

This alternative is not particularly appealing, however, given the financial condition of public libraries today. While demands for services increase and require the adoption of sophisticated technology for library operations including greater access to information for users, budgets are not increasing proportionally. In fact, libraries in many cities are open only three to four days per week. Also, purchasing power has seriously declined as inflation has negatively affected library budgets. These forces work against readers which, in effect, work against writers.[43]

The remaining possibility for funding PLR legislation in the United States is government funding. In each of the foreign countries with PLR schemes, a government created a fund from which payments to authors were disbursed.[44] In this country, funding for the arts and humanities has never been great and it seems to be on the decline at present. It is unlikely that in an era of huge deficits, the federal government will create a fund to pay writers for book loans. This is, however, what the Mathias bill proposed — that PLR payments come from congressional appropriations or from a specially created national trust.[45] To the contrary, most public library funding comes from the local sources rather than the federal government.[46] If locally funded public library loans harm an author's economic return on a book, should not a fund for reimbursement for these loans come from local sources? Although this is the logical conclusion, local funding authorities have little in the way of new resources to appropriate to libraries.

C. Authors Who Qualify

Several important issues factor into the determination of which authors will qualify for PLR payments: (a) an author's own characteristics, (b) collaborative efforts, (c) citizenship of the writer and (d) the scope of the definition of "author." No matter which basis is used for calculating the benefits, the largest payments will undoubtedly go to the most widely read authors, i.e., writers of best-sellers who least need the additional money PLR will generate. Authors of scholarly works clearly will not receive as much compensation as writers of popular fiction. The authors of scholarly works cannot expect large revenues either from sales royalties or from PLR payments because the highly specialized nature of the works they produce. On the other hand, the scholarly writer usually has other employment and the writing endeavor is often supported by her institution.[47]

If one of the goals of a PLR system is to reward little-known but worthy authors, it may be difficult to accomplish. It already has been established that under any system, popular authors will recover more than others. Great Britain placed a ceiling on the amount of payment any one author could receive, thus preserving the integrity of the scheme for lesser known writers.[48] Some countries use PLR as a method for encouraging native talent and promoting a national culture.[49]

The U.S. copyright law broadly defines the term "author" as the originator of a copyrightable work.[50] This includes photographers, translators, cartographers, artists, illustrators, etc. Should these "authors" be considered writers for PLR purposes? There is no agreement on this issue among PLR countries,[51] but any scheme considered for the United States should determine at an early stage who will qualify as an "author" for PLR purposes.

A second issue is whether publishers should receive payments under the PLR scheme. If one accepts the idea that publishers also suffer economic loss from library circulations, then publishers should be entitled to some of the payment.[52] In the United Kingdom, PLR is assignable; thus, a publisher can require an assignment of PLR or take it into account when negotiating authors' contracts.[53] Publishers are included in PLR in Australia and West Germany, for example, but excluded in Scandinavia and New Zealand.[54]

Should authors who collaborate to produce works be included? If so, how should payments be apportioned? What percentage of the book must an author have contributed to be considered a co-author? In Britain, strict rules were devised limiting the number of co-authors, including illustrators, to three while co-authors of dictionaries and encyclopedias were eliminated altogether in the original scheme.[55] Another important issue is whether authors must be citizens to qualify for PLR payments.[56] In some countries only nationals are eligible for payments. Others recognize "landed immigrants," i.e., those domiciled in the country with the intention of staying, in addition to citizens for eligibility in the PLR scheme. In Germany, however, foreign authors also are eligible to receive PLR payments.[57]

D. What Libraries Included

The focus has been on public libraries in PLR debates throughout the world. "Public" has meant the cross section of the population that uses public libraries, but there is no reason why the term could not be applied logically to any group of users served, no matter how specialized.[58] If the money to fund PLR is to come from users, then including as many libraries as possible will augment the benefit to writers. If academic and special libraries are included, then payments to authors of specialized scholarly works would be increased. One drawback to such a scheme is that the problems of data collection would be exacerbated if additional types of libraries were to be added.[59]

E. Which Books Qualify

Any PLR system would require that books be registered with some sort of a responsible agency in order to qualify for disbursements. To simplify matters, PLR could be limited to works currently registered for copyright. This would eliminate the necessity of a separate registration.[60] If copyright registration is used, then presumably the right to receive payments would automatically descend to heirs upon the author's death extending for the next 50 years.[61] If copyright registration is not used as a PLR registration, then the period for payments also must be determined. The period could be limited to any number of years after first publication — the life of the author or for any number of years after his death.

The types of books that qualify for PLR payments vary from country to country. In some schemes, all types of books are included; in others, reference books are excluded. Obviously, the problems of measuring on-premise use of books is difficult and discourages any attempt to apply PLR to reference books. Home loans are much easier to count because of traditional circulation records maintained by libraries.[62] The ease of counting the loans probably should not be determinative in selecting what types of works to include.

Page limitations for inclusions are present in all schemes. Although page limitations vary, 32 to 48 pages appears to be the norm for prose. Poetry and drama usually must be at least 24 pages in length to be eligible for PLR payments.[63] The British statute originally required that a page consist of at least one-half text to count in the page requirement and end pages were excluded. The result was that many children's books and other illustrated works did not qualify. Under an amendment effective May 1, 1983, all pages except blank ones count in the minimum page requirement.[64]

Should only print media be included for PLR purposes? The first nations to adopt PLR considered only book loans as eligible. As authors tend to choose nonprint media as a possible format for their expression, the less logical the limitation to books appears. Nonbook material is included in the Federal Republic of Germany, and there is an effort in Australia to expand PLR to include phonorecords and tapes.[65] As public libraries loan videotapes and computer software, should these works qualify for PLR payments? What about text preserved on floppy disks?

F. What Measure to Use

There are three methods of measuring use of books in libraries, but none of them is entirely accurate. Actual number of loans can be compiled or books purchased can be counted at the time of acquisition. Stock held at the end of a year also serves as a calculation method. Each of these methods is currently being used in at least one PLR country.

Counting the number of loans produces the most relevant measure since it is strictly related to use. The disadvantage of this approach, however, is that it requires recording a huge number of transactions. Several countries base their systems on total loans: Sweden, Finland, Great Britain and Germany.[66]

In Denmark and Iceland, actual holdings are counted in an annual census. This census places quite a burden on libraries and there have been some complaints and recommendations for alternative methods of calculation. Alternatively, a sampling of holdings is used in Australia and New Zealand.[67] The disadvantage to such a system is that counting books in stock results not in measuring actual use but availability for use. In such a scheme, seldom-used books will score as high as popular ones, so scholarly books are treated more favorably than they are under a loan-based system.

An acquisition-based system counts books only at the time they are purchased. The simplicity of this method makes it attractive, but the stock may not remain constant with lost and missing books. An easy way to perform the count is to insert a tear-out label in each book that contains the International Standard Book Number (ISBN). When a library purchases the book, the label is removed and sent to the administering agency which maintains the records. The disadvantage of this system is that it measures initial availability rather than actual use. Also, an acquisitions count has no application for books already in the collection.[68] If the United States were to adopt a PLR, perhaps a combination of these three systems would produce the most accurate results.

In any of these systems, a sampling could be used in lieu of using 100 percent of the libraries. The advantages of sampling are obvious relative to administrative costs and inconvenience to libraries. If user fees are to be used, then full data from all included libraries would be necessary making the calculation very complicated. Thus, a sampling is attractive. Even for a sampling, a computer system would be essential, both for collecting the data and for transmitting it to the agency, regardless of the counting method adopted.[69]

If PLR is funded through fees charged to users for loans, then using a sampling presents problems. Libraries would have to determine eligible books and collect fees for each loan. Perhaps a season ticket of sorts could be developed to reduce collections by libraries. Clearly, this would be the most expensive method of calculation since administrative costs in libraries would be greatly increased. Also, it might serve to discourage readers or influence them to read only noneligible books.[70]

G. *Who Administers PLR*

Under any system for calculating PLR fees, some administrative clearinghouse or agency is needed. It must register eligible authors and their works unless copyright registration is somehow substituted. If PLR were created as a right under copyright in this country, then perhaps the Copyright Office might administer PLR. In most countries, however, a separate agency has been created.

The agency should be independent and dedicated to the fair administration of the PLR scheme. The agency would be responsible not only for registration of works and authors but also for carrying out the authors' instructions about disbursements. The agency also would collect data from libraries regarding loans, stock and/or acquisitions. Finally, the agency would collate and aggregate these data, determine the amount to be paid to each registered writer and make disbursements.[71]

Obviously, a staff to process the PLR applications, gather the information and analyze the data is necessary. Some staff would need to have computer programming backgrounds to facilitate the work of the agency. Additionally, staff would be responsible for monitoring the PLR scheme, generating statistical reports, etc.[72] Administering a PLR scheme is expensive, and administrative costs generally have run between 5 and 10 percent of total income in PLR countries.[73]

H. *Disbursements*

Disbursements may present the most difficult problem of all. Thousands of libraries loan millions of books each year. For example, in the Netherlands, a small country by U.S. standards, there are approximately 100 million loans per year. Even if the data collection system for determining loans, stock, etc., is in place, establishing how much each author should fairly receive is quite difficult.[74] Although this issue is not as pressing as others to the library community, it is a very important issue should the United States adopt a PLR scheme. Since disbursement issues may affect or even determine the type of PLR scheme recommended, librarians should not dismiss this as an issue requiring less attention.

In some countries, disbursements have been made to an authors union similar to performing rights societies such as ASCAP.[75] That group then assumes part of the administrative burden for disbursements. Blanket licensing was examined as an alternative disbursement method in Great

Britain. Under blanket licensing, an author assigns her PLR rights to a collecting agency which then issues an annual license to each library. The collecting agency gathers PLR payments from libraries, then handles disbursements to writers.[76] This fits the ASCAP model. In the United States there is no all-encompassing authors union to facilitate disbursements,[77] although perhaps one could be formed primarily to handle PLR payments to writers.

Norway uses a similar system in which all funds from the government grant are paid to a collecting society. There are no individual disbursements at all; the money is used for scholarships and other social purposes. In some countries such as Finland, fees are split between authors and a social fund.[78] Direct payments to authors and publishers from the administering agency occur in Australia,[79] Great Britain[80] and Sweden.[81] There probably are increased administrative expenses to be borne by the agency whenever direct disbursements are made to individual authors.

Disbursements also would be affected by establishing minimum and maximum payment limits. Great Britain set such a limit in its system where royalties are limited to 5,000 pounds. This prevents authors of best-sellers from depleting the fund.[82]

IV. THEORETICAL ISSUES

A. *Purpose of PLR*

The theoretical issues concerning PLR relate to its purpose as legislated in a particular jurisdiction. There are three theories for PLR: copyright, cultural enhancement and social security. The central question for each country is whether the PLR scheme fulfills the purpose for which it was adopted. This still is an open question in all twelve PLR nations. There is no evidence that any writer ever produced a work because the country in which he resides had a PLR scheme.[83] There also is no evidence concerning whether PLR has succeeded in encouraging and protecting the national literature of a country.

Likewise, there is no uniform agreement on the theoretical underpinnings of PLR. In the United States, it appears most likely that PLR would evolve from copyright concepts. Under the Copyright Act, authors (or copyright holders) are given certain specific rights relative to the works they create.[84] The Act enumerates exceptions to the exclusive rights of the author such as library reproduction, fair use and exemptions for certain performances and displays in classrooms in nonprofit educational institutions.[85] Use of copyrighted works, therefore, is governed by the Act. If the purpose of PLR is to reimburse authors for the use made of their books through library loans, then "use" for PLR purposes properly is a copyright issue.

Inclusion under copyright has its problems, however, although the major one — the restrictive first sale doctrine — recently has been amended. As mentioned earlier, the Record Rental Amendment,[86] enacted in September 1984, represents the first exception to the first sale doctrine. This was followed by the Computer Software Rental Amendments Act,[87] also mentioned previously. PLR, if enacted under copyright, would represent yet another exception to the first sale doctrine. In other countries, the central legal issue in the development of PLR has been its relationship to copyright. In nine of the twelve PLR countries, proponents won the PLR battle but failed to have it included as a part of the scheme of copyright protection.[88]

Another copyright analogy used in the PLR debate concerns the public performance right. Composers of music are entitled to receive royalties each time their works are performed publicly[89] in addition to the royalties they receive from the sale of copies of the composition usually in sheet music form. In order to be accessible, music must be played publicly and the composer is then compensated. By analogy, an unread book has as much value as unheard music. Thus, the act of borrowing followed by reading might be thought of as a compensable event.[90]

The major disadvantage to enacting PLR as a part of the copyright statute is that because of treaty obligations, foreign authors must be treated as if they were nationals.[91] This means that PLR payments also would be made to foreign writers. Germany, the only country to embody PLR in its copyright statute, does make payments to foreign authors.[92] There are difficulties inherent with disbursements to foreign writers, but such authors could be required to register just as national authors in order to receive PLR payments. This would eliminate the difficulties of locating foreign writers. As an alternative, payment could be made to a collection society in the foreign author's country.

B. Cultural Enhancement

If the goal of PLR is to protect and encourage vernacular literature, then PLR should not be enacted under copyright since foreign authors normally are not involved in protecting national literature. Particularly in countries with small populations, writers are seen as national resources, pillars of national culture. Authors are not seen as "just another job classification of uncertain financial value like trapeze artists."[93]

Two of the earliest PLR countries, Denmark and Sweden, had as a purpose of their schemes protection of a native literature. These countries have small populations, a living language and a literary heritage; however, they also are very dependent on translations of works from other languages. Both Denmark and Sweden saw PLR as a means to encourage and protect their literary heritage by compensating authors who created works in these languages. Thus, the country's culture also would be protected and extended.[94] In countries where the protection of vernacular literature is viewed as essential, it may even outweigh librarians' objections to PLR.[95] In the early 1980s, before Canada had adopted a specific PLR scheme, the Canadian Library Association ceased opposing PLR in recognition of the cultural contribution of Canadian writers.[96]

This desire to protect national literature is of much less concern in the United States where American culture is not threatened with domination by a larger neighbor.[97] The number of translations of foreign-produced works published in the United States is far below those in European countries. In 1978 the United States published only one-fifth as many translations as did either Great Britain or West Germany, and one-sixth as many as those published in France.[98] Thus, larger countries apparently have less to fear about loss of cultural identity than do smaller ones which import so many foreign works.[99]

On the other hand, protection of authors for cultural enhancement should not be entirely dismissed. It has been suggested that as a form of PLR affirmative action, this country might consider support for ethnic and minority literature through a lending right.[100]

C. Social Security

In some PLR countries inadequate earnings of authors and the desire to support their work have been motivating factors. Social security is used as a general term for this concept. The desire to support authors financially through social security is especially important in Scandinavia where economic regulations have long been aimed at income equalization.[101]

Arguably, PLR payment to authors is not charity but is a right, especially if that right is created by statute. Throughout PLR countries, however, the financial condition of writers as a group has been raised. This initiates the impetus toward calling PLR social security legislation. The fact that PLR funds tend to come from government grants further stimulates the social security model. Perhaps the strongest evidence for considering PLR as social security, however, comes from countries like Norway where a portion of PLR funds are used for old age support and travel grants. In Finland, a percentage of the PLR funds are distributed to elderly and/or indigent writers.[102]

American authors might agree that government grants to support literature and the arts have been neither substantial nor particularly effective. Government does not appear to know how to support authors since they are individuals rather than institutions or organizations such as museums or symphonies. Some writers fear government control whenever there is government support, but PLR could eliminate this concern in large measure because of the intercession of an administering agency, etc.

An important issue under the social security model is how to ensure that needy authors receive the largest PLR payments. It cannot be done under any loan-based scheme since the number of loans would determine the amount of payments, and best-sellers circulate more than other books loaned in public libraries. The writers of best sellers are more likely to be able to live off royalties from book sales than are lesser known authors whose works enjoy only moderate or limited success. Even placing a ceiling on recoveries under PLR does not totally solve this problem.[103]

In 1983, the first year in which PLR disbursements were made in Great Britain, payments were made to 6,113 writers. Only 46 received the maximum 5,000 pounds; this was less than one percent of the total. Over 71 percent of the eligible authors received less than 100 pounds.[104] Novelist Jeffrey Archer earned the maximum and announced that he would donate it to charity. Romance writers Ursula Bloom and Barbara Cartland also earned near the maximum. Nonfiction writers did not fare nearly so well. Robert Gittings, noted Keats scholar, had six eligible books yet received less than 250 pounds in PLR payments.[105] In 1987, payments were made to 12,260 authors. Fifty-seven received the maximum 5,010 pounds, while 75 percent of the eligible authors received less than 100 pounds.[106] The British experience certainly indicates the likelihood that authors of potboilers will reap the greatest benefit. Yet it is the writers of scholarly works most U.S. citizens would most like to assist.

V. INTERNATIONAL INFLUENCE

Although several signatory nations to the Berne Convention[107] have PLR schemes, the Convention currently requires payment to authors after the initial sale only for cinematographic works (and works adapted for such works). After years of debate, in 1989 the United States joined the Convention. There appears to be little chance that the Convention will soon mandate any absolute right of authors to share in the proceeds of all subsequent sales of their works; such a proposal was handily rejected at the 1967 Stockholm Diplomatic Conference for the revision of the Berne Convention and has not been actively considered since that time.[108]

Despite this, much attention has been paid recently to two abbreviated versions of the right of distribution: the right of rental and the right of public lending of a work. While there seems to be an emerging consensus on the need to recognize a rental right, there is less support for the public lending right. At a 1992 meeting of a Berne Convention committee, delegations were widely divided on the issue of a public lending right. Most members opposed the inclusion of that right in the protocol, despite its endorsement in the 1991 meeting of the committee. In general, the committee seemed to feel that the issue remained insufficiently clear and required more study before any action should be taken.[109]

The chair summarized the second session of the committee in the following manner:

> (T)here was agreement that a possible protocol should provide for a right of rental; at the same time, the proposal for the recognition of a right of public lending had not received sufficient support. The scope of the right of rental and the definition of rental should be further clarified, and specific exceptions should be considered in respect of computer programs included in products.[110]

By the third meeting, it was "clear now that there would be no support for any provision on traditional public lending rights."[111]

VI. CONCLUSION

It is difficult to predict the future of PLR in the United States Clearly, the interest stimulated by the Authors Guild is unlikely to dissipate and guild members probably will be able to generate some congressional support. Despite one's view about whether PLR should be enacted in this country, if it is again seriously considered, a study commission should be created and an examination of PLR should be conducted. ALA supported the study in 1984 although any examination of PLR is bound to produce widely differing opinions with divisive results between authors and librarians.

The reasons for creating PLR abroad may not exist in this country. Although the Authors Guild survey[112] reflected low incomes for the majority of American authors, other social conditions present in PLR countries do not exist in the United States. The degree of social welfare is far lower in this country than in Scandinavia or the British Commonwealth nations. There is little need to further a national language or literature as the United States clearly dominates world markets with books, motion pictures and recorded music.

Whatever form an American PLR would take, it is clear that at least public libraries would be intimately involved. Because administrative costs associated with a library's participation, including staff time and computer equipment and time will be involved, librarians legitimately are concerned that PLR would hurt budgets and increase workloads.

Authors who seek the support of librarians for PLR should be aware that this is asking librarians to assume an altruistic view. What do librarians have to gain? Nothing unless PLR stimulates book production, an unlikely occurrence. Instead, the effect on libraries will be detrimental. Increased work and expense will be involved even if PLR funding comes from another source. Writers might be more persuasive in obtaining the support of librarians if some specific advantage to librarians could be identified. For authors, no additional work is required to profit from PLR except that the author would have to register to be eligible for payments. For librarians, on the other hand, a great deal of work would be required with no corresponding benefit to them, their institutions, the profession or the general public.[113]

According to Robert Wedgeworth, formerly of ALA, libraries currently support authors in a variety of ways. First, libraries represent a significant market for book sales. They sponsor awards for writers with accompanying publicity about their work. Public libraries offer public readings and other programs designed to give visibility to authors and express the importance of books. Because their interests are so closely allied, librarians can be expected to support authors and authorship, but not to the point of diminishing library budgets. Further, librarians will defend vigorously the principles, benefits and operations of free public libraries.[114]

Free public libraries are a universally accepted social institution that create what has been called a "public library right."[115] Worldwide, many librarians reject PLR because they feel it threatens the legitimacy of public library operations. There is some justification for the fear that PLR reduces library funding. In both Denmark and Britain, after government grants were set aside for PLR, library funding was reduced.[116] Perhaps the decentralization of public library funding in the United States will reduce this concern if the federal government funds any PLR effort.

There are a number of alternatives to PLR which would have the same benefit for writers, i.e., increasing their incomes. Some of these alternatives are more attractive to the library community than others. For example, a time limit could be placed on libraries, requiring them to refrain from purchasing books until two years after first publication. Most of the sales potential of a book is realized during the first two years.[117] Were this alternative selected, some exceptions for reference and informational works would have to be made. Perhaps the purchasing hiatus should be limited to best-sellers or to fictional works. Of course, these are the authors who earn the greatest royalties from sales and who would receive the largest PLR payments under a loan-based scheme anyway. If this alternative were selected and reference and other fact-based works such as scientific and technical books were included, libraries would be far less able to support research efforts throughout the country.

Another alternative is to extend the variation in price on various publications. The library rate for many subscriptions already is higher than the rate for individual ones. More titles could be included in this pricing structure. The effect of this alternative would be to intensify the informational function of libraries over the recreational, and libraries would pay a premium for likely use of works by a variety of users.[118] Obviously, any price increases will negatively impact library budgets; the pricing of videotapes offers some precedence, however. A library pays one price for a single copy and a higher price for the copy along with the right to make a limited number of additional copies. Admittedly, this involves the reproduction right, but, clearly, it also affects the amount of use a videotape will receive.

It has been suggested that tax relief for writers might be an appropriate response to the income deprivation they suffer as opposed to any PLR program. Royalty payments up to some maximum limit might be excluded from taxation. Again, this would benefit authors who derive sufficient income from their writings to pay income taxes, and there are plenty of authors who do not earn enough from their writing endeavors to pay taxes at all. (Of course, this group probably would not benefit much from a PLR scheme either).[119] Certainly, this alternative appeals to librarians as there is no inconvenience to them nor are users involved in the payment to authors.

Another possibility is to develop a royalty payment for the commercial lending of books. Many libraries purchase duplicate copies of current best-sellers for loan and impose a small fee for the loan transaction. A percentage of the loan fee could be returned to authors as a royalty.[120] The difficulty with this alternative is that a collection and disbursement agency would be needed, although it could be modeled after the Copyright Clearance Center. Using commercial lending as the PLR count would require a minimum of work on the part of libraries for recording fees. These loans likely represent only a small percentage of library loans, however. There is some precedence in the Record Rental and Computer Software Amendments where royalties must be paid for the commercial renting of records. Nonprofit libraries are excluded, however.[121] Again, best-selling authors would be the primary beneficiaries since the works included in these "loan for a fee" systems are current best-sellers.

Writers' economic status could be bolstered through direct bargaining with publishers or by government involvement. Book publishing always has been a buyers market, so this alternative is not likely to produce much in the way of results,[122] especially for lesser known authors. It does, however, offer the traditional method of producing author income. Enhanced writer revenue from publishers could result in increased book prices to libraries and other buyers, which is a disadvantage to this alternative.

A final alternative is to offer rewards for cultural contributions such as the Nobel Prize or Pulitzer Award. This would require a body responsible for value judgments on the quality of works, but organizations currently offering literary awards do this now. In fact, library holdings could even be used as one measure of cultural value.[123] This again would involve some work by libraries, but it would be considerably less burdensome than surveying complete holdings, counting loans, etc. Probably the greatest concern about such an alternative to a PLR scheme would be the fear that a government agency determining which works are meritorious reeks of government control.

If the United States again considers a PLR, then Congress must first identify the problem. If indeed the problem is inadequate income for authors, then various methods of correcting the problem, including PLR should be studied. Because PLR would affect libraries so substantially, librarians should be conscious of proposals regarding public lending rights and maintain an activist role.

ENDNOTES - CHAPTER 10

1   An earlier version of this chapter appeared as an article in 41 Ark. Libr. 7 (1984).

2   *See* Thomas Stave, *Pay As You Read: The Debate Over Public Lending Right*, Wilson Libr. Bull., Oct., 1987, at 22-23, noting that Belgium may become thirteenth country to institute PLR.

3   Jack R. Hart, *Public Lending Right: The American Author's Viewpoint*, 29 Libr. Trends 613, 616 (1981) [hereinafter Hart].

4   Jennifer M. Schneck, Note, *Closing the Book on the Public Lending Right*, 63 N.Y.U. L. Rev. 878, 892-93 (1988) [hereinafter Schneck]. Benefits based on library loans most clearly satisfies the objective of PLR. *Id.* at 892.

5   John Y. Cole, *Public Lending Right: A Symposium at the Library of Congress*, 42 Libr. of Cong. Info. Bull. 427, 427 (1983) [hereinafter Cole].

6   Angela C. Million & J. S. Healey, *Public Lending Right: An Overview*, 21 Pub. Libr., Sept. 1982, at 26, 27 [hereinafter Million].

7   Thomas Stave, *Public Lending Right: A History of the Idea*, 29 Libr. Trends 569, 572 (1981) [hereinafter Stave].

8   S. 2192, 98th Cong., 1st Sess. (1983).

9   129 Cong. Rec. S17059-60 (daily ed. Nov. 18, 1983) (statement of Sen. Mathias).

10   S. 658, 99th Cong., 1st Sess. (1985).

11   131 Cong. Rec. S2922 (daily ed. Mar. 14, 1985) (statement of Sen. Mathias).

12   17 U.S.C. + 109(a) (1988).

13   Record Rental Amendment of 1984, Pub. Law No. 98-450, 98 Stat. 1717, + 2, (codified at 17 U.S.C. + 109(b) (1988)).

14   *See* Daniel Y. Mayer, Note, *Literary Copyright and Public Lending Right*, 18 Case W. Res. J. Int'l. L. 483, 497-98 (1986) (discussing Record Rental Amendment in the Context of PLR).

15   Computer Software Rental Amendments Act of 1990, Pub. Law No. 101-650, 104 Stat. 5134, + 802, (codified at 17 U.S.C. 109(b)(2)(B) (Supp III. 1991)).

16   17 U.S.C. + 109(b) (Supp. III 1991).

17   S. 658, 99th Cong., 1st Sess. (1985).

18   *Id.*

19   *Id.*

20   S. 658, 99th Cong., 1st Sess. (1985).

21   American Library Association Resolution, adopted by the ALA Council, Jan. 11, 1984, in Wash., D.C., Council Doc. No. 31.9.

22   *Id.*

23   Letter from Eileen Cook, ALA Washington Representative, to Laura N. Gasaway (Sept. 17, 1984).

24   John Sumison, *PLR in Practice: A Report to the Advisory Committee* 134-49 (1988) [hereinafter Sumison].

25   Cole, *supra* note 5, at 429.

26   H.R. 4850, 93rd Cong., 1st Sess. (1973).

27   *Public Lending Right Advocated*, 54 Wilson Libr. Bull. 491 (1980).

28   Raymond G. Astbury, *The Situation in the United Kingdom*, 29 Libr. Trends 661, 664 (1981) [hereinafter Astbury].

29   Milo G. Nelson, *An Idea Whose Time to Debate, Has Come*, 58 Wilson Libr. Bull. 165, 165 (1983) [hereinafter Nelson].

30   Ernest A. Seeman, *A Comparative Look at Public Lending Right from the U.S.A.*, in *Public Lending Right: Report of an ALAI Symposium and Additional Materials* 137, 146 (H. Jehoram ed., 1983) [hereinafter Seeman].

31   Richard Hartzell, *Public Lending Right Stirs Debate*, 58 Wilson Libr. Bull. 567, 568 (1984) [hereinafter Hartzell].

32   Howard Fields, *Senate Bill to Study U.S. Public Lending Right*, 224 Pub. Wkly., Oct. 21, 1983, at 15.

33   Hart, *supra* note 3, at 617.

34   *Id.* at 616-17.

35   Million, *supra* note 6, at 28.

36   Nelson, *supra* note 29.

37   S. 658, 99th Cong., 1st Sess. (1985).

38   Cole, *supra* note 5.

39   Schneck, *supra* note 4, at 884-86.

40   *Id.* at 886-87.

41   Astbury, *supra* note 28, at 680.

42   Arthur C. Jones, *Practical and Economic Considerations*, 29 Libr. Trends 597, 608 (1981) [hereinafter Jones].

43   Cole, *supra* note 5.

44   *Id.*

45   S. 658, 99th Cong., 1st Sess. (1985).

46   Cole, *supra* note 5, at 429.

47   Seeman, *supra* note 30, at 144-45.

48   *Public Lending Right Update*, 22 Pub. Libr. 30, 31 (1983) [hereinafter Update].

49   *See* Million, *supra* note 6, at 29 (discussing PLR schemes of Australia, New Zealand and Sweden).

50   Melvill B. Nimmer & David D. Nimmer, 1 *Nimmer on Copyright* + 5.01 (rev. ed. 1993).

51   Schneck, *supra* note 4, at 888.

52   Seeman, *supra* note 30, at 161.

53   Gerald Dworkin, *Public Lending Right - The U.K. Experience*, 13 Colum.-VLA J.L. & Arts 49, 54-55 (1988) [hereinafter Dworkin].

54   Ernest A. Seeman, *A Look at the Public Lending Right*, 30 Copyright L. Symp. 71, 119 (1983).

55   Update, *supra* note 48. *See* Dworkin, *supra* note 53, at 51 (discussing eligibility of co-authors).

56   *See* Schneck, *supra* note 4, at 898-99 (noting the impact of international treaties on PLR benefits to foreign authors).

57   Stave, *supra* note 7, at 578.

58   Jones, *supra* note 42, at 597.

59   *Id.* at 601.

60   Ole Koch, *Situation in Countries of Continental Europe*, 29 Libr. Trends 641, 642, 646 (1981) [hereinafter Koch]. On the other hand, registration is voluntary in the United States and the Berne Convention insists requirement of any formalities.

61   17 U.S.C. + 302(a) (1988).

62   Jones, *supra* note 42, at 601.

63   Update, *supra* note 48.

64   *Id. Update on: Public Lending Right Scheme Amendments*, 22 Pub. Libr. 115, 115 (1983).

65   Million, *supra* note 6, at 28.

66   Koch, *supra* note 60, at 642-54.

67   *Id.*

68   Jones, *supra* note 42, at 602-03.

69   *Id.* at 604-05.

70   *Id.* at 607-08.

71   *Id.* at 606.

72   Million, *supra* note 6, at 27-28.

73   Seeman, *supra* note 30, at 160.

74 Jaap H. Spoor, *Problems and Possibilities of Public Lending Right*, in *Public Lending Right: Report of an ALAI Symposium and Additional Materials*, 57, 60 (H. Jehoram ed., 1983).

75 American Society of Composers, Authors and Publishers. See Chapter 4 for a discussion of licensing agencies and other collectives.

76 Astbury, *supra* note 28, at 673.

77 Hart, *supra* note 3, at 623.

78 Koch, *supra* note 60, at 646, 649-50.

79 Henning Rasmussen, *Public Lending Right: Situation in New Zealand and Australia*, 29 Libr. Trends 687, 701-02 (1981).

80 Ian Trewin, *Letter from London*, Pub. Wkly., Oct. 8, 1982, at 14, 16.

81 Koch, *supra* note 60, at 646-47.

82 Hartzell, *supra* note 31, at 567.

83 Cole, *supra* note 5, at 430.

84 17 U.S.C. + 106 (1988). These rights are reproduction, distribution, adaptation, performance and display.

85 *Id*. ++ 107-18 (1988).

86 Record Rental Amendment of 1984, Pub. L. No. 98-450, 98 Stat. 1717, + 2, (codified at 17 U.S.C. + 109(b) (1988)).

87 Computer Software Rental Amendments Act of 1990, Pub. L. No. 101-650, 104 Stat. 5134, + 802, (codified at 17 U.S.C. + 109(b)(2)(B) (Supp. III 1991)).

88 Dennis Hyatt, *Legal Aspects of the Public Lending Right*, 29 Libr. Trends 583, 583 (1981).

89 17 U.S.C. + 106(4) (1988).

90 Seeman, *supra* note 30, at 163-64.

91 The U.S. statute already affords equal treatment for foreign works. *See* 17 U.S.C. + 104(b) (1988).

92 Koch, *supra* note 60, at 952-54.

93 Seeman, *supra* note 30, at 143.

94 Jones, *supra* note 42, at 598-99.

95 George Pitternick, *Points of View of Librarians: Alternatives to PLR*, 29 Libr. Trends 627, 628 (1981) [hereinafter Pitternick].

96 Million, *supra* note 6, at 30.

97 *Id*.

98 Hartzell, *supra* note 31, *quoting* an unnamed UNESCO publication.

99 Seeman, *supra* note 30, at 143.

100 Million, *supra* note 6, at 30.

101 Stave, *supra* note 7, at 574.

102 Koch, *supra* note 60, at 645, 650.

103 Hart, *supra* note 3, at 617-18.

104 Sumison, *supra* note 24, at 8.

105 Hartzell, *supra* note 31, at 567.

106 Sumison, *supra* note 24, at 8.

107 Berne Convention for the Protection of Literary and Artistic Works, July 24, 1971, 828 U.N.T.S. 221. For further discussion of the Berne Convention see Chapter 8.

108 World Intellectual Property Organization, Report of the Committee of Experts on a Possible Protocol to the Berne Convention for the Protection of Literary and Artistic Works, 1st Session, Geneva, Nov. 4-8, 1991, Doc. No. BCP/CE/I/3 (Oct. 8, 1991), at 21. *See also* Jorg Reinbothe & Silke Von Lewrinski, *The EC Directive on Rental and Lending Rights and on Piracy* (1993).

109 World Intellectual Property Organization Report of the Committee of Experts on a Possible Protocol for the Protection of Literary and Artistic Works, 2d Session, Geneva, Feb. 10-18, 1992, Doc. No. BCP/CE/11/1, at 16. At the Second Session of the Committee of Experts it was felt that "any provision on the public lending right ... would be premature."

110   *Id.*
111   World Intellectual Property Organization, Report of the Committee of Experts on a Possible Protocol for the Protection of Literary and Artistic Works, 3d Session, Geneva, Part III, June 21, 25, 1993, Doc. No. BCP/CE/III/2-III, at 7.
112   Nelson, *supra* note 29, and text accompanying notes 29-30.
113   Million, *supra* note 6, at 29.
114   Cole, *supra* note 5 at 429.
115   Stave, *supra* note 7, at 570 *quoting* George Pitternick.
116   Pitternick, *supra* note 95, at 627, 633-34.
117   *Id.* at 638.
118   *Id.*
119   *Id.*
120   Million, *supra* note 6, at 30.
121   *See* text accompanying notes 86-87.
122   Pitternick, *supra* note 95, at 638-39.
123   *Id.* at 639.

CHAPTER 11

CONCLUSION

Copyright law and practice have changed significantly in recent years, and continued change is a certainty. Copyright developed around the invention of the printing press and current technological developments make it increasingly difficult for copyright to expand to new uses of copyrighted works. Copyright has been able to encompass new types of works fairly easily, but the electronic era with increased ease of copying will be the greatest challenge yet. Scholars continue to speculate about whether there will be such fundamental changes in the way that the world interacts with copyrighted materials and whether copyright will become a useless concept. There is also concern about whether an author will lose complete control of his work once it is available electronically. Integrity of data, whether the concept of "authorship" will change significantly, and how to limit access to authorized users are all issues that remain to be solved. These issues are also fundamental to libraries and to the exchange of information; thus, librarians and information specialists have a stake in being involved in any congressional discussions concerning these matters.

No one knows exactly what changes the electronic environment will bring, but it is clear that copyright will be at the heart of any solutions developed to facilitate the free flow of information. Concepts such as fair use and library reproductions of copyrighted works are likely to continue to be important to society. And librarians must act to secure those rights for the research community.

Librarians must become knowledgeable about copyright and stay current on developments. Professional reading should include articles and books about copyright law in addition to more job-specific types of readings. Another way to stay current is to attend professional meetings where copyright issues are discussed. Watch for announcements and news notes from the copyright committees of the Special Libraries Association and other library organizations. Attend meetings of those committees when possible.

All organizations that use copyrighted works in their research endeavors should have written copyright policies. If a librarian's organization does not have a company-wide policy, then she should work to develop one for the library or information center. Few corporate attorneys are knowledgeable about copyright and even fewer can combine copyright knowledge with an understanding of libraries and library operations. Thus, librarians can do much to facilitate the development of a copyright policy by preparing a draft policy to submit to corporate counsel. The draft policy should address the following issues at a minimum:

1. Fair use;
2. The limits of reproduction that a user can expect from the library;
3. Any limitations on the way certain works are to be used;
4. Arrangements for seeking permission to use works in excess of fair use (such as CCC membership);
5. How, for what activities and to whom royalty payments will be submitted,
6. License agreements for software, databases, CD-ROM materials and any other works;
7. To whom questions should be addressed (probably a librarian and then legal counsel); and
8. How the policy will be enforced.

Once the draft policy is ready, submit it to corporate counsel for comment and approval. The policy should not apply solely to the library but should be company-wide. It is not enough for the company to adopt a policy, however. The company must publicize the existence of the policy widely and regularly; it should make the policy a part of any employee handbook. Further, if the policy is to be effective, the company must take disciplinary action against employees who violate the policy. In other words, a company must discipline anyone who operates outside of the copyright policy just as it would discipline employees who violate any other company policy.

Some librarians and information professionals find themselves in a difficult position when they are directed by management to infringe copyright. Individuals are charged with obeying the law. Librarians should make management aware that the activity they have suggested or directed is contrary to the copyright law of the United States If management still chooses to go forward with the activity, the librarian may want to put his objection in writing. Not to defend the company or managers who think "if I want it, then it must be fair use," but there are many practices that are not clear violations of the law. Companies make risk assessment decisions every day about all kinds of situations that may involve technical violations of a law. The way a company should approach copyright law is no different.

If a librarian is unsure about whether some practice is permissible under the law, there are several places where help is available. The first stop certainly should be the corporate counsel. There are copyright committees of various library organizations that may be able to answer questions along with appropriate licensing groups such as the Copyright Clearance Center. Seeking permission from the copyright holder is always an alternative when it is not clear whether the practice may be an infringing one. For further help, a bibliography of books and articles on copyright appears as Appendix 9. Sample permission letters are in Appendix 8.

Librarians and information professionals must take an active role in exercising users' rights under the copyright law. First, know your rights and then use them! Then communicate what you know about copyright. Write articles for professional journals; submit articles and editorials to subject-matter journals; write columns on copyright compliance for in-house newsletters. Take the opportunity to speak about copyright at employee meetings and to appear on programs at professional associations. One does not have to be an expert on copyright to lead a discussion of particular problems of compliance or issues of interest to a special librarians.

Copyright law is of vital interest to libraries and librarians. It is not so complicated that only an attorney can be knowledgeable about the law. It is the duty of librarians and information professionals to learn about copyright and to educate their users and each other.

APPENDIX A

ANSWERS TO QUESTIONS POSED IN CHAPTER 1

The questions posed in Chapter 1 are answered below briefly. References to relevant chapters appear at the end of each answer.

QUESTION 1. A research librarian is asked by an advertising executive to locate and photocopy a magazine advertisement and a cartoon that first appeared in 1939. The executive is both writing an article on the history of advertising psychology and preparing a speech for a group of local public relations professionals. Are the ad and cartoon still protected by copyright or are they now in the public domain? May the executive reproduce the ad and/or the cartoon in his book without the permission of original copyright holder? May he reproduce the ad on a transparency to use to illustrate his speech?

Answer: In accordance with the Copyright Act of 1909, works created prior to the effective date of the 1976 Copyright Act, January 1, 1978, had to be registered with the U.S. Copyright Office in order to receive protection. It may be presumed that the advertisement was registered; if it were not, the work would have entered the public domain prior to 1978, thereby rendering a further discussion of copyright protection moot.

If the ad and cartoon were registered at the time of its publication, they would have been protected for 28 years. At the end of that period, the owner of the copyright had the option to renew for a second 28-year term. If the copyright was not renewed, it would have entered the pubic domain in 1967. The passage of the 1976 Act would not have altered its status.

Had it been renewed for a second 28-year term in 1967, the copyright would have been in effect in 1978. Under the 1976 Act, a copyright endures for the life of the author plus 50 years. For works such as anonymous, pseudonymous, posthumous and corporate-authored works, the term is 75 years from the date of first publication. The 75-year term also applies to subsisting works, as in the example in question. Thus, works in their second term receive an additional 47 years of protection for a total of 75 years from the original date of publication. So, copyright in the advertisement would be effective under the 1976 Act until the year 2014.

The exclusive rights granted to the copyright holders include the rights to reproduce the work and to authorize derivative works. The executive would need to secure permission from the owner before using the copyrighted advertisement and cartoon in a book. Making a transparency for use in a speech arguably could be a fair use; however, it should be easy to obtain the permission and it is unlikely the owner would require royalties for such use.

See Chapter 2 for further information.

QUESTION 2. A music librarian is asked by a vocal artist to copy a rare recording of arias by Maria Callas so the performer can study at home. What if the recording were a new release on Classic Compact Discs which was readily available in any record store? Does it matter if the student wanted to copy only a few bars to study a difficult phrase?

Answer: Sounding recordings are one of the eight categories of copyrightable works. While underlying musical compositions on sound recordings have long been copyrightable works, it was not until the advent of a commercially viable recording industry in the early twentieth century that the economic ramifications of denying protection in the recordings as well became evident. The copyrights on most classical operas have long since passed into public domain, although newer arrangements and other variations may be copyrighted as derivative works. Individual performances preserved as recordings are eligible for copyright, and the reproduction of such recordings is infringement.

In cases where the copyright on the underlying musical composition remains current, permission of the owner of the copyright in the musical composition may be required prior to public performance. Many modern composers list their works with a performing rights society, such as ASCAP or BMI. These societies police the performance of registered musical works and collect royalties which are later distributed to copyright owners. The name of the performing rights society that lists the musical work appears on the recording's packaging. Note that there are no performance rights in sound recordings. Thus, any royalties are paid to the composer.

Under most circumstances, the owner of copyright in a recording has the exclusive right to reproduce and distribute copies of the recording and to prepare derivative works such as new arrangements. It is not an infringement of the owner's exclusive rights, however, for another individual to use the copyrighted work in the question if the use is a fair use. Examples of fair use included in the statute are criticism, comment, news reporting, teaching, scholarship or research. The court applies the four criteria listed in section 107 of the statute to determine whether a particular use is a fair use and balances the factors to determine whether the benefit to be gained by allowing the use free of the material outweighs the owner's loss of royalties.

In the case of the Callas recording, if the recording were newly released and easily available, reproducing a copy of the entire work from a library master has the effect of robbing the copyright holder of a sale. The potential market effect would be devastating if such behavior were widespread. Scholarship is one of the uses normally found to be fair, however. The guidelines on the educational use of music govern the situation of the performer's study of the Callas arias.

If the performer wants to copy only a few bars for study at home, the argument for allowing this reproduction as a fair use becomes even stronger. First, the purpose is scholarly, which is an acceptable basis under the statute for a fair use. Second, the amount of the recording the singer wants to copy is probably not substantial enough to justify a sale. It is unlikely she will buy a recording in order to secure only a couple of bars; therefore, copying the portion she requested will not discourage a sale and will have no effect on the recording's market. Third, no profit can be gained from the copy of so insignificant a portion of a recording. Even if the recording is still readily available for sale, this scenario may well be one in which a fair use exemption to the Copyright Act is justified.

A more difficult situation is presented in the case of a rare recording. If the whole work is to be copied, the substance of the work will be taken, which usually requires the payment of royalties. The rarity of the recording, however, implies that it is no longer on the market. If the recording remains under copyright, then the owner should receive royalties for reproductions that are not fair use.

Despite of the potential infringement caused by a library patron's use, section 108 the Act grants a lending organization some modicum of insulation from liability. Unfortunately, musical works are excluded from the section 108 exemptions except in very narrow circumstances.

See Chapters 2, 3, 5 and 7 for additional information.

QUESTION 3. A motion picture archivist is asked to screen *The Chimes at Midnight* for a university Shakespeare society. Are royalties owed? What if the society wants to charge admission?

Answer: Copyright protection in motion pictures grants its owner the bundle of exclusive rights provided in section 106 including the performance right. The issue in this question is whether the situation is one in which performing or displaying a work is permissible without either paying royalties or infringing the copyright on the work.

Section 110 of the Act states it is not an infringement to perform a copyrighted film in a classroom or place of instruction "of a nonprofit educational institution" for the purpose of instruction, so long as the screening is not for profit and the film copy was lawfully obtained. If screening a motion picture based on a Shakespeare play for students interested in Shakespearean studies is considered "instruction," if teacher and students are present in the same place, and if the performance takes place in a room where instruction normally occurs, the showing of the film without charge to viewers likely is an exempted performance. To qualify for the exemption, however, the educational institution must be nonprofit.

If admission is charged for the performance of the film, the situation is somewhat more complicated. Section 110 also states that a performance of "a non-dramatic literary or musical work" for a private audience will continue to be exempt from royalty payments if any proceeds gained are used "exclusively for educational, religious or charitable purposes and not for private financial gain." It is nowhere stated that similar allowances are made for motion picture screenings. Therefore, it may be presumed that any charge levied in connection with a screening of a copyrighted dramatic film does not qualify for the exemption for and royalties are warranted.

See Chapter 5.

QUESTION 4. An aerospace librarian has completed a complicated database search for researchers in the company. Is downloading on a disk permitted? May the librarian retain the search results for later use and update the results on a regular basis?

Answer: The issue in this question is whether downloading data obtained from a database inquiry may be an exempted use. Technically, a database is a compilation which is defined in section 101 as "a work formed by the collection and assembling of preexisting materials that are selected coordinated, or arranged in such a way that the resulting work as a whole constitutes an original work of authorship." Thus, the compilation may be subject to copyrights from two differing sources: copyrights on the individual materials contained in the database, if any, and copyright on the compilation itself. Still, this particular use would be allowed if the use were determined to be a fair one. To make this determination, a helpful technique is to determine whether making a print copy from the database is a fair use.

Downloading to a disk a copy of the search results to give to a user is equivalent to printing a copy for her and is permissible unless the license agreement that covers the subscription to the database prohibits it. Retention of the search results for repeated use presents another problem, however.

Many databases are funded on a per search charge as opposed to a single purchase price. Any avoidance of the per search fee is an infringement absent specific permission in the license agreement. So, retaining results for reuse is infringement. If the database subscription is on a flat monthly or annual rate, then reusing the results may be permitted. The librarian could retain his search strategy, in either event.

See Chapter 6 on computers, databases and electronic publishing for further explanation.

QUESTION 5.   A school librarian wants to make a videotape to interest younger students in reading and wishes to include slides of illustrations by Dr. Seuss and Maurice Sendak. May she make these slides without infringing the artists' rights? What if she reproduces them to create a pamphlet for a first-grade teacher's use as a primary reader?

Answer: The use of children's book illustrations either in a videotape or in a pamphlet raises the issue of fair use in making adaptations of a copyrighted work. If the proposed use of the illustrations in the texts constitute fair use, her use will not infringe the owner's rights in the illustrations.

Section 110 of the Act permits a teacher in a nonprofit school to display or perform any copyrighted work so long as the copy used for the performance or display is a lawfully acquired copy and the display is within the normal course of instruction. Making slides of copyrighted illustrations to display to a class is permitted. Likewise, including the slides on a videotape for display in the classroom probably is not an infringement.

Adding the videotape to the library collection in order to retain it for later use, however, probably is an infringement. The teacher should obtain permission from the copyright holder to convert the format of the book illustrations to videotape to retain it for requested use.

When a teacher becomes an "author" by writing a pamphlet, he may not copy illustrations for inclusion without permission of the copyright holder. Also, he must give credit to the source for the original work.

See Chapters 2 and 5 for additional information.

QUESTION 6.   A pharmaceutical company purchases an original work from a famous sculptor to place in its reception hall. After several years, the sculpture is moved to the library. In order to fit the sculpture in the available space, the librarian has it removed from its base. Does this infringe the sculptor's copyright?

Answer: The focus of copyright is to ensure the owner the maximum economic benefit from a work issued for public sale, but until 1990 that protection did not extend beyond the first sale of works of visual arts. Once an individual purchased a work of fine art, he could destroy it, alter the work, etc.

In 1990, the Visual Artists Rights Act was enacted as section 106A of the Act. It ensures artists, sculptors and photographers some control over a work after its sale in order to prevent purchasers from damaging the artist's reputation by altering or mutilating the work. Works of visual art are defined as works which exist in a single copy or in 200 or fewer signed and numbered reproductions. Further, the amendment grants to creators of works of visual art the rights of integrity and attribution but only for the lifetime of the artist. The right of attribution gives artists the right to have any publicly displayed work attributed to them. The right of integrity means an artist can prevent the alteration, distortion, mutilation or other distortion of any publicly displayed work. These rights vest in works of visual art created after 1990 even if the artist is not the owner of the copyright in the work.

In the question involving a sculpture in the corporate lobby, assuming the work was created after 1990 and that the artist is still alive, the integrity right attaches. It is unlikely that the movement of the statue would infringe the artist's right of integrity, however, unless the placement was considered necessary to the artistic effect. The Act states that the modification of a work of art for the purpose of conservation or public presentation does not infringe the right. Thus, the law places some limitations on an artist's ability to claim her work was mutilated.

Whether the removal of the base from the sculpture constitutes a distortion of the work is a subjective question. There is no definition in the Act clarifying the meaning of "distortion, mutilation or modification," but it is unlikely that the removal of the work from its base is mutilation.

See Chapter 2 for more details.

QUESTION 7.   The librarian in a technical library is asked to photocopy material from a technical report published under the auspices of NASA. May she do so? Does it make a difference that the report has been republished by a commercial publisher?

Answer: U.S. federal documents are not eligible for copyright protection, and there is a strong likelihood that the report is a government document. Therefore, the librarian could freely reproduce the work even in multiple copies. Traditionally, any work produced under a government contract or grant was ineligible for copyright protection, and this is still true in the main. In the past few years, however, the terms of some federal contracts permit an author to hold copyright in works produced under a grant. Only by examining the terms of the contract can one ascertain the status of works produced under the contract or grant.

In all likelihood, the publication under NASA's auspices is a U.S. federal document and thus is excluded from copyright protection. Even for commercially published works which contain government-produced information, that information is within the public domain and may be reproduced. The commercial publisher may claim copyright in the work as a whole, and a librarian would not be free to copy an entire work to which value has been added (such as indexes, bibliographies, tables of cases, and the like). The copyright does not cover any government-produced information included in the commercially published work, however, and it could be reproduced.

See Chapter 2 for further information.

QUESTION 8.   An archivist fears that the public lending of sound recordings of the poets Dylan Thomas and T. S. Eliot reading their own works will endanger the recordings because of the fragility of the long-playing phonorecords. Instead, he retains the albums as masters and copies the recordings onto cassette tapes for lending. Has he infringed copyright? What if the recordings are still available on the original label, but at a very high price due to their historical value?

Answer: The issue involved is to what extent a library or archive is allowed to reproduce items in its collection for the purpose of creating a master file for preservation. The Act provides the right to create one copy or phonorecord if the copy is not made for commercial advantage, and if the collection is open to the public research, and the copy contains notice of a copyright.

Under section 108(b), a library may copy an ***unpublished*** work for the purpose of preservation. The facts included in the question are silent as to whether the sound recordings are published. Assuming such recordings are published, the archive has no right to reproduce the work even for the purpose of preservation. In fact, major preservation projects such as the Library of Congress's videodisk project obtain permission from copyright owners in order to reproduce the work even for preservation. Most copyright holders are willing to grant such permission when only one copy is produced. If multiple copies are to be produced and marketed, the owner probably will require the payment of royalties.

Section 108(c) permits the copying of published works to replace a lost, damaged, stolen or deteriorating copy after the library has made a reasonable effort to obtain an unused copy at a fair price. Sound recordings are included in this exemption. Creating a master copy in anticipation of destruction is not permitted, however.

See Chapter 3 on section 108.

APPENDIX B

ANSWERS TO QUESTIONS POSED IN CHAPTER 6

The questions posed in Chapter 6 are answered briefly below. References to relevant chapters appear at the end of each answer.

QUESTION 1. A library purchases a network version of Microsoft Word to load on the library's local area network (LAN) for in-house library use. May a library patron use the word processing program on the network?

Answer: Section 109 of the Act permits nonprofit libraries to lend software to patrons for nonprofit purposes provided the software is appropriately labeled with the warning of copyright law as promulgated by the Register of Copyrights. Instead of loaning software directly to patrons, libraries often load software onto a network in the library or in a computer lab. Provided a program is appropriately licensed to run on machines attached to a network and the license is not restricted to staff use, libraries are free to permit a patron may use the word processing program in-house.

Existing licensing practices may raise new issues in the network environment. Ever since the development of networks in the 1970s software publishers and vendors have struggled with the application of license agreements in the networked environment. When a library permits a patron to use software a library has loaded on a network, problems arise with providing the copyright warning. To comply with the spirit of the Act, a library should place the notice on the machines themselves, whether the machines stand alone or are connected to a network, or the library could place the warning of the menus or the screen following the menu. See Chapter 6 for the text of the mandated warning.

For-profit libraries may not loan software to users without permission of the copyright holder or without a site license authorizing the loan.

QUESTION 2. A library has Quattro Pro loaded on the LAN available to library users. May a patron bring in a blank disk and copy the program for his own personal use?

Answer: The owner of a copy of the computer program and the owner of the program (the copyright owner) are distinct. Section 117 of the Act grants to the "owner of a copy of a computer program" the right to make a copy as an "essential step" in using the program and to make a copy for archival purposes.

Although the library is the owner of the copy of the program, the library's use of the software is further limited by section 109 of the Act which amended the first sale doctrine. A nonprofit library may lend software to patrons for nonprofit uses so long as the software packages bear the software warning. Similarly, a library can make programs available through a LAN for use by patrons. A library may not make a copy of the software for subsequent use by patrons, however, unless its license agreement so permits.

The right to make lawful copies of software is set out in section 117 and is applicable only to the owner of the copy, not an independent user. A patron's right to copy, if any, must be found in section 107, fair use. It is doubtful that a court, applying the four fair use factors, would find a patron's copying of a library's software program to be a fair use.

First, the purpose of the use may be educational or scholarly. If not, and the purpose is commercial, the first factor weighs against a fair use finding. Second, the nature of the work, as software based on algorithms, is nonetheless sufficient by creative as opposed to factual to warrant a finding in favor of the publisher. Third, the amount and substantiality will weigh against the user because he copied the entire program. Finally, the market may be affected because the patron copied the software. She could have relied on using the library's copy for subsequent use. Whether the vendor actually would have made another sale is debatable. A patron may not copy for his own use a program such as Quattro Pro licensed to run on a network available to users.

QUESTION 3.   A one-volume book arrived in the library with a computer disk to accompany it. May the library make a backup copy of the disk?

Answer: Section 117(2) of the Act states that it is not an infringement to make a copy of a software program for archival purposes. Recognizing the risk of electronic mutilation or destruction that can occur in utilizing a program, Congress intended to protect the owner of the copy as distinguished from the owner of the copyright. A library, therefore, can make an archival copy.

The Act does not distinguish between types of software: software that operates machines; software that permits applications such as word processing or electronic spreadsheets; or software that does not stand alone but rather accompanies a book or video. Therefore, a library may make a backup copy of a disk purchased together with a book or other information to be used by the library itself or circulated to patrons. Only one user at a time may use the software, however. The archival (backup) copy may not be used simultaneously with the original copy.

QUESTION 4.   In order to utilize a piece of expensive software purchased by the library, a librarian must load it onto the hard disk of her personal computer. Is this permitted under the law?

Answer: Among the exclusive rights of copyright holders defined in section 106 of the Act are the rights of reproduction, distribution, adaptation, performance and display. These rights are limited by sections 107 to 118 that permit owners of software to use their copies in certain ways. More specifically, section 117 authorizes the "owner of a copy of a computer program" to copy or make an adaptation under two circumstances. A copy owner may make an authorized copy of a program when doing so is "an essential step in the utilization of a computer program in conjunction with a machine ..." Using any software requires making a copy. For a computer to execute a program from memory the program must first have been copied by the user into the machine's memory. Some programs do this automatically when the software instructions interact with computer's instructions. Two copies exist: one copy is in the internal memory of the machine; the other copy is on the hard disk or on a diskette. The internal copy is not an infringing copy because it was made as an essential step in using the program. Section 117(1) of the Act grants the librarian the right to copy the program to the hard disk of her machine when doing so is "an essential step" in the utilization of the program with her machine.

QUESTION 5.   Must the library invest in the more expensive network version of software in order to load it onto the computer laboratory's network?

Answer: License agreements as contracts determine what use the copy owner can make of the licensed program including who may use the program, on what machines and for how long.

Among issues addressed in licenses are payment terms, delivery of software, warranties, upgrades and downloading provisions. Licenses typically address the number of users and the number of machines on which software may be used. The most common license is authorization for use on a single machine.

Today, companies are more frequently negotiating site licenses with software publishers which grant a user through the license the right to make a specified number of copies for use at a specific location or to authorize a specified class of individuals as users, for instance, employees. These types of network licenses provide greater flexibility for users and simplify distribution. The library can purchase a license for the number of individual machines on which it wishes to run a specific program even though the machines can also run programs loaded on the network. A library may also purchase concurrent use licenses that permits a specified number of users simultaneous access to a program. Generally, concurrent use licenses are less expensive than network versions of a program.

QUESTION 6. The library has purchased two copies of a software upgrade. The Acquisitions Department asks to use the old version since the administrative office will be using the upgrade. Should the library permit this use?

Answer: Under the terms of a software license the copy owner is authorized to use the licensed version. Normally a license dictates that a company or individual cannot use both an earlier version and the current version of a computer program. Absent permission from the software publisher, a library should not permit a department to use an old version of a software program to which it has recently purchased an upgrade. For example, a library purchasing an upgrade of WordPerfect from version 5.2 to 6.0 has not purchased an additional copy but rather has upgraded the earlier version probably at a reduced price.

Some companies like WordPerfect Corporation (WPC), for example, through its School Software Donation Program, however, permit donation of old versions to a qualified primary or secondary school (kindergarten through grade 12) and WPC will issue a valid license to the school.

QUESTION 7. A library patron wants to donate several computer disks that contain copies of freeware. May the library accept the gift?

Answer: Many creators who develop computer software make no claim of copyright for the programs they create. Everyone may freely use and reproduce this type of software (freeware) without infringement. A library may accept a gift of freeware and make it readily available to patrons or permit staff to use this gift. A library, however, should be cautious about accepting copies of shareware which is a type of software initially distributed without charge with the understanding that if the program is retained and used distributors generally anticipate compensation.

QUESTION 8. May a reference librarian download bibliographic citations from a variety of databases to create a bibliography which will become a section of the library's monthly newsletter?

Answer: There are two sources of copyright law that may govern the downloading of bibliographic information and its subsequent reuse. The first is the U.S. Supreme Court's analysis concerning compilations and the second is the merger doctrine.

While underlying facts cannot be copyrighted, the selection and arrangement of facts in a database may be sufficient to entitle the database to protection as a compilation. The merger doctrine applies to the situation where an idea can be described only in a limited number of ways.

When the idea and the expression of the idea, for instance an alphabetically arrangement of data, can be expressed only in such limited ways, the idea and expression are said to have merged and there can be no protected interest.

Unless restricted by a license, when a librarian downloads information and supplies a single copy of that information to a user, the issue of merger is less important because the librarian supplied the search results as intended. When a librarian intends to distribute copies of a bibliography created by a librarian, the merger doctrine becomes relevant. The arrangement of data may be merged with the expression that there can be no copyright to be infringed. If the bibliography includes abstracts protected by copyright, the abstracts should be removed before a librarian circulates the bibliography she prepared. For an additional charge, some vendors will license libraries to distribute abstracts widely.

QUESTION 9.   Users have access to the library's LAN and can use all programs loaded on the LAN. Should the library check the network periodically to ensure that no patron has added other software to the network?

Answer: With the growth of computer networks including local area networks (LANs), wide area networks (WANs) and metropolitan area networks (MANs) a library should periodically check the network to determine that network use is still in compliance with its various licenses. Since often the use of software is licensed for concurrent use, a library needs to establish some type of monitoring system to ensure that no more than the authorized number of simultaneous users have access to the program. If the library has agreed to other terms by signing a license agreement, the library must monitor the system to ensure that the library is in compliance with other terms in the contract. Further, a library should check periodically to determine whether a patron has added software to a library's equipment that may not be authorized for such use.

QUESTION 10.   After a complicated computer database search, the librarian retains copies of both the results and his search strategy. May he reuse either or both?

Answer: Whether a librarian may reuse copies of search results depends on whether the database is funded on a per use charge or a flat basis. Reusing the search results would be infringement if the database is funded by a per use charge. A library is expected to conduct and pay for a separate search for each individual use for this type of database. Conversely, reusing search results is permitted if the database is funded on a flat rate basis. The fee for searching the database is the same regardless of the number of times it is searched. Because of constant updating of databases, however, librarians should be cautious about reusing search results that might be outdated. Retaining search strategies for subsequent reuse is permitted because the search statement is the librarian's own work product.

QUESTION 11.   The library purchases a copy of software. May it modify the program for its own internal use?

Answer: Section 117 deals specifically with computer programs and authorizes the copy owner of a program to make a copy or to adapt the program under two circumstances. The first instance permits the copy owner to make a copy when doing so is an essential step to running the program. A common occurrence requires conversion from one disk size to another and a copy owner is permitted to make a noninfringing copy of a program when such conversion is required. Copying is permitted also to convert a program from one computer language to another. Similarly, section 117 would permit the addition of features not present. This right is personal and not commercial. Therefore, the adaption should be used only in-house. Under no circumstances should the library attempt to market the product or to distribute it to others. When modifications are permitted, a library should recognize that existing warranties on the underlying program are not extended automatically to the adaptation.

QUESTION 12. A library staff member has requested permission to duplicate a copy of a program owned by the library to use on her home computer so she may work at home. Should the library approve this request?

Answer: License agreements essentially are contracts between a user of the software and owners such as Microsoft or vendors such as Dialog of a copyrighted work. These agreements are governed by state law and not the Copyright Act. Parties to a contract may agree to limit the user's rights afforded by the Act or broaden the rights a user has under the law. Licenses limit users' rights to copy and access to databases and downloading or they may specify legitimate uses of information. Unless the making of a copy is required as an "essential step" in using the software or is so permitted by the terms of the license, generally a library should not permit a staff member to copy software owned by the library to use on her home computer to facilitate work at home. WordPerfect under some circumstances permits owners of copies of the software to make a copy for home use of a legally licensed copy. Under no circumstances should the copying be allowed unless the license permits a home use copy or if the library has obtained permission from the copyright holder to permit reproduction of a copy for home use. When copying is permitted for home use, common sense dictates that the licensed office copy cannot be used by one person and the home use copy by another at the same time.

QUESTION 13 Instead of producing a printout to record the results of a computer scarch, a librarian downloads the results to a disk. May he give the results to the user in this format?

Answer: Downloading restrictions are among the most common restrictions included in a license agreement for use of an online database. If it is permissible for a library to supply a printout of the search results to a user, a library may give the user a disk containing the information or send the information by e-mail as equivalents of a printout. Some vendors, however, restrict the number of lines or entries that may be downloaded and libraries should check the terms of the license agreement to determine permissible limits of downloading. Other vendors permit downloading of "insubstantial portions" of a database but offer no further guidance as to what the supplier believes constitutes an insubstantial portion.

Many corporate libraries have negotiated agreements that permit them to download primarily bibliographic entries with abstracts, but in some cases textual files, and to load them on a company server for distribution and use by multiple users within the company.

APPENDIX C

AGREEMENT ON GUIDELINES FOR CLASSROOM COPYING IN NOT-FOR-PROFIT EDUCATIONAL INSTITUTIONS WITH RESPECT TO BOOKS AND PERIODICALS

The purpose of the following guidelines is to state the minimum standards of educational fair use under Section 107 of H.R. 2223. The parties agree that the conditions determining the extent of permissible copying for educational purposes may change in the future; that certain types of copying permitted under these guidelines may not be permissible in the future; and conversely that in the future other types of copying not permitted under these guidelines may be permissible under revised guidelines.

Moreover, the following statement of guidelines is not intended to limit the types of copying permitted under the standards of fair use under judicial decision and which are stated in Section 107 of the Copyright Revision Bill. There may be instances in which copying which does not fall within the guidelines stated below may nonetheless be permitted under the criteria of fair use.

GUIDELINES

I. Single Copying for Teachers

A single copy may be made of any of the following by or for a teacher at his or her individual request for his or her scholarly research or use in teaching or preparation to teach a class:

A. A chapter from a book;
B. An article from a periodical or newspaper;
C. A short story, short essay, or short poem, whether or not from a collective work;
D. A chart, graph, diagram, drawing, cartoon, or picture from a book, periodical, or newspaper.

II. Multiple Copies for Classroom Use

Multiple copies (not to exceed in any event more than one copy per pupil in a course) may be made by or for the teacher giving the course for classroom use or discussion; provided that:

A. The copying meets the tests of brevity and spontaneity as defined below; and,
B. Meets the cumulative effect test as defined below; and,
C. Each copy includes a notice of copyright.

Definitions

Brevity

(i) Poetry: (a) A complete poem is less than 250 words and if printed on not more than two pages or, (b) from a longer poem, an excerpt of not more than 250 words.

(ii) Prose: (a) Either a complete article, story, or essay of less than 2,500 words, or (b) an excerpt from any prose work of not more than 1,000 words or 10 percent of the work, whichever is less, but in any event a minimum of 500 words.
[Each of the numerical limits stated in "i" and "ii" above may be expanded to permit the completion of an unfinished line of a poem or of an unfinished prose paragraph.]

(iii) Illustration: One chart, graph, diagram, drawing, cartoon, or picture per book or per periodical issue.

(iv) "Special" works: Certain works in poetry, prose, or in "poetic prose" which often combine language with illustrations and which are intended sometimes for children and at other times for a more general audience fall short of 2,500 words in their entirety. Paragraph "ii" above notwithstanding such "special works" may not be reproduced in their entirety; however, an excerpt comprising not more than two of the published pages of such special work and containing not more than 10 percent of the words found in the text thereof, may be reproduced.

Spontaneity

(i) The copying is at the instance and inspiration of the individual teacher, and

(ii) The inspiration and decision to use the work and the moment of its use for maximum teaching effectiveness are so close in time that it would be unreasonable to expect a timely reply to a request for permission.

Cumulative Effect

(i) The copying of the material is for only one course in the school in which the copies are made.

(ii) Not more than one short poem, article, story, essay or two excerpts may be copied from the same author, nor more than three from the same collective work or periodical volume during one class term.

(iii) There shall not be more than nine instances of such multiple copying for one course during one class term.

[The limitations stated in "ii" and "iii" above shall not apply to current news periodicals and newspapers and current news sections of other periodicals.]

III. Prohibitions as to I and II Above

Notwithstanding any of the above, the following shall be prohibited:

A. Copying shall not be used to create or to replace or substitute for anthologies, compilations, or collective works. Such replacement or substitution may occur whether copies of various works or excerpts therefrom are accumulated or reproduced and used separately.

B. There shall be no copying of or from works intended to be "consumable" in the course of study or of teaching. These include workbooks, exercises, standardized tests and test booklets, and answer sheets and like consumable material.

C. Copying shall not:

   (a)   substitute for the purchase of books, publishers' reprints or periodicals;
   (b)   be directed by higher authority;
   (c)   be repeated with respect to the same item by the same teacher from term to term.
   (d)   No charge shall be made to the student beyond the actual cost of the photocopying.

Agreed March 19, 1976

    Ad Hoc Committee on Copyright Law Revision:
        By Sheldon Elliott Steinbach.

    Author-Publishers Group:
        Authors League of America:
        By Irwin Karp, Counsel.

    Association of American Publishers, Inc.:
        By Alexander C. Hoffman,
        Chairman, Copyright Committee.

APPENDIX D

GUIDELINES FOR EDUCATIONAL USE OF MUSIC

The purpose of the following guidelines is to state the minimum and not the maximum and standards of educational fair use under section 107 of H.R. 2223. The parties agree that the conditional purposes may change in the future; that certain types of copying permitted under these guidelines may not be permissible in the future, and conversely that in the future other types of copying not permitted under these guidelines may be permissible under revised guidelines.

Moreover, the following statement of guidelines is not intended to limit the types of copying permitted under the standards of fair use under judicial decision and which are stated in Section 107 of the Copyright Revision Bill. There may be instances in which copying which does not fall within the guidelines stated below may nonetheless be permitted under the criteria of fair use.

A. Permissible Uses

1. Emergency copying to replace purchased copics which for any reason are not available for an imminent performance provided purchased replacement copied shall be substituted in due course.

2. For academic purposes other than performance, single or multiple copies of excerpts of works may be made, provided that the excerpts do not comprise a part of the whole which would constitute a performable unit such as a selection, movement or aria, but in no case more than 10 percent of the whole work. The number of copies shall not exceed one copy per pupil.

3. Printed copies which have been purchased may be edited or simplified provided that the fundamental character of the work is not distorted or the lyrics, if any, altered or lyrics added if none exist.

4. A single copy of recordings of performances by students may be made for evaluation or rehearsal purposes any may be retained by the educational institution or individual teacher.

5. A single copy of a sound recording (such as a tape, disk, or cassette) of copyrighted music may be made from sound recordings owned by an educational institution or an individual teacher for the purpose of constructing aural exercises or examinations and may be retained be the educational institution or individual teacher. (This pertains only to the copyright of the music itself and not to any copyright which may exist in the sound recording.)

B. Prohibitions

1. Copying to create or replace or substitute for anthologies, compilations, or collective works.
2. Copying of or from works intended to be "consumable" in the course of study or of teaching such as workbooks, exercises, standardized tests, and answer sheets and like material.
3. Copying for the purpose of performance, except as in A(1) above.
4. Copying for the purpose of substituting for the purchase of music, except as in A(1) and A(2) above.
5. Copying without inclusion of the copyright notice which appears on the printed copy.

APPENDIX E

INTERLIBRARY LOAN GUIDELINES

Guidelines for the Proviso of Subsection 108(g)(2)

1. As used in the proviso of subsection 108(g)(2), the words "... such aggregate quantities as to substitute for a subscription to or purchase of such work" shall mean:

    (a) with respect to any given periodical (as opposed to any given issue of a periodical), filled requests of a library or archive (a "requesting entity") within any calendar year for a total of six or more copies of an article or articles published in such periodical within five years prior to the date of the request. These guidelines specifically shall not apply, directly or indirectly, to any request of a requesting entity for a copy or copies of an article or articles published in any issue of a periodical, the publication date of which is more than five years prior to the date when the request is made. These guidelines do not define the meaning, with respect to such a request, of "... such aggregate quantities as to substitute for a subscription to [such periodical]".

    (b) With respect to any other material described in subsection 108(d), (including fiction and poetry), filled requests of a requesting entity within any calendar year for a total of six or more copies or phonorecords of or from any given work (including a collective work) during the entire period when such material shall be protected by copyright.

2. In the event that a requesting entity -

    (a) shall have in force or shall have entered an order for a subscription to a periodical, or

    (b) has within its collection, or shall have entered an order for, a copy or phonorecord of any other copyrighted work, material from either category of which it desires to obtain by copy from another library or archives (the "supplying entity"), because the material to be copied is not reasonably available for use by the requesting entity itself, then the fulfillment of such request shall be treated as though the requesting entity made such copy from its own collection. A library or archive may request a copy or phonorecord from a supplying entity only under those circumstances where the requesting entity would have been able, under the other provisions of section 108, to supply such copy from materials in its own collection.

    (c) No request for a copy or phonorecord of any material to which these guidelines apply may be fulfilled by the supplying entity unless such request is accompanied by a representation by the requesting entity that the request was made in conformity with these guidelines.

(d) The requesting entity shall maintain records of all requests made by it for copies or phonorecords of any materials to which these guidelines apply and shall maintain records of the fulfillment of such requests, which records shall be retained until the end of the third complete calendar year after the end of the calendar year in which the respective request shall have been made.

(e) As part of the review provided for in subsection 108(i), these guidelines shall be reviewed not later than five years from the effective date of this bill.

APPENDIX F

GUIDELINES FOR OFF-AIR RECORDINGS OF BROADCAST PROGRAMMING FOR EDUCATIONAL PURPOSES

In March 1979, Congressman Robert Kastenmeier, chairman of the House Subcommittee on Courts, Civil Liberties, and Administration of Justice, appointed a Negotiating Committee consisting of representatives of education organizations, copyright proprietors, and creative guilds and unions. The following guidelines reflect the Negotiating Committee's consensus as to the application of "fair use" to the recording, retention, and use of television broadcast programs for educational purposes. They specify periods of retention and use of such off-air recordings in classrooms and similar places devoted to instruction and for homebound instruction. The purpose of establishing these guidelines is to provide standards for both owners and users of copyrighted television programs.

1. The guidelines were developed to apply only to off-air recording by nonprofit educational institutions.
2. A broadcast program may be recorded off-air simultaneously with broadcast transmission (including simultaneous cable retransmission) and retained by a nonprofit educational institution for a period not to exceed the first forty-five (45) consecutive calendar days after date of recording. Upon conclusion of such retention period, all off-air recordings must be erased or destroyed immediately. "Broadcast programs" are television programs transmitted by television stations for reception by the general public without charge.
3. Off-air recordings may be used once by individual teachers in the course of relevant teaching activities, and repeated once only when instructional reinforcement is necessary, in classrooms and similar places devoted to instruction within a single building, cluster or campus, as well as in the homes of students receiving formalized home instruction, during the first ten (10) consecutive school days in the forty-five (45) day calendar day retention period. "School days" are school session days — not counting weekends, holidays, vacations, examination periods, and other scheduled interruptions — within the forty-five (45) calendar day retention period.
4. Off-air recordings may be made only at the request of and used by individual teachers, and may not be regularly recorded in anticipation of requests. No broadcast program may be recorded off-air more than once at the request of the same teacher, regardless of the number of times the program may be broadcasted.
5. A limited number of copies may be reproduced from each off-air recording to meet the legitimate needs of teachers under these guidelines. Each such additional copy shall be subject to all provisions governing the original recording.

6. After the first ten (10) consecutive school days, off-air recordings may be used up to the end of the forty-five (45) calendar day retention period only for teacher evaluation purposes i.e., to determine whether or not to include the broadcast program in the teaching curriculum, and may not be used in the recording institution for student exhibition or any other non-evaluation purposes without authorization.

7. Off-air recordings need not be used in their entirety, but the recorded programs may not be altered from their original content. Off-air recordings may not be physically or electronically combined or merged to constitute teaching anthologies or compilations.

8. All copies of off-air recording must include the copyright notice on the broadcast program as recorded.

9. Educational institutions are expected to establish appropriate control procedures to maintain the integrity of these guidelines.

APPENDIX G

LIBRARY RESERVE GUIDELINES

At the request of a faculty member, a library may photocopy and place on reserve excerpts from copyrighted works in its collection in accordance with guidelines similar to those governing formal classroom distribution for face-to-face teaching discussed above. This University [College] believes that these guidelines apply to the library reserve shelf to the extent it functions as an extension of classroom readings or reflects an individual student's right to photocopy for his personal scholastic use under the doctrine of fair use. In general, librarians may photocopy materials for reserve room use for the convenience of students both in preparing class assignments and in pursuing informal educational activities which higher education requires, such as advanced independent study and research.

If the request calls for only *one* copy to be placed on reserve, the library may photocopy an entire article, or an entire chapter from a book, or an entire poem.

The negotiated safe-harbor guidelines for classroom uses are in many ways inappropriate for the college and university level. "Brevity" simply cannot mean the same thing in terms of grade-school readings that it does for more advanced research. Because university professors were not specifically represented in the negotiation of the classroom guidelines, ALA published *Model Policy Concerning College and University Photocopying for Classroom Research and Library Reserve Use* (the "Model Policy").

In general with respect to classroom uses, the standard guidelines should be followed:

1. The distribution of the same photocopied material does not occur every semester.
2. Only one copy is distributed for each student.
3. The material includes a copyright notice on the first page of the portion of material photocopies.
4. The students are not assessed any fee beyond the actual cost of the photocopying.

Requests for multiple copies on reserve should meet the following guidelines:

1. The amount of material should be reasonable in relation to the total amount of material assigned for one term of a course taking into account the nature of the course, its subject matter, and level, 17 U.S.C. + 107(1) and (3).
2. The number of copies should be reasonable in light of the number of students enrolled, the difficulty and timing of assignments, and the number of other courses which may assign the same material, 17 U.S.C. + 107(1) and (3).
3. The material should contain a notice of copyright, *see*, 17 U.S.C. + 401.
4. The effect of photocopying the material should not be detrimental to the market for the work. (In general, the library should own at least one copy of the work.) 17 U.S.C. + 107(4).

From: *ALA Model Policy Concerning College and University Photocopying for Classroom, Research and Library Reserve Use*, March 1982.

APPENDIX H(1)

SAMPLE PERMISSION LETTER: PHOTOCOPIES

February, 1994

Permission Department
Publishing Company
1608 Avenue Of The Americas
New York, NY 10019

Dear Sir or Madam:

May I have permission to copy the following for use in my classes next semester:

Title: *Philosophical and Literary Essays*

Copyright: Publishing Co., 1986, 1991

Material to be duplicated: Essays 7, 12 and 22

Distribution: The material will be distributed to students in my classes; they will pay only the cost of photocopying.

Type of Reprint: Photocopy

Use: Supplementary teaching materials.

I have enclosed a self-addressed envelope for your convenience in replying to this request. Your response by _____(date) would be most appreciated.

Sincerely yours,

Faculty Member

I (We) grant permission to faculty member to reproduce the above material.

_____   _____
                                     Date
Please use thc following credit line.

_____

APPENDIX H(2)

SAMPLE PERMISSION LETTER: AUDIOVISUAL WORK

February, 1994

Permission Department
Video Productions Company
1313 Sunset Boulevard
Hollywood, CA 90017

Dear Sir or Madam:

The _____ library is interested in obtaining a copy of the videocassette identified on the attached purchase order. Please do not supply the named videocassette unless you grant the following rights to the library.

_____ To make an archival copy;

_____ To duplicate up to _____ copies of the work (we understand that the cost of the work may be greater than the price for one copy);

_____ To convert the format of the videocassette to whatever becomes the standard for video (the earlier format will be destroyed upon any coversion of format);

_____ To perform the audiovisual work publicly at no admission charge. This would include use in the evenings by employee groups on company premises, special showings during the lunch hour, use at corporate educational seminars, and the like.

_____ To use the videocassette for instructional broadcasting purposes.

Please sign this letter below as an indication that you have granted the permissions checked above and return the letter with the copy(ies) of the videocassette.

Sincerely yours,

Librarian

I (we) grant the permissions noted above to this library to facilitate use of the video material(s) identified on the attached purchase order.

_____   _____
                                                     Date

APPENDIX I

LIST OF CASE CITATIONS

ABKCO Music, Inc. v. Harrisongs Music, Inc., 722 F.2d 988 (2d Cir. 1983).

Allen-Myland v. Int'l Business Mach. Corp., 796 F. Supp. 520 (E.D. Pa. 1990).

Am. Geophysical Union v. Texaco, Inc., 802 F. Supp. 1 (S.D.N.Y. 1992).

Apple Computer, Inc. v. Franklin Computer Corp., 714 F.2d 1240 (3d Cir. 1983).

Arica Inst., Inc. v. Palmer, 970 F.2d 1067 (2d Cir. 1992)

Atari, Inc. v. JS&A Group, Inc., 597 F. Supp. 5 (N.D. Ill. 1983).

Atari Games Corp. v. Nintendo of Am., Inc., 975 F.2d 832 (Fed. Cir. 1992).

Atari Games Corp. v. Oman, 979 F.2d 242 (D.C.Cir. 1992).

Baker v. Selden, 101 U.S. 99 (1879).

Basic Books, Inc. v. The Gnomon Corp., 10 Copyright L. Dec. _ 25,145 (CCH) (D.Conn. 1981).

Basic Books, Inc. v. Kinko's Graphics Corp., 758 F. Supp. 1522 (S.D.N.Y. 1991).

Broadcast Music, Inc. v. Niro's Place, Inc., 619 F. Supp. 958 (N.D. Ill. 1985).

Canadian Admiral Corp., Ltd. v. Rediffusion Inc., et al [1954] Ex C.R. 382.

C.B.S. v. Ames Records, 2 W.L.R. 973 (1981).

Coditel v. Cone Vog Films S.A., Judgment of March 18, 1980, Case 62/79, E.C.R. 881 (1980).

Columbia Broadcasting Sys. Inc., v. Am. Soc'y of Composers, Authors and Publishers, 620 F.2d 930 (2d Cir. 1980), *cert. denied*, 450 U.S. 970 (1981).

Computer Assocs. Int'l, Inc. v. Altai, Inc., 982 F.2d 693 (2d Cir. 1992).

Copyright Clearance Ctr., Inc. v. Comm'r, 79 T.C. 793 (1982).

Deutsche Grammophon Gmb. H. v. Firma Pap Import, 1971 C. & M.L.R. 631.

Eckes v. Card Prices Update, 736 F.2d 859 (2d Cir. 1984).

Electra Records Co. v. Gem Electronics Distribs., 360 F. Supp. 821 (E.D.N.Y. 1973).

Federal Election Comm'n v. Int'l Funding Inst., 969 F.2d 1110 (D.C. Cir. 1992).

Feist Publications, Inc. v. Rural Telephone Services Co. Inc., 499 U.S. 340 (1991).

Financial Information, Inc. v. Moody's Investors Service, 808 F.2d 204 (2d Cir. 1986).

Folsom v. Marsh, 9 F. Cas. 342 (C.C.D. Mass. 1841).

Foresight Resources Corp. v. Pfortmiller, 719 F. Supp. 1006 (D. Kan. 1989).

Gershwin Publishing Corp. v. Columbia Artists Mgmt., Inc., 443 F.2d 1159 (2d Cir. 1971).

Harper & Row Publishers, Inc., v. Nation Enters., 471 U.S. 539 (1985).

Harper & Row Publishers, Inc. v. Tyco Copy Service, Inc., 10 Copyright L. Dec. (CCH) _ 25,230 (D. Conn. 1981)

Heim v. Universal Pictures Co., 154 F.2d 480 (2d Cir. 1946).

Herbert Rosenthal Jewelry Corp. v. Kalpakian, 336 F.2d 738 (1971).

Jones v. CBS, Inc., 733 F. Supp. 748 (S.D.N.Y. 1990).

Latern Press, Inc. v. Am. Publishers Co., 419 F. Supp. 1267 (D.C.N.Y. 1976).

Kamar Int'l., Inc. v. Russ Berrie & Co., 657 F.2d 1059 (9th Cir. 1981).

Key Publications, Inc., v. Chinatown Today Publishing Enters., Inc., 945 F.2d 509 (2d Cir. 1992).

Kregos v. Associated Press, 731 F. Supp. 113 (S.D.N.Y. 1990), *aff'd in part, rev'd in part* 937 F.2d 700 (2d Cir. 1991).

Ladbroke v. William Hill, 1 W.L.R. 273 (1964).

Los Angeles News Serv. v. Tullo, 973 F.2d 791 (9th Cir. 1992).

Meeropol v. Nizer, 560 F.2d 1061 (2d Cir. 1977), *cert. denied*, Nizer v. Meerpol, 434 U.S. 1013 (1977).

Moreau v. St. Vincent [1950] Ex. C.R. 198.

New Era Publications Int'l, ApS v. Henry Holt & Co., Inc., 873 F.2d 576 (2d Cir. 1989).

New Era Publications Int'l, ApS v. Carol Publishing Group, 904 F.2d 152 (2d Cir. 1990).

Nichols v. Universal Pictures, Corp., 45 F.2d 119 (2d Cir. 1930), *cert. denied*, 282 U.S. 902 (1931).

Nichols v. Universal Pictures, 45 F.2d 119 (2d Cir. 1930), *cert. denied*, 282 U.S. 902 (1930).

Pasha Publications, Inc. v. Enmark Gas Corp., 22 U.S.P.Q.2d (BNA) 1076 (N.D. Tex. 1992).

Pushman v. New York Graphic Society, Inc., 39 N.E.2d 249 (1942).

The Queen v. James Lorimer & Co. Ltd. [1984] 1 F.C. 1065.

RCA Records, et. al. v. All-Fast Systems, Inc., 594 F. Supp. 335 (S.D.N.Y. 1984).

Salinger v. Random House, Inc., 811 F.2d 90 (2d Cir. 1987).

Schiller & Schmidt, Inc. v. Nordisco Corp., 969 F.2d 410 (7th Cir. 1992).

Sega Enters., Ltd. v. Accolade, Inc., 977 F.2d 1510 (9th Cir. 1992).

Sony Corp. of Am. v. Universal City Studios, Inc., 464 U.S. 417 *rehearing denied*, 465 U.S. 1112 (1984).

Spectravest, Inc. v. Mervyn's, Inc., 673 F. Supp. 1486 (N.D. Cal. 1987).

Step-Saver Data Sys. v. Wise Technology, 939 F.2d 91 (3d Cir. 1991).

Television Digest, Inc. v. U.S. Telephone Ass'n, 47 Pat. Trademark & Copyright J. (BNA) 32 (1993).

T.J. Moore v. Accessories de Bureau de Quebec, Inc. [1973] 14 C.P.R. (2d) 113 (Fed. T.D.).

University of London Press, Ltd. v. University Tutorial Press, Ltd. [1916] 32 T.L.R. 698.

Vault Corp. v. Quaid Software, Ltd., 847 F.2d 255 (5th Cir. 1988).

Waterlow Directories v. Reed Information Services, Ltd., Chancery Division, Times, Oct. 11, 1990.

Wheaton v. Peters, 33 U.S. 591 (1834).

Williams & Wilkins Co. v. National Library of Medicine, 487. F. 2d 1345 (Ct. Cl. 1973), aff'd by an equally divided Court, 420 U.S 376 (1975).

Wright v. Warner Books, Inc., 953 F.2d 731 (2d Cir. 1989).

BIBLIOGRAPHY:
BOOKS AND CHAPTERS

American Library Association, *Model Policy Concerning College and University Photocopying for Classroom Research and Library Reserve Use*, Chicago: American Library Association, 1982.

Bonham-Carter, Victor, *Authors by Profession*. London: Society of Authors, 1978.

Brophy, Brigid, *A Guide to Public Lending Right*. Hampshire, England: Gower, 1983).

Cavendish, J. M. & Poole, Kate, *Handbook of Copyright in British Publishing Practice*. London: Cassell, 3d ed. 1993.

Chickering, Robert B. & Hartman, Susan. *How to Register a Copyright and Protect Your Creative Work: A Basic Guide to the Copyright Law and How It Affects Anyone Who Wants to Protect Creative Work*. New York: Scribner, 1987.

Churchill, Robert, *Golden Egg Production: The Goose Cries 'Foul'*, in John Sheldon Lawrence & Bernard Timberg, *Fair Use and Free Inquiry* 169. Norwood, NJ: Ablex, 1980.

Council of School Attorneys, *Copyright Law: A Guide for Public Schools*. Alexandria, VA: National School Boards Association, 1986.

Crews, Kenneth D., *Copyright, Fair Use and the Challenge for Universities: Promoting the Progress of Higher Education*. Chicago: Univ. Chicago Press, 1993.

Green Paper on Copyright and the Challenge of Technology, Com (88) final (June 7, 1988).

Goldstein, Paul, *Copyright: Principles, Law and Practice*. Boston: Little Brown, 3 vols. 1989.

Hammer, Richard R., *The Church Guide to Copyright Law*. Matthews, NC: Christian Ministry Resources, 2d ed. 1989.

Hancock, W.A., *User's Guide to Software Licenses*. Chesterland, OH: Business Laws, 1985.

Harris, Lesley Ellen, *Canadian Copyright Law*. Toronto: McGraw-Hill Ryerson, 1992.

Heller, James S. & Wiant, Sarah K., *Copyright Handbook*. Littleton, CO: Rothman, 1984.

Jacob, Robin & Alexander, Daniel, *A Guidebook to Intellectual Property: Patents, Trademarks, Copyright and Designs*. London: Sweet & Maxwell, 4th ed. 1993.

Johnson, Beda *How To Acquire Legal Copies of Video Programs: Resource Information*. San Diego: Video Resources Enterprises, 6th rev. ed. 1993.

Katsh, Ethan M., *The Electronic Media and the Transformation of Law*. New York: Oxford University Press 1989.

Kellner, Douglas, *Television Research and Fair Use*, in John Sheldon Lawrence & Bernard Timberg, *Fair Use and Free Inquiry* 90. Norwood, NJ: Ablex, 1980.

Kunzle, David, *Hogarth Piracies and the Origin of Visual Copyright*, in John Sheldon Lawrence and Bernard Timberg, *Fair Use and Free Inquiry* 21. Norwood, NJ: Ablex, 2d ed. 1989.

Latman, Alan, Gorman, Robert & Ginsburg, Jane C., *Copyright for the Nineties*. Charlottesville, VA: Michie, 3d ed. 1989.

Lawrence, John Sheldon & Timberg, Bernard, *Fair Use and Free Inquiry*. Norwood, NJ: Ablex, 2d ed. 1989.

Mast, Gerald, *Film Study and Copyright Law*, in John Sheldon Lawrence & Bernard Timberg, *Fair Use and Free Inquiry* 72. Norwood, NJ: Ablex, 1980.

Medical Library Association, *The Copyright Law and the Health Sciences Librarian*. Chicago: Medical Library Association, 1989.

Miller, Arther R., *Intellectual Property: Patents, Trademark, and Copyrights in a Nutshell*. St. Paul, MN: West, 1990.

Miller, Jerome K., *Applying the New Copyright Law: A Guide for Educators and Librarians*. Chicago: American Library Association, 1979.

Miller, Jerome K., *The Duplication of Audiovisual Materials in Libraries*, in John Sheldon Lawrence & Bernard Timberg, *Fair Use and Free Inquiry* 128. Norwood, NJ: Ablex, 1980.

Miller, Jerome K., *Using Copyrighted Videocassettes in Classrooms, Libraries, and Training Centers*. Friday Harbor, WA: Copyright Information Services, 2d ed. 1988.

National Commission of New Techological Uses of Copyrighted Works, *Final Report*. Wash., DC: GPO, 1978.

Nimmer, Melville B. & Geller, Paul Edward, *International Copyright Law and Practice*. New York: Matthew Bender, 1988.

Nimmer, Melville B. & Nimmer, David D., *Nimmer on Copyright*. New York: Matthew Bender, 4 vols. 1990.

Patterson, L. Ray & Linberg, Stanley W., *The Nature of Copyright: A Law of User's Rights*. Athens, GA: Univ. Georgia Press, 1991.

Patry, William F., *The Fair Use Privilege in Copyright Law*. Wash., DC: Bureau of National Affairs, 1985.

Patry, William F., *Latman's The Copyright Law*. Wash., DC: Bureau of National Affairs, 6th ed. 1986.

Ploman, Edward W. & Hamilton, L. Clark, *Copyright: Intellectual Property in the Information Age*. London: Routledge & Kegan Paul, 1980.

Powell, Jon T. & Gair, Wall, *Public Interest and the Business of Broadcasting*. New York: Quorum Books, 1988.

Reed, Mary Hutchins, *The Copyright Primer for Librarians and Educators*. Chicago: American Library Association, 1987.

Reinbothe, Jorg & Von Lewinski, Silke, *The EC Directive on Rental and Lending Rights and Piracy*. London: Sweet & Maxwell, 1993.

Report of the Register of Copyrights, *Library Reproduction of Copyrighted Works* (17 U.S.C. + 108). Wash., DC: GPO, 1988.

Ringer, Barbara A., *The Demonology of Copyright*. New York: Bowker, 1974.

Seeman, Ernest A., *A Comparative Look at Public Lending Right From the U.S.A.*, in *Public Lending Right: Reports of an ALAI Symposium and Additional Materials* 137. Deventer, Netherlands: Kluwer, 1983.

Seltzer, Leon E. *Exemptions and Fair Use in Copyright: The Exclusive Rights Tension in the 1976 Copyright Act*. Cambridge, MA: Harvard Univ. Press, 1987.

Sinofsky, Esther R., *Off-Air Videotaping in Education: Copyright Issues, Decision, Implications*. New York: Bowker, 1984.

Special Libraries Association, *Library Photocopying and the U.S. Copyright Law of 1976*. New York: Special Libraries Association, 1977.

Spoor, Jaap H., *Problems and Possiblities of Public Lending Right*, in Herman Cohen Jehoram, *Public Lending Rights: Report of an ALAI Symposium and Additional Materials* 57. Deventer, Netherlands: Kluwer, 1983.

Stewart, Stephen M., *International Copyright and Neighboring Rights*. London: Butterworths, 2d ed. 1989.

Sumison, John, *PLR in Practice: A Report to the Advisory Committee*. Stockton-on-Tees, Eng.: Registrar of Public Lending Right, 1988.

Talab, R. S., *Commonsense Copyright: A Guide to the New Technologies*. Jefferson, NC: McFarland, 1986.

Timberg, Bernard, *New Forms of Media and the Challenge to Copyright Law*, in John Sheldon Lawrence & Bernard Timberg, *Fair Use and Free Inquiry* 247. Norwood, NJ: Ablex, 1980.

Timberg, Sigmund, *A Modernized Fair Use Code for Visual, Auditory and Audiovisual Copyrights: Economic Context, Legal Issues and the Laocoon Shortfall*, in John Sheldon Lawrence & Bernard Timberg, *Fair Use and Free Inquiry* 311. Norwood, NJ: Ablex, 1980.

Whale, R. F., *Copyright: Evolution, Theory and Practice*. Totowa, NJ: Roman and Littlefield, 1972.

U.S. Copyright Office, *Compendium II; Compendium of Copyright Office Practices*. Wash., DC: GPO, 1984.

Vlcek, Charles W., *Adoptable Copyright Policy*. Wash., DC: Association for Educational Communications and Technology, 1992.

ARTICLES

Aiello, Vincent F., *Note: Educating Sony: Requiem for a 'Fair Use'*, 22 Cal. W.L. Rev. 159 (1985).

Anderson, Micheal G. & Brown, Paul F., *The Econimics Behind Copyright Fair Use: A Principled and Predictable Body of Law*, 24 Loy. U. Chi. L. J. 143 (1993).

Angel, Frank P., *France, More Ded Arts ... at Ded Laid: Also for Foreign Works in France*, 30 J. Copyright Soc'y 335 (1983).

Arden, Thomas P., *The Conflicting Treatments of Compilations of Facts under the United States and British Copyright Laws*, 19 AIPLA Q. J. 267 (1991).

Astbury, Raymond G., *The Situation in the United Kingdom*, 29 Libr. Trends 661 (1981).

Bennett, Scott, *Copyright and Innovation in Electronic Publishing: A Commentary*, 19 J. Acad. Libr. 87 (1993).

Berman, Bayard F. & Boxer, Joel E., *Copyright Infringement of Audiovisual Works and Characters*, 52 S. Cal. L. Rev. 315 (1979).

Besen, Stanley M., Kirby, Sheila N. & Salop, Steven C., *An Economic Analysis of Copyright Collectives*, 78 Va. L. Rev. 383 (1992).

Bianchi, David J., Brenna, Grant H. & Shannon, James P., *Comment, Basic Books, Inc. v. Kinko's Graphic Corp.: Potential Liability for Classroom Anthologies*, 18 J. of C. & U. L. (1992).

Blum, Debra E., *Use of Photocopied Anthologies for Courses Snarled by Delays and Costs of Copyright-Permission Process*, Chronicle of Higher Educ. Sept. 11, 1991, at A19.

Callison, Daniel, *Fair Payment for Fair Use in Future Information Technology Systems*, Educ. Tech., Jan. 1981, at 20.

Cole, John Y., *Public Lending Right: A Symposium at the Library of Congress*, 42 Libr. of Cong. Info. Bull. 427 (1983).

Cornish, Graham P., *The New United Kingdom Copyright Act and its Implications for Libraries and Archives*, 83 Law Libr. J. 51 (1991).

Deutsch, Sarah, *Fair Use in Copyright Law and The Nonprofit Organization: A Proposal for Reform*, 34 Am. U. L. Rev. 1327 (1985).

Duggan, Mary Kay, *Copyright Of Electronic Information: Issues and Questions* OnLine, May, 1991, at 20.

Dworkin, Gerald, *Public Lending Right - The U.K. Experience*, 13 Colum.-VLA J. L. & Arts 49 (1988).

Fields, Howard, *Senate Bill to Study U.S. Public Lending Rights*, 224 Pub. Wkly., Oct. 21, 1983, at 15.

Finkel, Evan, *Copyright Protection For Computer Software In The Nineties*, 7 Santa Clara Computer & High Tech. L. J. 201 (1991).

Fujitani, Jay M., *Comment, Controlling the Market Power of Performing Rights Societies: An Administrative Substitute for Antitrust Regulation*, 72 Cal. L. Rev. 103 (1984).

Gasaway, Laura N., *Audiovisual Material and Copyright in Special Libraries*, 74 Special Libr. 222 (1983).

Gasaway, Laura N., *Nonprint Works and Copyright in Special Libraries*, 74 Special Libr. 156 (1983).

Gasaway, Laura N., *Wide Impact Seen for Photocopying Case*, 15 Nat'l. L. J., Aug. 16, 1993, at 21.

Gilbert, Maria Franciose, *International Copyright Law Applied to Computer Programs in the United States and France*, 14 Loy. U. L. J. 105 (1982).

Ginsburg, Jane C., *Copyright Without Walls?: Speculations on Literary Property in the Library of the Future*, 42 Representations 53 (1993).

Goddard, Connie, *Textbook Authors Hear Kinko's View on Customizing*, Pub. Wkly, July 19, 1991, at 12.

Goldstein, Paul, *Commentary On "An Economic Analysis of Copyright Collectives,"* 78 Va. L. Rev. 413] (1992).

Gorman, Robert A., *The Feist Case: Reflections On A Pathbreaking Copyright Decision*, 18 Rutgers Computer & Tech. L. J. 731 (1992).

Greengrass, Alan R., *Databases and Their Off-Spring*, Bookmark, Winter, 1992, at 147.

Hansen, Linda, *The Half-Circled "C": Canadian Copyright Legislation*, 19 Gov't Publ. Rev. 137 (1992).

Hart, Jack R., *Public Lending Right: The American Author's Viewpoint*, 29 Libr. Trends 613 (1981).

Hartzell, Richard, *Public Lending Right Stirs Debate*, 58 Wilson Libr. Bull. 567 (1984).

Holland, Arnold J., *The Audiovisual Package: Handle with Care*, 23 Bull. Copyright Soc'y 104 (1974).

Holland, Bill & Terry, Ken, *RIAA Spearheads New Royalty Group: Alliance Will Collect Artists' Digital Fees*, Billboard, Feb. 13, 1993, at 1.

Hyatt, Dennis, *Legal Aspects of the Public Lending Right*, 29 Libr. Trends 583 (1981).

Iocupa, Nora Maija, *The Development of Special Provisions in International Copyright Law for the Benefit of Developing Countries*, 29 J. Copyright Soc'y 405 (1982).

Jenson, Mary Brandt, *Electronic Reserve and Copyright*, 13 Computers In Libr. 40 (1993).

Jenson, Mary Brandt, *Is The Library Without Walls on a Collision Course with the 1976 Copyright Act?* 85 L. Libr. J. 619 (1993).

Johnson, Mark, *The New Copyright Law: Its Impact on Film and Video Education*, 32 J. U. Film Ass'n 67 (1980).

Jones, Arthur C., *Practical and Economical Considerations*, 29 Libr. Trends 597 (1981).

Jones, Beryl R., *An Introduction to the European Economic Community and Intellectual Properties*, 18 Brook. J. Int'l L. 665 (1992).

Kasunic, Robert, *Fair Use and the Educator's Right to Photocopy Copyrighted Material for Classroom Use*, 19 J. of C. & U. L. 271 (1993).

Katsh, Ethan & Rifkin, Janet, *The New Media and a New Model of Conflict Resolution: Copying, Copyright, and Creating*, 6 Notre Dame J. of L. Ethics & Pub. Pol'y 49 (1992).

Kent, Marian, *From Sony to Kinko's: Dismantling the Fair Use Doctrine*, 12 J. L. & Com. 133 (1992).

Kies, Cosette, *Copyright Versus Free Access: CBS and Vanderbilt University Square Off*, 50 Wilson Libr. Bull. 242 (1975).

Kinko's Drops Course Packets to Focus on Electronic Services, Pub. Wkly. June 14, 1993, at 21.

Klaver, Franca, *The Legal Problems of Video Cassettes and Audiovisual Discs*, 23 Bull. Copyright Soc'y 152 (1976).

Koch, Ole, *Situation in Countries of Continental Europe*, 29 Libr. Trends 641 (1981).

Kravetz, Paul I., *Copyright Protection Of Screen Displays After Lotus Development Corporation v. Paperback Software International*, 4 DePaul Bus. L. J. 485 (1992).

Landes, William M. & Posner, Richard A., *An Economic Analysis of Copyright Law*, 18 J. Legal Stud. 325 (1989).

Landesman, Betty & Brugger, Judith M., *Everything You Always Wanted to Know about SISAC (Serials Industry Systems Advisory Committee)*, 21 Serials Libr. 211 (1991).

Leval, Pierre N., *Toward A Fair Use Standard*, 103 Harv. L. Rev. 110 (1990).

Levin, Martin B., *Soviet International Copyright: Dream or Nightmare*, 31 J. Copyright Soc'y 99 (1983).

Litman, Jessica, *Copyright and Information Policy*, 55 Law & Contemp. Probs. 185 (1992).

McAnanama, Judith, *Copyright Law: Libraries and Their Uses Have Special Needs*, 6 Intell. Prop. J. 225 (1991).

McCartney, Sheila J., *Moral Rights Under the United Kingdom's Copyright, Designs and Patents Act of 1988*, 15 Colum.-VLA J. L. & Arts 205 (1991).

McDowell, Edwin, *The Media Business; Royalties From Photocopying Grow*, N.Y. Times, June 13, 1988, at D9.

MacPherson, Lillian B., *The State of Copyright Legislation in Canada and its Impact on Libraries*, 1 Commonwealth L. Libr. (1992).

Magnuson, Jan, *Duplicating AV Materials Legally*, 13 Media & Methods 52 (1977).

Marik, Katherine, *The Visual Artists Rights Act of 1990; The United States Recognizes Artists and Their Rights*, 8 Ent. & Sports Law. 7 (1991).

Martin, Scott M., *Photocopying and the Doctrine of Fair Use: The Duplication of Error*, 39 J. Copyright Soc'y 345 (1992).

Mayer, Daniel Y., *Note, Literary Copyright and Public Lending Right*, 18 Case W. Res. J. Int'l. L. 483 (1986).

Merry, Susan, *Canada Calling*, 15 SpeciaList, June, 1992, at 1.

Miller, Authur R., *Copyright Protection for Computer Programs, Databases, and Computer-Generated Works: Is Anything New Since CONTU?*, 106 Harv. L. Rev. 977 (1993).

Million, Angela C. & Healey, J.S., *Public Lending Right: An Overview*, 21 Pub. Libr., Sept. 1982, at 26.

Mills, Mary L., *Note and Comment, New Technology And The Limitations of Copyright Law: An Argument For Finding Alternatives To Copyright Legislation In An Era Of Rapid Technological Change*, 65 Chi.-Kent L. Rev. 307 (1989).

Motyka, Carol, *U.S. Participation in the Berne Convention and High Technology*, 39 Copyright L. Symp. (ASCAP) 105 (1992).

Nelson, Milo G., *An Idea Whose Time to Debate, Has Come*, 58 Wilson Libr. Bull. 165 (1983).

Note: Toward a Unified Theory of Copyright Infringement for an Advanced Technological Era, 96 Harv. L. Rev. 450 (1982).

Ogle, Melinda, *Not by Books Along: Library Copying of Nonprint Copyrighted Material*, 70 Law Libr. J. 153 (1977).

Olsson, Jennifer T., *Rights In Fine Art Photography: Through a Lens Darkly*, 70 Tex. L. Rev. 1489 (1992).

Pitternick, George, *Points of View of Librarians: Alternatives to PLR*, 29 Libr. Trends 627 (1981).

Plumleigh, Michael, *Digital Audio Tape: New Fuel Stokes the Smoldering Home Taping Fires*, 37 U.C.L.A. L. Rev. 733 (1990).

Pubnet Permissions [Electronic System to Facilitate Copyright] for Course Antholgies Set for Test Launch, 238 Pub. Wkly., Mar. 1, 1991, at 24.

Raskind, Leo J., *Protecting Computer Software in the European Economic Community: The Innovative New Directive*, 18 Brook J. Int'l L. 729 (1992).

Rasmussen, Henning, *Public Lending Right: Situation in New Zealand and Australia*, 29 Libr. Trends 687 (1981).

Report of the Register of Copyrights, *Technological Alterations to Motion Pictures and Other Audiovisual Works ...*, 10 Loyola Ent. L. J. 1 (1990).

Ringer, Barbara & Flacks, Louis I., *Applicability of the Universal Copyright Convention to Certain Works in the Public Domain in Their Country of Origin*, 27 Bull. Copyright Soc'y 157 (1980).

Riordan, Virginia, *Copyright Clearance Center, 1988: A Progress Report*, 15 Serials Libr. 43 (1988).

Rogers, Edward S., *Copyright and Morals*, 18 Mich. L. Rev. 390 (1920).

Roit, Natasha, *Soviet and Chinese Copyright: Ideology Gives Way to Economic Necessity*, 6 Loy. Ent. L. J. 53 (1986).

Samuelson, Pamela & Glushko, Robert J., *Intellectual Property Rights for Digital Library and Hypertext Publishing Systems*, 6 Harv. J. of L. & Tech. 237 (1993).

Savage, Diane, *Law of the LAN*, 9 Santa Clara Computer & High Tech. L. J. 193 (1993).

Schaper, Louise Levy and Kawecki, Alicja T., *Towards Compliance: How One Global Corporation Complies with Copyright Law*, Online, Mar. 1991, at 15.

Schneck, Jennifer M., *Note, Closing the Book on the Public Lending Right*, 63 N.Y.U. L. Rev. 878 (1988).

Scott, Michael D. & Talbott, James N., *Interactive Multimedia What Is It, Why Is It Important and What Do I Need to Know About it?* 11 Ent. & Sports Law. 13 (1993).

Seeman, Ernest A., *A Look At The Public Lending Right*, 30 Copyright L. Symp. (ASCAP) 71 (1983).

Somorjai, John E., *The Evolution of a Common Market: Limits Imposed on the Protection of National Intellectual Property Rights in the European Economic Community*, 9 Int'l Tax & Bus. Law 431 (1992).

Stave, Thomas, *Pay As You Read: The Debate Over Public Lending Right*, 62 Wilson Libr. Bull., Oct., 1987, at 22.

Stave, Thomas, *Public Lending Right: A History of the Idea*, 29 Libr. Trends 569 (1981).

Stover, June M., *Copyright Protection for Computer Programs in the United Kingdom, West Germany, and Italy: A Comparative Overview*, 7 Loy. L.A. Int'l & Comp. L. J. 278 (1984).

Tanenbaum, William A. & Wells, William K. Jr., *Multimedia Works Requre Broad Protection*, 16 Nat'l L. J., Nov. 1, 1991, at S11.

Thomas, Gloria Jean, *Copyrights: The Law, the Teacher, and the Principal* 1992 B.Y.U. J. L. & Educ. 1.

Timburg, Sigmund, *A Modernized Fair Use Code for the Electronic Age As Well As the Guttenberg Age*, 75 Nw. U. L. Rev. 193 (1980).

Valauskas, Edward J., *Copyright: Know Your Electronic Rights!* 117 Libr. J. 40 (1992).

Vandebeek, Victor, *Realizing the European Community Common Market by Unifying Intellectual Property Law: Deadline 1992*, 1990 B.Y.U. L. Rev. 1605.

Vaver, David, *Copyright Phase 2: The New Horizen*, 6 Intel. Prop. J. 37 (1991).

Voss, Christopher, *The Legal Protection of Computer Programs in the European Economic Community*, 11 Computer/L. J. 441 (1992).

Waldron, Jeremy, *From Authors To Copiers: Individual Rights and Social Values in Intellectual Property*, 68 Chi.-Kent L. Rev. 841 (1993).

Wall, Raymond A., *Copyright: The New Act of 1988*, 21 Law Libr. 18 (1990).

Wargo, Natalie, *Copyright Protection for Architecture and the Berne Convention*, 65 N.Y.U. L. Rev. 403 (1990).

Winick, Raphael, *Copyright Protection For Architecture After The Architectural Works Copyright Protection Act of 1990*, 41 Duke L. J. 1598 (1992).

Wright, Brad, *Note, Changing the Standard for Computer Software Copyright Infringement: Computer Associates Int'l v. Altai*, 13 Geo. Mason U. L. Rev. 663 (1992).

Wyunbrandt, Robert A., *Musical Performances in Libraries: Is a License from ASCAP Required?* 24 Pub. Libr. 224 (1990).

Zimmerman, Barbara, *The Trouble with Multimedia: Copyright Clearance and the Uncertain Future*, AV-Video, Jan., 1993, at 46.

INDEX

A

AAP POSITION PAPER
Library copying, 56
ABSTRACTION OF WORKS
Substantial similarity test in infringement action, 25
ABUSE OF JUDICIAL PROCESS
Infringement defense, 32
ACCESS TO WORK
Infringement action, elements of, 24, 25
ACTION OR SUIT
See REMEDIES FOR INFRINGEMENT
ADAPTATIONS
See DERIVATIVE WORKS
ADONIS
Computer works, licensing of, 75, 76
ADVISORY COMMITTEE ON COPYRIGHT REGISTRATION AND DEPOSIT ACCORD
Registration, proposal to eliminate, 24
AGENCIES AND COLLECTIVES FOR LICENSING
See LICENSING AND LICENSING AGENCIES AND COLLECTIVES
AGREEMENTS
Agreement on Guidelines for Classroom Copying, Appx C, 28, 57, 72, 144–146
Copyright act, license agreement as superseding, 52, 132
Government works for hire as copyrightable (ed: xref if able to build up work for hire), 19
Library copying, 52
Manuscripts, archival copying of, 47
Writing requirements for agreements, 21, 92
ALA
Copyright notice format, 45
Fair price for replacement, 47
Reserves, model policy, Appx G, 148
Unsupervised copying notice, 51
ALLIANCE OF ARTISTS AND RECORD COMPANIES (AARC)
Music licensing by, 74
AMERICAN GEOPHYSICAL UNION V TEXACO
Fair use and, 29, 44, 45, 55, 60–62
ANNUAL AUTHORIZATION SERVICE (AAS) BLANKET LICENSES
Copyright Clearance Center (CCC), 69
ANTHOLOGIES AND COURSE PACKS
Kinko's case as affecting, 71–73, 147
Prohibited use, 146
APPLICATIONS FOR COMPUTERS
Software or applications for computers. *See under* COMPUTERS AND COMPUTER-GENERATED WORKS
ARCHITECTURAL WORKS
Protected status, 18
ARCHIVES
See LIBRARIES AND ARCHIVES
ARRANGEMENTS OF MUSICAL WORKS
Public domain works, copyrightability of compilations of, 22
ARTICLES IN JOURNALS
See BOOKS, JOURNALS, AND OTHER LITERARY WORKS
ASCAP (AMERICAN SOCIETY OF COMPOSERS, ARTISTS, AND PUBLISHERS)
Music licensing agencies and collectives, 74
ASCAP (AMERICAN SOCIETY OF COMPOSERS AND PUBLISHERS)

Blanket license for excepted uses, 25, 26
ASSIGNMENT OR OTHER TRANSFER OF RIGHTS IN WORK
Main discussion, 21
Author's or artist's transfer, generally, 21
Copyright as transferred with ownership, 21
Dramatization, assignment for, 4
Exclusive rights of owner, 20
Formal requirements for executing, 20, 21
Music licensing collectives, collection of royalties by, 74
Severability of rights, 20
Termination of transfer, 21
ATTORNEYS' FEES
Infringement damages as including award of, 24, 33
Registration as prerequisite to recovery of attorneysfees in civil action, 24
AUDIO HOME RECORDING ACT OF 1992
DAT recordings, payment of royalties on, 93
AUDIOTAPES
Audio Home Recording Act of 1992, 93
DAT (digital audio tape) recordings, scheme for paying royalties on, 93
Electra Records Co. v Gem Electronics Distributors and sound recordings, 92, 93
First sale doctrine where user makes copies of loaned sound recordings, 94
Library taping of sound recordings, 91–94
RCA Records, et al. v All-Fast Systems and, 93
AUDIOVISUAL WORKS
Main discussion, 81–106
See also particular kinds, e.g., FILMS AND FILMSTRIPS, VIDEOTAPES, etc.
Copyrightable works, list, 18
Exclusive rights of authors or artists, 20
Fair use doctrine, main discussion, 83–87
Library copying. *See under* LIBRARIES AND ARCHIVES
Permission letter, form of, 244
AUTHOR OR ARTIST
See also EXCLUSIVE RIGHTS OF OWNER; OWNER OF COPYRIGHT
Authorship, defined, 3
British copyright, waiver of author's rights under, 189
European Economic Community or EU, software rights in, 172
Ownership as vested in, 21
Public Lending Right (PLR), compensation under, 199 et seq.
Unknown author, duration of copyright as affected by, 21
AUTOMATIC RENEWAL OF COPYRIGHT
Copyrights eligible for, 22

B

BAKER V SELDEN
Merger doctrine created by, 129
BASIC BOOKS, INC. V KINKO'S
Professor publishing as affected by, 71–73, 147
BERLIN ACT OF 1908
History of copyright law, international aspects of, 9
BERNE CONVENTION AND UNION
Main discussion, 162–166
EU members as members of, 168
First publication rule and, 165
Foreign works as copyrightable under Copyright Act, 19

History of copyright law, 9
Implementation Act of 1988, United States as treaty signatory, 9
Integrity protection, 164
Minimum procedural rights under Berne Convention, 163–165, 167
National treatment, 163–165
New technologies and, 165
Notice requirement as affected by, 23
Paternity protection, 164
Reciprocity of protection among members, 165
Registration, 163, 164
United States as signatory, 9, 163, 165
United States copyright law conformed to, 1
Universal Copyright Convention (UCC) as complementing, 166, 167

BETAMAX CASE
See SONY CORPORATION OF AMERICA V UNIVERSAL STUDIOS

BIBLIOGRAPHIC DATA
Databases and, 130, 131

BIOGRAPHIES
Fair use, 26
Harper and Row v The Nation Magazine, 30

BLANKET LICENSE
Excepted uses, 26

BLANKET LICENSE OR PERMISSION
Annual Authorization Service (AAS) of Copyright Clearance Center (CCC), 69
Library duplication of recordings from archive or master copy, 91, 92

BMI (BROADCAST MUSIC, INC)
Music licensing agencies and collectives, 74

BOOKS, JOURNALS, AND OTHER LITERARY WORKS
Agreement on Guidelines for Classroom Copying in Not-For-Profit Educational Institutions with Respect to Books and Periodicals, Appx C, 28, 57, 72, 144–146
Classroom copying, 28
Copyrightability, 18
Copyright Clearance Center (CCC) licensing copying of, 68 et seq.
Factual works, 29, 72
Libraries *See under* LIBRARIES AND ARCHIVES
Software as literary work, 18

BRITISH COPYRIGHT
Main discussion, 187–193
Contributory or secondary infringement, 190
Copyrightable works, 188–190
Copyrights, Designs, and Patents Act (CDPA), generally, 187 et seq.
Crown Copyright, 189, 190
Defenses to infringement action, 191, 192
Duration of copyright, 188, 189
Exclusive rights of owners, 188
Fair dealing, 192, 193
History of, 4–6
Infringement, 190–192
Library uses, 192, 193
Licensing and library uses, 192, 193
Moral rights, 189
Originality requirement, 188
Photographs, 189
Secondary or contributory infringement, 190
Software, originality principle, 172
Videocassette recording, 190
Waiver of author's rights, 189

BRUSSELS ACT OF 1948
History of copyright law, international aspects of, 9

BURDEN OF PROOF
Independent creation defense, difficulty of proving, 31
Infringement action, plaintiff's burden of proving elements, 24
Presumptions. *See* PRESUMPTIONS, INFERENCES, AND PRIMA FACIE EVIDENCE
Profit loss, owner's burden of proving, 33
Registration as prima facie evidence of copyright, 23

C

CANADIAN COPYRIGHT
Main discussion, 181–187
Copyrightable works, 182–184
Educational or scholarly purpose, 185, 186
Historical background, 181
Infringement, 184, 185
International copyright compared, 182
Libraries and archives, 186, 187
Publication not required, 182
Remedies for infringement, 184, 185

CARL/UNCOVER
Copyright Clearance Center (CCC) and, 71

CCC
See COPYRIGHT CLEARANCE CENTER (CCC)

CD-ROM
Copyrightable works, 18
Licensing, 75
Notice of copyright, 131

CD (SOUND RECORDINGS)
See PHONORECORDS AND CD

CERTIFICATE OF REGISTRATION
Issuance, 23
Validity of copyright without, 23

CERTIFICATE OF TRANSFER OF RIGHTS IN WORK
Necessity of, 21

CHIPS FOR COMPUTERS
Misuse defense and, 32

CHOREOGRAPHIC WORKS
Copyrightable works, list, 18

CIRCUMSTANTIAL EVIDENCE
Infringement action, proving elements of, 24, 25

CITIZENS AND ALIENS
Copyrightability of foreign works as affected by citizenship of owner, 19

CLASSROOM COPYING
See also EDUCATIONAL OR SCHOLARLY USE
Cumulative effects test, 28
Fair use, 28
Guidelines, Appx C, 28, 57, 72, 144–146
Libraries as subject to guidelines for, 57
Single copying of chapter, 28

COLLABORATIVE WORKS
Infringement action by collaborator, good faith requirement as to, 32

COLLECTIVES
See LICENSING AND LICENSING AGENCIES AND COLLECTIVES

COLLECTIVE WORKS
Compilation as including, 18

COMMERCIAL PURPOSE OR ADVANTAGE
CARL/UnCover, 71
Criminal action for infringement, 34
Direct or indirect commercial advantage, 44, 45
Exceptions as affected by, generally, 25
Fair use as affected by, 26–31

First sale doctrine
  International copyright, 170, 171
  Software, 123, 124
  Sound recordings, 94
For-profit libraries. *See* FOR-PROFIT LIBRARIES
Home videotaping, 83–87
Infringement damages as including loss of profit, 33
Kinko's case and fair use, 71–73, 147
Library copying, 44, 45, 55, 56
Loss of profit, infringement damages for, 33
Market effect of copying. *See* MARKET EFFECT OF COPYING
Personal use as distinguished by lack of, 25
Software adaptation, 120
Software leasing prohibited, 123, 125
Sony Corporation of America v Universal Studios, Inc., 83–86
Tape recordings, copying, 28, 83–86, 92–94
COMMON LAW COPYRIGHT
  Copyright Act of 1976 as preempting, 22
COMPILATIONS
  Databases. *See* DATABASES
  Defined, 18
  Public domain works, copyrightability of compilations of, 22
  Tables of contents copyrightable as, 48
COMPUTERS AND COMPUTER-GENERATED WORKS
  Authorship, computer generated works as within definition of, 3
  Compiler or assembler described, 117
  Cost of coping as exacerbating copying problem, 144
  Databases. *See* DATABASES, ELECTRONIC
  Hardware and software distinguished, 117
  I/C chips and misuse defense, 32
  Library practice. *See under* LIBRARIES AND ARCHIVES
  Licensing agencies, 75
  Misuse defense to infringement action, 32
  Networks. *See* NETWORKS
  Software or applications
    Adapted programs, vending, 120
    Alternative forms of protection proposed, 114
    Archival copies, 119, 120
    Backup copies, proper making and use of, 119
    British copyright, originality principle, 172
    Commercial leasing prohibited, 123, 124
    Computer Software Directive of European Economic
      Community or EU, 171–173
    Computer Software Rental Amendment Act, 125, 126
    Copying inherent in use, 119
    Copyrightability, 4, 18, 118
    Copyright protection, adequacy of, 114
    Defined, 116–118
    Exclusive rights of owner, 118
    Executing licensing agreements, 122, 123
    First sale doctrine, 123, 124
    Freeware defined, 118
    Hardware distinguished, 117
    International copyright, 161, 167, 171–173
    Languages, conversion between, 119, 120
    Library practices. *See under* LIBRARIES AND ARCHIVES
    Licensing and licensing agreements, 52, 114, 120–123
    Literary works as including, 118
    Multimedia production, 101
    Network licenses, 121, 122
    Object code and source code described, 117
    Ownership, 118, 119
    Pirating software, libraries's difficult position vis a vis, 114
    Programming described, 117
    PROLOK case, 119
    Section 117 and, 118–120
    Sega case, 120
    Shareware defined, 118
    Shrinkwrapped licensing, 122
    Site licenses, 52, 121
    Source code, copying, 120
    Use copy, 119
    Warning notice of copyright, 126, 127
COMPUTER SOFTWARE RENTAL AMENDMENTS ACT
  Nonprofit library exemption, 125, 126
CONCEPT
  Copyrightable work, unexpressed idea as, 3–4
CONCLUSION TO SUBJECT MATTER
  Main treatment, 217, 218
CONFLICT OR CHOICE OF LAW
  Foreign works, law governing copyrightability of, 19
  International copyright, 166, 168
CONTINUING INFRINGEMENT
  Limitation of actions, 31
CONTRACTS
  See AGREEMENTS; LICENSES
CONTRIBUTORY OR SECONDARY INFRINGEMENT
  British copyright, 190
CONTU
  Interlibrary loan guidelines, 53–55
  Software copyrightability, 118
CONVENTIONS AND TREATIES
  Berne Convention. *See* BERNE CONVENTION AND UNION
  Universal Copyright Convention. *See* UNIVERSAL COPYRIGHT
    CONVENTION
CONVERSION OF FORMATS
  Library copying. *See under* LIBRARIES AND ARCHIVES
COPIES AND COPYING
  See REPRODUCTION OF WORK
COPYRIGHTABLE WORKS
  Main discussion, 17–19
  See also FIXATION
  British copyright, 188–190
  Canadian copyright, 182–184
  Databases, 128
  Databases, electronic, 18, 128–130
  Defined, 3, 17, 18
  Fair use doctrine as applicable to, 29
  Fraudulent works, 18
  Government works, 19
  Idea unexpressed, 3–4
  International copyright, 19, 160, 161
  Merger doctrine as affecting copyrightability, 130
  Microforms, 18
  Object code in computer programming, 117
  Obscene or profane works, 18
  Pantomime, list of copyrightable works, 18
  Software, 4, 18, 118
  Tables of contents, 48
COPYRIGHT ACT OF 1909
  Correlation of rights under 1976 Copyright Act, 8, 22
  History of copyright law, 8
  Renewal of works copyrighted under, necessity, 22
COPYRIGHT ACT OF 1976
  For particular provisions, see specific subject headings throughout
    this index
  Main discussion, 17–34
  Common law copyright as preempted by, 22

Computer Software Rental Amendment Act, 125, 126
Duration of copyright increased, 21
Fair use doctrine, 27, 31
Grandfather clause applicable to copyrights prior to effective date of, 8
Libraries and. *See* LIBRARIES AND ARCHIVES
Notice requirements, liberalization of, 23
Record rental amendment, 94
Registration requirement abolished, 23
Unpublished works and fair use, amendment affecting, 31
COPYRIGHT CLEARANCE CENTER
Blanket license, 25, 26
COPYRIGHT CLEARANCE CENTER (CCC)
Main discussion, 68–73
Annual authorization service (AAS) blanket licenses, 69
CARL/UnCover and, 71
Course packs, CCC as taking over responsibility after Kinko decision, 73
Educational copying, licensing for, 70, 71
Fair use, 69, 70
International organizations, coordination with, 74
Journals, licensing copying of, 68 et seq.
Licenses, types of, 69, 70
Nonprofit organization CCC as, 68
Registration of organizations, 68, 69
Royalty fees, collection and distribution of, 68–70
Serial issue coding, 69
SIID code as replacing CCC code, 69
Surveys by, 69, 70
Transaction reporting service (TRS), 69
COPYRIGHTS, DESIGNS, AND PATENTS ACT (CDPA)
British copyright, 187 et seq.
COSTS OF ACTION
Infringement action, award of costs in, 33
COURSE PACKS AND ANTHOLOGIES
Guidelines prohibiting, 146
Kinko's as affecting, 71–73, 147
CRIMINAL PROSECUTION
Infringement, 34
CROWN COPYRIGHT
British copyright, 189, 190

D

DAMAGED WORKS
Library copying to replace, 47
DAMAGES FOR INFRINGEMENT
Actual damages, 33
Attorney's fee awards, 33
Innocent infringement as mitigating, 32
Statutory damages for infringement, 33
DATABASES, ELECTRONIC
Main discussion, 127–132
Compilation, database as, generally, 128
Copyrightability, 18, 128–130
Defined, 128
Downloading restrictions, 131, 132
Feist Communications Inc. v Rural Telephone Co., Inc. as affecting, 128, 129
Internet, 133
Libraries. *See under* LIBRARIES AND ARCHIVES
Merger doctrine, 128–130
Notice of copyright, 131
Per search basis of charging, 128

DAT (DIGITAL AUDIO TAPE) RECORDINGS
Royalties, scheme for payment of, 93
DATE
See TIME OR DATE
DEATH OF AUTHOR
Duration of copyright as affected by, 21
DECOMPILING SOFTWARE PROGRAMMING
European Economic Community or EU, 173
DEFENDANTS
Infringement action, compulsory joinder in, 24
DEFENSES TO INFRINGEMENT
British copyright, 191, 192
Fair use. *See* FAIR USE
Independent creation, 31
Limitation of actions, 31
Main discussion, 31, 32
DERIVATIVE WORKS
Abstracting works, substantial similarity, 25
Anthologies created for educational or scholarly purpose, 146
Compilations, 18, 22
Copyrightability, 18, 22
Dramatization, assingment of copyright for, 4
Exclusive rights of owner, 20
Fair use doctrine, 28
Library copying, 48
Public domain works, copyrightability of adaptations or compilations of, 22
Software, adapted, 120
Software adaptations, permissibility of, 120
Transformative or productive use, fair use and, 28
DESTRUCTION OF WORKS
Infringement remedies, 33
Library replacement of tapes, 91
DIPLOMATIC OR CONSULAR OFFICIALS
Transfer of owner's rights, issuance of certificate of, 21
DIRECTORIES
Copyrightability of databases, 129
DISPLAY OF WORK
See PERFORMANCE OR DISPLAY OF WORKS
DISTRIBUTION OF WORKS
Circulation by library. *See* LIBRARIES AND ARCHIVES
Commercial distribution. *See* COMMERCIAL PURPOSE OR ADVANTAGE
Databases, electronic, 130–132
Exclusive rights of owner, 20, 25
First sale doctrine
Software, 123, 124
Sound recodings, 94
Personal use, tests for, 25
Single copy rule, 44–46, 48–51, 53
Software
Adapted software, 120
First sale doctrine, 123, 124
DIVISIBILITY OF RIGHTS IN WORK
See SEVERABILITY OR DIVISBILITY OF RIGHTS IN WORK
DOCUMENT DELIVERY SERVICES
Copyright Clearance Center (CCC) as, 71
Libraries. *See* LIBRARIES AND ARCHIVES
DOWNLOADING RESTRICTIONS
Databases, 131, 132
DRAMATIC WORKS AND DRAMATIZATIONS
Copyrightable works, 4, 18
Library copying, 56

DURATION AND RENEWAL OF COPYRIGHT
Main discussion, 21, 22
Assignment or transfer of rights, time of reversion to author after, 21, 22
British copyright, 188, 189
Copyright Act of 1976, renewal of earlier copyrights, 8, 22
International copyright, 21, 159, 167, 188, 189
Unpublished works, 22

E

EDUCATIONAL OR SCHOLARLY USE
Anthologies and course packs
Guidelines as prohibiting, 146
Kinko's case as affecting, 71–73, 147
Prohibited, 28
Articles. See lines throughout this topic
Brevity, 28
Canadian copyright, 185, 186
Exclusive rights exception, generally, 3, 25
Face-to-face teaching exemption, 75
Fair use doctrine, generally, 26–29, 72
Films, 75, 95
Guidelines for educational use and copying
Acceptance of, 146
Graphic works, 89
Musical works, Appx D
Off-air recordings of broadcasts, Appx F, pp 239 et seq.
Requirements, generally, Appx C, 28, 57, 72, 144–146
International copyright, 168
Kinko's case and fair use, 71–73, 147
Library copying. *See under* LIBRARIES AND ARCHIVES
Licensing for educational copying
main discussion, 71–73
See also under LIBRARIES AND ARCHIVES
Copyright Clearance Center (CCC), 70, 71
Fair use as to factual works, 72
Kinko's, Basic Books v, 71–73, 147
Multiple copying for classroom use, guidelines for, Appx C, 28, 57, 72, 144–146
Single-article rule
Library copying, 48
teachers, 145
Software lending between institutions, faculty, and students, 126
Spontaneity, 28
EEC
See EUROPEAN ECONOMIC COMMUNITY OR EUROPEAN UNION
ELECTRA RECORDS CO. V GEM ELECTRONICS DISTRIBUTORS
Sound recordings and, 92, 93
ELECTRONIC MEDIA
See COMPUTERS AND COMPUTER-GENERATED WORKS
ELECTRONIC PUBLISHING OF JOURNALS
Libraries and, 132, 133, 150, 151
E-MAIL
Libraries and, 133
ENCYCLOPEDIA BRITANICA EDUCATIONAL CORP. V COOKS
Videotaping and, 99, 100
EUROPEAN ECONOMIC COMMUNITY OR EUROPEAN UNION
Main treatment, 168
Author's rights in software, 172
Computer Software Directive, 171–173
Conflict or choice of laws problem, 168
Decompiling software programming, 173

Exclusive rights of software author, 172
Exhaustion doctrine, 170, 171
First sale doctrine, 170, 171
Gray market goods, 170
Historical development, 168
Industrial and commercial property as protected, 169, 170
Programming translations of software, 173
Proportionality principle, 171
Software and, 171–173
Structure of, generally, 169
Territoriality principle, 171
Trade restriction bar, 168, 169
EVIDENCE
Burden of proof. *See* BURDEN OF PROOF
Presumptions. *See* PRESUMPTIONS, INFERENCES, AND PRIMA FACIE EVIDENCE
EXCERPTS
Library copying, 48, 49
EXCLUSIVE RIGHTS OF OWNER
Main discussion, 19–21
Adaptation, 20
British copyright, 188
Distribution, 20
International copyright, 161, 162, 172
Limitations on exclusive rights, main discussion of, 25–31; *See also* FAIR USE DOCTRINE
Performance or display, 20
Personal use, 25
Reproduction, 20
Severability or divisibility of rights, 20, 21
Software, 115, 118
EXHAUSTION DOCTRINE
European Economic Community or EU, 170, 171
EXPIRATION OF COPYRIGHT
See DURATION AND RENEWAL OF COPYRIGHT

F

FACE-TO-FACE TEACHING EXEMPTION
Motion picture use, 75
FACTUAL WORKS
Databases as copyrightable works, 128
Fair use, 29, 72
FAIR PRICE OF REPLACEMENT FOR LOST OR DAMAGED WORK
Library copying, 47
FAIR USE
Main discussion, 26–31
American Geophysical Union v Texaco and, 29, 44, 45, 55, 60–62
Amount and substantiality of portion used, 29, 30, 85, 143
Audiovisual works, 83–87
Balancing factors to determine fair use, 3, 27, 28
Biographies, 26
Brevity, 28
British copyright and fair dealing, 192, 193
Classroom copying, 28
Commercial purpose as affecting fair use, 26–28, 30, 31
Defined, 26, 27
Derivative works, 28
Educational uses, generally, 26–28, 72, 144 et seq.
Factors in determining, 26–28
Factual works as subject to, 29
First Amendment rights and, 30
Freedom of speech and, 30

Harper and Row v The Nation magazine as affecting, 30, 31, 141, 142
International copyright, 160
Kinko's case and fair use, 71–73, 147
Libraries and. *See under* LIBRARIES AND ARCHIVES
Market effect test, 30, 31, 86, 143
Meeropol v Nizer substantiality doctrine as developed by, 30
Nature of the work, 29, 143
Originality as affecting, 29, 30
Out-of-print works as subject to, 29
Potential market as affecting aplicability of, 30, 31, 86, 143
Purpose and character of use, 141, 142
Relationship to section 108, 31
Software, section 117 as allowing fair use of, 117–120
Sony Corporation of America v Universal Studios, Inc. and, 28, 83–86
Spontaneity, 28
Statutory criteria for fair use, 27
Substantiality of use as affecting, 29, 30, 85, 143
Transformative or productive work, 28
Unpublished works, 29, 31, 141–144
Value of the work test, elements of, 30, 31
Videotaping at home, 83–87
FAXING
See LIBRARIES AND ARCHIVES
FEE-BASED SERVICES
Commercial enterprises, generally. *See* COMMERCIAL PURPOSE OR ADVANTAGE
Library copying, 44, 45, 55, 56
FEIST COMMUNICATIONS, INC. V RURAL TELEPHONE CO., INC.
Databases, electronic, as affected by, 128, 129
FILMS, INC
Licensing collectives, 75
FILMS AND FILMSTRIPS
For matters relating to audiovisual works, generally, *See* AUDIOVISUAL WORKS
Copyrightable works, list, 18
Exclusive rights of author or artist, 20
Libraries and. *See under* LIBRARIES AND ARCHIVES
Licensing agencies, 75
Motion picture licensing corporation (MPLC), 75
Piracy, FBI warning, 94, 95
Videotape duplication. *See* VIDEOTAPES, VIDEOCASSETTES, AND VIDEODISCS
FINES AND PENALTIES
Criminal action for infringement, 34
Deposit of work, author's failure to make, 24
FIRST AMENDMENT RIGHTS
Fair use and, 30
FIRST PUBLICATION RULE
Berne Convention and Union, 165
FIRST SALE DOCTRINE
European Economic Community or EU, 170, 171
Software, 123, 124
Sound recordings, 94
FIXATION
Defined, 3
International copyright, 158, 159
Registration requirement, fixation as replacing, 23
Software, 118
Videotapes compared with films, 96
FOREIGN WORKS
See also INTERNATIONAL COPYRIGHT

Copyrightability under Copyright Act, 19
Transfer of owner's rights, issuance of certificate of, 21
FORMS
Permission letters, 243, 244
FOR-PROFIT LIBRARIES
Commercial advantage, direct or indirect, 44
Copying by, 44, 45, 55, 56
Criminal penalties for infringement, 34
Open to the public or available to researchers, section 108 requirement, 45
Section 108 of Copyright Act and, 44, 45
Software circulation, 126
FRAUDULENT WORKS
Copyrightable works, 18
Criminal prosecution, 34
FREEDOM OF SPEECH
Fair use and, 30
FREEWARE
Defined, 118

G

GOOD FAITH
Infringement defenses, 32
Unsupervised copying, library's liability, 52
GOVERNMENT WORKS
Copyrightability, 19, 167
Universal Copyright Convention, 167
GRAPHIC WORKS
Pictorial or graphic works. *See under* LIBRARIES AND ARCHIVES
GUIDELINES FOR USE AND COPYING
ALA copyright notice guidelines, 45
ALA fair price guideline and copying for replacement, 47
ALA guideline for unsupervised copying, 51
Classroom copying, Appx C, 28, 57, 72, 144–146
Graphic works, educational copying, 89
Interlibrary loans, Appx E, 47, 50, 53–55
Libraries
  ALA guidelines, 45, 47, 51
  interlibrary loans, Appx E, 47, 50, 53–55
  reserves, Appx G, 148
  Reserves, copying for, 57
Music, educational copying, Appx D
Off-the-air recording for educational purpose, Appx F
Videotapes, educational copying, 98, 99

H

HARPER AND ROW V THE NATION MAGAZINE
Fair use and, 30, 31, 141, 142
Fair use doctrine and, 30, 31, 141, 142
HISTORY OF COPYRIGHT LAW
Main discussion, 4–10
Berne Convention, 9
British copyright law, brief history of, 4–6
Canadian copyright, 181
CCC (Copyright Clearance Center), historical context, 68
Computer Software Amendments Act of 1990, 124
Fair use doctrine, 26, 27
Libraries, 1976 Act as including, 43
1909 Copyright Act, 8
1976 Copyright Act, historical context of, 17
Notice of copyright, 23
Photocopier, effect of advent of, 67, 68
United States copyright law, main discussion of history of, 6–8

HOME DUPLICATION
Sound recordings circulated from libraries, 94
Videotaping. *See under* VIDEOTAPES, VIDEOCASSETTES, AND VIDEODISCS, 83–87
HOUSE REPORT 94–1476
Library copying, 43 et seq.

I

IDEA OR CONCEPT
Copyrightable work, unexpressed idea as, 3–4
IMPLEMENTATION ACT OF 1988
Berne Convention, United States as signatory of, 9
IMPOUNDMENT OF WORKS
Infringement remedies, 33
INDEPENDENT CONTRACTORS
Government works for hire as copyrightable works, 19
INDEPENDENT CREATION DEFENSE
Infringement, 31
INFERENCES
See PRESUMPTIONS, INFERENCES, AND PRIMA FACIE EVIDENCE
INFORMATION AND RETRIEVAL SYSTEMS
See COMPUTERS AND COMPUTER-GENERATED WORK
INFRINGEMENT
Main discussion, 24–34
Abuse of process defense, 32
Access to work as element of cause of action, 24, 25
Attorneys' fee awards, 24, 33
British copyright, 190–192
Canadian copyright, 184, 185
Collateral estoppel defense, 32
Criminal prosecution for, 34
Damages. *See* DAMAGES FOR INFRINGEMENT
Defenses
Main discussion, 31, 32
Death of author, presumption of, 21
Defined, 24
Elements of cause of action, 24
Equitable defenses, 32
Fair use. *See* FAIR USE DOCTRINE
Good or bad faith, defenses involving, 32
Independent creation defense, 31
Innocent infringement
Defense to infringement action, 32
Discretion of court, 32
FBI warning as affecting, 95
Notice abolishment as affecting remedies against, 23
Limitation of actions, 31
Misuse, defense of, 32
Notice abolishment as affecting remedies for, 23
Presumptions, inferences, and prima facie evidence
Author's death as complete defense, 21
Elements of infringement action, inferences establishing, 24, 25
Ownership, prima facie showing of, 24
Registration, infringement remedies as limited by lack of, 23, 24, 33
Remedies. *See* REMEDIES FOR INFRINGMENT
Res judicata (matters decided by court) defense, 32
Similarity
Independent creation defense, 31
Substantial similarity below in this group
Software, whether or not to sue for infringement of, 116
Substantial similarity
Abstract similarity, 25
Evidence establishing, 25

Infringement action, elements of, 24, 25
Ordinary observer test for abstract similarity, 25
Striking similarity test, 25
Unclean hands defense, 32
INHERITANCE
Owner's rights in work, 21
INJUNCTIONS
Infringement remedies, 33
Videotaping of television program, 99, 100
INNOCENT INFRINGEMENT
See INFRINGEMENT
INTEGRATED CIRCUIT CHIPS
Misuse defense to infringement, 32
INTENT OR MOTIVE
Innocent infringement. *See* INFRINGEMENT
Personal use exception as affected by intent to distribute, 25
Profit motive. *See* COMMERCIAL PURPOSE OR ADVANTAGE
Single copy rule, duty of library to detect multiple copy intent, 46
INTERLIBRARY LOANS AND REQUESTS
Guidelines, Appx E, 47, 50, 53–55
INTERNATIONAL COPYRIGHT
Main discussion, 157–174
Berne Convention. *See* BERNE CONVENTION AND UNION
Canada. *See* CANADIAN COPYRIGHT
Copyrightable works, 160, 161
Duration and renewal, 21, 159, 167, 188, 189
England. *See* BRITISH COPYRIGHT
European Economic Community or European Union. *See* EUROPEAN ECONOMIC COMMUNITY OR EUROPEAN UNION
Exclusive rights of owner, 161, 162, 172
Fair use, 160
Fixation, 158, 159
Licensing organizations, 73–75
Moral rights, 160, 164
Originality, 158, 159, 172
Patent law distinguished, 159
Registration, 160, 163, 164
Software, 161, 167, 171–173
United States copyright law and, 163–165
Universal Copyright Convention. *See* UNIVERSAL COPYRIGHT CONVENTION
INTERNET
Libraries and, 133
INTESTATE SUCCESSION
Owner's rights, transfer of, 21

J

JOINDER OF PARTIES
Infringement action, 24
JOURNALS
See BOOKS, JOURNALS, AND OTHER LITERARY WORKS

K

KINKO'S CASE
Anthologies and course packs, 71–73, 147

L

LIBRARIES AND ARCHIVES
Abstracts, databases and, 130
ALA. *See* ALA
American Geophysical Union v Texaco, fair use and, 44, 45, 55, 60–62

Articles from journals, copying, 48, 49
Section 107 and relationship to section 108, 57–59
Audiovisual works
  Main discussion, 90–102
  DAT (digital audio tape) recordings, 93
  Films and filmstrips, 94–96
  Multimedia computer works, 101, 102
  Packaging of audiovisual works in–house, 100, 101
  Performance in libraries, 103–105
  Phonorecords and CDs, 90–94
  Section 108 and, 52, 56, 82, 99
  Slides and transparencies, 89
  Sound recording duplication, 90–94
  Television shows, videotaping, 97, 98
  Videotapes, videocassettes, and videodiscs, main discussion, 96–100
Authority to copy, generally, 49, 58, 59
Bibliographic data in electronic databases, 130, 131
Blanket permission to duplicate recordings from archive or master copy, 91, 92
Books, journals, and other literary works
  Main discussion, 48, 49, 54, 56
  Electronic format, conversion to, 43, 50
  Electronic journals, 132, 133, 150, 151
  Public Lending Right (PLR), books qualifying for, 204, 205
  Section 108 as to literary or dramatic works, 56
  Tables of contents, copying, 48
  Titles of journals, copying, 48
British copyright, 192, 193
Section 108
  main discussion, 43 et seq.
  American Geophysical Union v Texaco, 44, 45
  Articles or small excerpts, generally, 48
  Audiovisual works, 52, 56, 82
  Dramatic works, 56
  Electronic journals and, 133
  Electronic reserves and, 151
  Excerpt copying, generally, 48
  Fair use as applicable to libraries, 31
  Graphic works, 89
  License agreement expanding or restricting scope of, 52, 132
  Musical works, 88
  Notice of copyright requirements, 45, 46
  Open to the public or available to researchers, section 108 requirement, 45
  Packaging of audiovisual works by libraries, 101
  Photocopying of printed works, generally, 43 et seq.
  Replacement, copying for, 47
  Section 107 and, 57–59
  Slides or transparencies, 90
  special libraries, 44
  Unpublished works, copying for archives, 46–47
Canadian copyright, 186, 187
CARL/UnCover, faxing and, 71
Catalog, availability on campus network, 132
CD-ROM, unsupervised copying, 51
CDs. Phonorecords and CDs, below in this topic
Circulation. See lines throughout this topic
Commercial purpose or advantage, 44, 45, 55, 56
Computers and computer-generated works
  Main discussion, 113–134
  Conversion from print to electronic format, 43, 50, 89, 90
  Databases. See Databases, electronic, below in this topic
  Internet, 133

Journals, electronic, 132, 133
License agreements, 114
Multi-media productions by libraries, 101
Reserves, electronic, 150, 151
Scanning data into computer file, 43, 50, 89, 90
Software. See software or applications, below, in this topic
Transparencies or photographs, conversion and storage, 89, 90
Computer Software Rental Amendments Act, 125, 126
Contractual obligations, 52
CONTU guidelines for interlibrary loans, 53–55
Conversion of formats
  Film to videotape, 95
  Print to electronic, 43, 50
  Sound recordings, 91, 92
  Transparencies or photographs, digitizing, 90
  Videotape formats, conversion between, 97
Copyright Act of 1976
  Libraries as within purview of, generally, 43
  Section 108. See Section 108, below in this topic
Section 117 of 1976 Copyright act, fair use of software under, 118
Damaged works, copying to replace, 47
Damages in infringement actions against, 33
Databases, electronic
  See also Online databases in this topic
  Abstracts in, 130
  Campus network, availability through, 132
  Catalog, availability on campus network, 132
  Charging for use, 128, 131
  Downloading, 131
  Library access, generally, 114
  Licensing and licensing agreements, 130, 132
  Local mounting, 131
  Merger doctrine as affecting distribution, 130
  Network availability, 132
  Problems libraries face regarding use, 128
  Searching for users, 131
DAT (digital audio tape) recordings, taping, 93
Derivative works, tables of contents of, 48
Destruction of audiotapes, problem of replacing, 91
Digitizing transparencies or photographs, 90
Document delivery services
  CARL/Uncover, 56
  Fee-based services, 55, 56
  Single copy rule, duty of library to detect conduct violating, 46
  University library, direct or indirect commercial advantage, 44
Dramatic works, section 108 as limited to, 56
Duplication or reproduction of works. See lines throughout this topic
Educational or scholarly purpose
  Databases, network availability, 132
  Library use as synonymous with, 94, 103
  Performance of works in libraries, 103–105
  Sound recordings, taping, 93
  Television videotaping, 98, 99
Electronic capture of work, 43, 50, 89, 90
Electronic copies as permitted under section 108, 43
Electronic journals, 132, 133, 150, 151
E-mail, 133
Entire work, copying, 49–51
Entire work, library's authority to copy, 49
Exception from copyright protection, generally, 44, 51, 52
Excerpts, copying, 48
Excerpts, copying small, 48, 49
Fair price of replacement for lost or damaged work, 47

Fair use
   See also Section 108 in this topic
   Benefit to libraries, 31, 32
   Copying and, generally, 44, 45, 52, 56, 57–62
   Electronic journals and, 133
   Films or filmstrips, 95
   Musical works, 88
   Pictorial or graphic works, 88, 89
   Section 108, relationship to, 31
   Slides or transparencies, 90
   Software, 118
   Sound recordings, 93, 94
Faxing
   CARL/UnCover, 71
   Entire work, 50
   Notice or warning and, 49
Fee-based services, 44, 45, 55, 56
Films and filmstrips
   Graphic works, microfilming, 90
   Multimedia production, license to use film clips for, 102
   Packaging of audiovisual works in-house, 100, 101
   Performance in libraries, 104
   Videotaping or other duplication, 94, 96
Five-year reports to register, 57
Fixation of videotapes compared with films, 96
For-profit libraries. *See* FOR-PROFIT LIBRARIES
Good faith as affecting liability for excessive unsupervised
   copying, 52
Graphic works. *See* Pictorial or graphic works, below in this topic
Guidelines
   Classroom copying and graphic works, 89
   CONTU guidelines for systematic reproduction, 53–55
   interlibrary loans and requests, Appx E, 47, 50, 53–55
   Reserves, ALA guidelines for, 148
   Television programs, videotaping, 98, 99
Home duplication of circulated sound recordings, 94
House report 94–1476, 43 et seq.
Interlibrary loans
   Articles and small excerpts copying, 48
   Copy requests, 47, 50
   Entire work, copying, 50
   Guidelines, Appx E, 53–55
   Section 108 as satisfied by interlibrary lending from for-profit
     library, 45
Internal use of library, copying for, 46–48, 96
Internet, 133
Journals. *See* Books, journals, and other literary works, above in
   this topic
Library of Congress and registration, 24
Licensing or permission
   Abstracts, distribution through databases, 130
   British copyright, 192, 193
   Campus network, database available through, 132
   Copying, generally, 52
   Copyright Act, licensing agreement as superseding, 52, 132
   Database use and distribution, 114, 130–132
   Downloads of databases, 131
   Entire work, copying, 49
   Expanding or restricting section 108 use by license agreement,
     52, 132
   Lending software, 126
   Multimedia production by libraries, 101, 102
   Performance in libraries, 104
   Software or applications, 52, 114, 122, 123, 126
   Sound recording duplication from master or archive copy, 91, 92

Listserv, 133
Literary works. *See* Books, journals, and other literary works, above
   in this topic
Local mounting of databases, arranging for, 131
Lost or stolen works, copying to replace, 46, 47
Manuscripts, archival copying, 47
Market effect of copying or duplicating, 56, 95, 96
Master or archival copies
   See also Reserves, below in this topic
   Film or videotape, 96, 97
   Sound recordings, 91
Merger doctrine as shielding libraries from liability vis a vis
   abstracts, 130
Microforms, 90
Multimedia productions by libraries, 101, 102
Multiple copy requests, duty of library to detect, 46
Musical works
   Copying or duplicating, 88, 90–94
   Performance in libraries, 103–105
   Recordings. *See* Phonorecords and CDs, below in this topic
Network availability of databases, 132, 133
Newsletters
   Created from database downloads, 131
   Copying or duplicating, 53
News programs, copying or duplicating, 52, 98
Nonprint works, main discussion, 88–102
Notice or warning of copyright, 23, 45–51, 94, 95, 108, 109, 126,
   127
Off-the-air recording, Appx F, 97, 98
Online databases
   Electronic journals, 132, 133, 150, 151
   Internet, 133
   Merger doctrine as affecting, 130
   Use, generally, 131
Patrons, copying for or by, 48–51
Performance or display works, 89, 103–105
Permission. *See* Licensing or permission, above in this topic
Phonorecords and CDs
   Copying or duplication, 46, 88, 90–94
   Fair use, 31
   Performance in libraries, 104
   section 108 as including, 43
Photographs, 89, 90, 102
Pictorial or graphic works
   Main discussion, 88, 89
   Exemption for library use, generally, 82
   Literary work, copying of graphics included in, 56
   Photographs, 89, 90, 102
   Slides and transparencies, 90
Preservation of works
   Damaged works, 47
   Films, 96
   Lost works, 46, 47
   Master or archival copy of nonprint works, 91, 96, 97
Profit motive, 44, 45, 55, 56
Public Lending Right. *See* PUBLIC LENDING RIGHT
Public Lending Right (PLR), 199 et seq.
Public library requirement, 45
Qualifying libraries, 44–46
Reasonable effort to purchase work, 47, 49
Relationship between section 107 and section 108, 57–59
Reports to register of copyrights, 57
Reproduction or duplication of work. *See* lines throughout this
   topic

Reserves
  Main discussion, 148–151
  ALA model policy, Appx G, 148
  Classroom copying guidelines as governing copying for, 57
  Electronic reserves, 150, 151
  Single copy rule as affecting copying for, 46
Role of librarian, generally, 1–2
Routing of tables of contents copies to users, 48
Royalties
  Electronic reserves, 150, 151
  Film duplication and, 97
  Sound recording duplication, royalty pool for, 93
  Systematic reproduction, 55
  Television programs, videotaping, 99, 100
Scanning of works into computer documents, 43, 50, 89, 90
Scholarly use. See Educational or scholarly use, above in this topic
SDI and systematic copying, 53
Senate report 94–473, 43 et seq.
Single-article rule for copying, 48
Single copy rule, 44–46, 48–51, 53, 58, 151
Site license to copy software, 52
Slides and transparencies, 89, 90
Software and applications
  Computer Software Rental Amendments Act, 125, 126
  Lending regulations, 125
  Problems libraries face in protecting copyright, 114
  Scenarios raising copying issues, 115, 116
  Site licenses, 126
Software or applications
  Main discussion, 113 et seq.
  Circulation controversy, 125, 126
  Copyrightability, 118
  Defined, 116–118
  Library exemption, 126
  Licensing and licensing agreements, 52, 114, 122, 123
  Multimedia production by libraries, 101
  Reserves, electronic, 150, 151
  Warning notice of copyright, 126, 127
Sound recordings
  Audiotapes, 91–94
  Audiotapes, main discussion, 88, 90–94
  Conversion of formats, 91, 92
  DAT (digital audio tape) and royalties, 93
  Duplication, 92, 93
  Educational or scholarly use, taping for, 93
  Fair use and, 93, 94
  Home duplication of circulated recordings, 94
  License or permission, 91, 92
  Master or archival copy, 91
  Phonorecords and CDs, 88, 90–94
  Royalties, 93
Special libraries, section 108 and, 44
Stolen or lost works, copying to replace, 46, 47
Substantiality of use
  Entire work, authority to copy, 49
  Excerpts, small, 48, 49
  Film duplication for internal use, 96
  Slides and transparencies, 90
Synchronization license for multimedia works, 102
Systematic copying, 53–56
Tables of contents, copying, 48
Taping of sound recordings, 91–94
Taping of video works. See Videotapes, videocassettes, and
  videodiscs, below in this topic

Telephone orders for copies as requiring warning, 49
Texaco case (American Geophysical Union v Texaco), fair use, 44,
  45, 55, 60–62
Titles of journals, copying, 48
Unpublished works, copying, 46, 47, 51
Unsupervised copying machines, 51
Used book search prior to copying, 49
User liability for excessive copying, 51, 52
Users, copying for, 48–51
Vicarious liability for patron's excessive copying, 51, 52
Videotapes, videocassettes, and videodiscs
  Main discussion, 96–100
  Conversion of formats, 95, 97
  Educational or scholarly use, guidelines for, 98, 99
  Films, taping, 94–96
  Fixation compared with films, 96
  Master or archival copies, 96, 97
  Multimedia production by library, permission to use video clips
    for, 102
  News broadcasts, taping, 98
  Television programs, taping, 97–100
Warning or notice of copyright, 45, 46, 48–51, 94, 95
Williams & Wilkins Co. v National Library of Medicine, fair use,
  59–61
Written and signed agreement for permission to duplicate
  recordings, 92
LIBRARY OF CONGRESS
  Registration of copyrights, 24
LIBRARY RESERVES
  See under LIBRARIES AND ARCHIVES
LICENSING AGENCIES AND COLLECTIVES
  Alliance of Artists and Record Companies (AARC), music licensing
    by, 74
  ASCAP (American Society of Composers, Artists, and Publishers),
    25, 26, 74
  BMI (Broadcast Music, Inc), 74
  Copyright Clearance Center. See COPYRIGHT CLEARANCE
    CENTER
  Films, Inc., 75
  International organizations, list, 73–75
  Motion Picture Licensing Corporation, 75
  Music collectives, 74
  Pubnet as licensing agency, 71
LICENSING OR PERMISSION
  Agencies and collectives. See LICENSING AGENCIES AND
    COLLECTIVES
  Blanket license. See BLANKET LICENSE
  British copyright, 192, 193
  Databases, 132
  Exceptions to exclusive rights of owner, necessity of license, 25
  Exclusive use exceptions, licensing of use as within, 25
  Films, duplication of, 96
  Form of permission letter, sample, 243, 244
  Library copying. See under LIBRARIES AND ARCHIVES
  Software, main discussion of licensing of, 120–123
  Universal Copyright Convention, 167, 168
LIMITATION OF ACTIONS
  Criminal prosecution for infringement, 34
  Infringement actions, 31, 34
LISTSERV
  Libraries and, 133
LITERARY WORKS
  See BOOKS, JOURNALS, AND OTHER LITERARY WORKS
LIVES IN BEING
  Duration of copyright, 21

LOSS OF PROFIT
Infringement damages as including, 33
LOST WORKS
Library copying, 46, 47

M

MANUSCRIPTS
Libraries, archival copying by, 47
MARKET EFFECT OF COPYING
Fair use doctrine, 30, 31, 86, 143
Home videotaping and Sony case, 86
Library copying, 56, 95, 96
MASTER OR ARCHIVAL COPIES
Film or videotape, library copying, 96, 97
Sound recordings, library copying, 91
MEEROPOL V NIZER
Fair use doctrine, substantiality of use as affecting, 30
MERGER DOCTRINE
Abstracts and, 130
Baker v Selden, 129
Bibiliograhic data and, 130
Database copyrightability, 128–130
Defined, 129
METHOD OF OPERATION
Copyrightability, 3, 4
MICROFORMS
Copyrightable works, 18
Library copying or reproduction, 90
MISUSE DEFENSE TO INFRINGEMENT
Restraint of trade compared, 32
MITIGATION OF DAMAGES
Innocent infringement, 32
MONOPOLIES AND RESTRAINTS OF TRADE
Copyright distinguished, 2–3
Misuse defense to infringement, 32
MORAL RIGHTS
British copyright, 189
International copyright, 160, 164
MOTION PICTURES
See FILMS AND FILMSTRIPS
MOTIVE
See INTENT OR MOTIVE
MULTIMEDIA
Library productions, 101, 102
MULTIPLE COPYING FOR CLASSROOM USE
Guidelines for, Appx C, 28, 57, 72, 144–146
MUSICAL WORKS
Copyrightable works, list, 18
Guidelines for education use of music, Appx D
Libraries and. *See* LIBRARIES AND ARCHIVES
Licensing of musical works, collectives for, 74
Public domain works, copyrightability of arrangements of, 22
Recordings. *See* PHONORECORDS AND CD

N

NATIONAL ASSOCIATION OF COLLEGE STORES
Licensing agreement with, 71
NATIONAL SCIENCE FOUNDATION
Copyrightable works, 19
NATIONAL TECHNICAL INFORMATION SERVICE (NTIS)
Copyrightability of government works made for hire, 19
NETWORKS
Database availability on, 132, 133
Internet, libraries and, 133

Pubnet as licensing agency, 71
Software licenses, 121, 122
NEWSLETTERS
Database downloads creating, 131
Copying or duplicating, 53
NEWS PROGRAMS AND BROADCASTS
Library copying or reproduction, 52, 98
NONFICTION
Fair use, 29, 72
NONPRINT WORKS
See particular kinds, e.g., AUDIOVISUAL WORKS, MOTION
PICTURES, PHONORECORDS AND CD, PICTORIAL OR
GRAPHIC WORKS, etc.
Library copying, main discussion, 88–102
NONPROFIT ORGANIZATIONS
Copyright Clearance Center (CCC) as, 68
Educational institutions. *See* EDUCATIONAL OR SCHOLARLY
USE
Exclusive use, exceptions to, 25
Public libraries. *See* LIBRARIES AND ARCHIVES
NOTARY PUBLIC
Transfer of owner's rights, issuance of certificate of, 21
NOTICE OR WARNING
Main discussion, 23
Abolished by 1976 Copyright Act, 23
Berne Convention adoption as abrogating notice requirement, 23
Classroom copying, 28
Copyright Act of 176 as eliminating requirement of, 23
Crime of removing copyright label, 34
Death of author as affecting duration of copyright, 21
Film piracy, FBI warning, 94, 95, 104
Format of, 23, 45, 46
Good faith users as affected by, 23
Innocent infringement as affected by, 23
Library copying, 45, 46, 48–51, 94, 95
Section 108 and, 45, 46
Software, 126–127
Term of art, notice as, 23

O

OBSCENE OR PROFANE WORKS
Copyrightability, 18
ONGOING INFRINGEMENT
Limitation of actions, commencement of, 31
ONLINE DATABASES
See also DATABASES, ELECTRONIC
Internet, 133
Libraries and. *See under* LIBRARIES AND ARCIIIVES
ORDINARY OBSERVER TEST
Substantial similarity test in infringement action based on
abstraction of work, 25
ORIGINALITY
See also COMPILATIONS; DERIVATIVE WORKS; Substantial
similarity under INFRINGEMENT
British copyright, 172, 188
Copyrightable works, 18
Database creation, 129
Defined, 3
Fair use doctrine as affected by, 29, 30
Independent creation defense, 31
International copyright, 158, 159, 172
OUT-OF-PRINT WORKS
Fair use doctrine as applicable to, 29

OWNERSHIP OF WORK
  Main discussion, 21
  Alienability. *See* ASSIGNMENT OR OTHER TRANSFER OF
    RIGHTS IN WORK
  Exclusive rights. *See* EXCLUSIVE RIGHTS OF OWNER
  Infringement action, standing to bring, 24
  Registration as proof of, 23
  Software, 118, 119

P

PANTOMIMES
  Copyrightable works, list, 18
PARIS ACT OF 1971
  History of copyright law, international aspects of, 9, 165
PARTIES TO ACTIONS
  Infringement suits, 24
PENALTIES AND FINES
  Criminal prosecution for infringement, 34
  Deposit of work with register, failure to make, 24
PERFORMANCE OR DISPLAY OF WORKS
  Educational use, 25
  Exclusive rights of author or artist, 20
  Film collectives, 75
  Libraries, performances in, 89, 103–105
  Music, 74, 103–105
  Pantomime as copyrightable work, 18
  Public performance or display defined, 20
PERIODICALS.
  See BOOKS, JOURNALS, AND OTHER LITERARY WORKS
PERMANENT INJUNCTION
  Infringement remedies, 33
PERMISSION
  See LICENSING OR PERMISSION
PERSONAL OR PRIVATE USE
  Exclusive rights, 25
  Home videotaping. *See under* VIDEOTAPES, VIDEOCASSETTES,
    AND VIDEODISCS
  Software, 120
  Teachers, single copying for, 145
PHONORECORDS AND CD
  Electra Records Co. v Gem Electronics Distributors and, 92, 93
  Fair use, 31
  Leasing commercially, prohibition of, 123
  Libraries and, 44, 88, 90–94, 104
  Record rental amendment of copyright act of 1976, 94
  Registration, deposit of copies as required for, 24
PHOTOCOPYING
  Copies and copying. *See* REPRODUCTION OF WORKS
PHOTOGRAPHS
  British copyright, 189
  Library copying or reproduction, 89, 90, 102
PICTORIAL, GRAPHIC, AND SCULPTURAL WORKS
  Copyrightable works, list, 18
PICTORIAL, GRAPHIC, OR SCULPTURAL WORKS
  Library copying. *See under* LIBRARIES AND ARCHIVES
  Photographs, 89, 90, 102, 189
  Visual Artists' Rights Act of 1990, 9
PIRACY OF WORKS
  See also INFRINGEMENT
  FBI warning as to, 94, 95
  Software, libraries' difficult position vis a vis, 114
PLAGIARISM
  See INFRINGEMENT

PLR
  See PUBLIC LENDING RIGHT (PLR)
POETRY
  Classroom copying, 28
  Fair use, 28, 29
PRESERVATION OF WORKS
  Library copying. *See under* LIBRARIES AND ARCHIVES
PRESIDENTIAL PROCLAMATION
  Foreign work as copyrightable by, 19
PRESUMPTIONS, INFERENCES, AND PRIMA FACIE EVIDENCE
  Certificate of transfer as prima facie evidence of transfer, 21
  Death of author, duration of copyright as affected by presumption
    of, 21
  Infringement. *See* INFRINGEMENT
  Notice requirement abolishment as affecting user's assumptions
    about protected status, 23
  Registration of copyright as prima facie evidence of copyright, 23
PROCEDURE, PROCESS, SYSTEM, OR METHOD OF OPERATION
  Copyrightability, 3, 4
PRODUCTIVE OR TRANSFORMATIVE USES
  Fair use doctrine and, 28
PROFANITY
  Copyrightability, 18
PROFESSOR PUBLISHING
  Kinko's case as affecting, 71–73, 147
PROFIT MOTIVE
  See COMMERCIAL PURPOSE OR ADVANTAGE
PROGRAMS FOR COMPUTERS
  Software or applications for computers. *See under* COMPUTERS
    AND COMPUTER-GENERATED WORKS
PROPORTIONAL AMOUNT AND SUBSTANTIALITY USED
  Fair use doctrine, 29, 30, 85
PUBLIC DISTRIBUTION
  See DISTRIBUTION OF WORKS
PUBLIC DOCUMENTS OF UNITED STATES
  Copyrightability, 19
PUBLIC DOMAIN
  Databases, electronic, 127, 128
  Expiration of copyright as effecting
  Government works as copyrightable, 19
  Notice absence as indicating, 23
  Notice requirement abolishment as affecting user's assumptions as
    to, 23
PUBLIC LENDING RIGHT (PLR)
  Main discussion, 199–212
  Administration of PLR, 206
  Books qualifying, 204, 205
  Compensation of authors for circulated works, 199
  Cultural enhancement issue, 208
  Definition, 199
  Disbursements to authors, 206, 207
  Financial burden, allocation of, 202
  Historical background, 199–202
  Hypothetical questions, 207
  International participation or influence, 199, 209
  Libraries included in PLR, 204
  Obligor (who pays), 202
  Purposes, 202, 207, 208
  Qualifying authors, 203, 204
  Social security issue, 208
  Study commission, 200, 201
  Theoretical issues, 207
  Use, keeping track of, 205, 206
PUBLIC LIBRARIES
  See LIBRARIES AND ARCHIVES

PUBLIC PERFORMANCE
   See PERFORMANCE OR DISPLAY OF WORKS
PUBNET
   Licensing agency, pubnet as, 71
PURPOSE OF COPYRIGHT
   Main discussion, 4

R

RCA RECORDS, ET AL. V ALL-FAST SYSTEMS
   Tape recordings and, 93
RECORDINGS
   See PHONORECORDS AND CD; SOUND RECORDINGS;
      VIDEOTAPES, VIDEOCASSETTES, AND VIDEODISCS
REGISTRATION
   Main discussion, 23, 24
   Abolishment of registration requirements by 1976 Copyright Act,
      23
   Attorney's fees in infringment action as dependent on, 33
   Certificate of registration, 23
   Copyright Clearance Center (CCC), 68, 69
   Deposit as affecting validity of, 24
   Infringement remedies as limited by lack of, 23, 24, 33
   International copyright, 160, 163, 164
   Probative value of, 23
REMEDIES FOR INFRINGEMENT
   Attorneys' fee awards, 24, 33
   Canadian copyright, 184, 185
   Costs of action, award of, 33
   Criminal prosecution, 34
   Damages. *See* DAMAGES FOR INFRINGEMENT
   Impoundment, 33
   Injunction, 33
   Innocent infringers, notice as affecting remedies against, 23
   Registration as affecting scope of, 23, 24
RENEWAL OF COPYRIGHT
   See DURATION AND RENEWAL OF COPYRIGHT
REPRODUCTION OF WORK
   Educational copying. *See* CLASSROOM COPYING
   Exclusive rights of author or artist, 20
   Libraries. *See* LIBRARIES AND ARCHIVES
   Permission letter for photocopies, form of, 243
   Taping. *See* TAPE RECORDING OR DUPLICATION
RESERVES
   See under LIBRARIES AND ARCHIVES
RES JUDICATA (MATTERS DECIDED BY COURT) DEFENSE
   Infringement, 32
RESTRAINING ORDERS
   Infringement remedies, 33
RESTRAINT OF TRADE
   Copyright distinguished from monopoly, 2, 3
   Misuse defense to infringement, 32
REVERSION OF RIGHTS IN WORK AFTER TRANSFER
   Time of, 21
REVOCATION
   Transfer of owner's rights, 21
ROME ACT OF 1928
   History of copyright law, international aspects of, 9
ROYALTIES
   See also PUBLIC LENDING RIGHT
   ASCAP, collection by, 74
   Blanket license, 26
   BMI, collection by, 74
   Copyright Clearance Center (CCC) collecting and distributing, 68–
      70
   Film duplication, price as determining right of, 96

First sale doctrine and tape duplication of sound recordings, 94
Library copying. *See under* LIBRARIES AND ARCHIVES
License as affecting manner of paying, 26
Out-of-print works, 29

S

SAFE HARBOR GUIDELINES
   Educational or scholarly purpose, copying for, Appx C, 28, 144–
      146
SALE OF RIGHTS IN WORK
   See ASSIGNMENT OR OTHER TRANSFER OF RIGHTS IN
      WORK
SCANNING
   Library copying or reproduction, 43, 50, 89, 90
SCHOLARLY USE
   See EDUCATIONAL OR SCHOLARLY USE
SDI
   Systematic copying, 53
SECONDARY OR CONTRIBUTORY INFRINGEMENT
   British copyright, 190
SECTION 108 OF 1976 COPYRIGHT ACT
   Library copying. *See under* LIBRARIES AND ARCHIVES
SEMICONDUCTOR CHIPS
   I/C chips, 32
SENATE REPORT 94–473
   Library copying, 43 et seq.
SERIAL ISSUE CODING
   Copyright Clearance Center (CCC), 69
SEVERABILITY OR DIVISIBILITY OF RIGHTS IN WORK
   Assignment or other alienation of rights, 20
   Copyright as separate from ownership, 21
SHAREWARE
   Defined, 118
SHRINKWRAPPED LICENSING
   Software, 122
SIGNATURES
   Assignment or other transfer of rights in work, 21
SIID CODE
   Copyright Clearance Center (CCC) code as replaced by, 69
SINGLE-COPY RULES
   Library copying, 44–46, 48–51, 53
   Teachers' personal use, single copy for, 145
SITE LICENSES
   Software, 52, 121
SLIDES AND TRANSPARENCIES
   Library copying or reproduction, 89, 90
SOFTWARE OR APPLICATIONS FOR COMPUTERS
   See under COMPUTERS AND COMPUTER-GENERATED
      WORKS
SONY CORPORATION OF AMERICA V UNIVERSAL STUDIOS,
      INC.
   Fair use doctrine, 28, 83–86
   Home videotaping and fair use doctrine, 83–86
SOUND RECORDINGS
   Audio Home Recording Act of 1992 and DAT (digital audio tape),
      93
   CDs. *See* PHONORECORDS AND CD
   Copyrightable works, list, 18
   First sale doctrine, 94
   Library copying. *See under* LIBRARIES AND ARCHIVES
   Tape recording of, 93, 94
STANDING (ELIGIBILITY OR QUALIFICATION) TO SUE
   Infringement action, 24

STATE GOVERNMENTS
    Copyrightability of works of, 19
    Infringement by officer or employee of, 24
STATUTES OF LIMITATION
    Criminal prosecution for infringement, 34
    Infringement actions, 31, 34
STOCKHOLM ACT OF 1967
    History of copyright law, international aspects of, 9
STRIKING SIMILARITY TEST
    Substantial similarity test in infringement action, 25
SUBSTANTIALITY OF USE
    Fair use doctrine as affected by, 29, 30, 85, 143
    Library copying. *See under* LIBRARIES AND ARCHIVES
    Meeropol v Nizer, 30
    Sony Corporation of America v Universal Studios, Inc., 85
SUBSTANTIAL SIMILARITY
    See INFRINGEMENT
SUMMARY AND CONCLUSION TO SUBJECT MATTER
    Main treatment, 217, 218
SURVEYS
    Copyright Clearance Center (CCC), 69, 70
SYNCHRONIZATION LICENSE
    Multimedia works, 102
SYSTEMATIC COPYING
    Library copying, 53–56
SYSTEMS
    Computer systems. *See* COMPUTERS AND COMPUTER-
        GENERATED WORKS
    Copyrightability of procedure, process, system, or method, 3, 4

T

TABLES OF CONTENTS
    Library copying, 48
TAPE RECORDING OR DUPLICATION
    Audiotapes. *See* AUDIOTAPES
    Off-air recordings for educational use, guidelines, Appx F, pp 239 et
        seq.
    Sony case, home recording, 28, 83–86
    Videotapes. *See* VIDEOTAPES, VIDEOCASSETTES, AND
        VIDEODISCS
TELEPHONE DIRECTORIES
    Copyrightability of databases, 129
TELEPHONE ORDERS FOR COPIES
    Library copying as requiring notice or warning of copyright, 49
TELEVISION PROGRAMS
    Guidelines for off-air broadcasting recordings for educational use,
        Appx F
    Library videotaping, 97, 98
TEMPORARY INJUNCTION
    Infringement remedies, 33
TERRITORIALITY PRINCIPLE
    European Economic Community or EU, 171
TESTAMENTARY TRANSFERS
    Transfer of owner's rights by, 21
TEXACO, AMERICAN GEOPHYSICAL UNION V
    Fair use doctrine, 29, 44, 45, 55, 60–62
TEXTBOOKS
    Fair use for educational purpose, 29
TIME OR DATE
    Deposit of copies with Library of Congress, 24
    Duration of copyright. *See* DURATION AND RENEWAL OF
        COPYRIGHT
    Limitation of actions, 31, 34
    Registration as prerequisite to infringement remedies, 24

Statutes of limitation, 31, 34
    Transfer of rights in work, duration of, 21
TITLE 17 OF UNITED STATES CODE
    Copyright law embodied in, 1
TRANSACTION REPORTING SERVICE (TRS)
    Copyright Clearance Center (CCC), 69
TRANSFER OR RIGHTS IN WORK
    See ASSIGNMENT OR OTHER TRANSFER OF RIGHTS IN
        WORK
TRANSFORMATIVE OR PRODUCTIVE USES
    Fair use doctrine, 28
TREATIES AND CONVENTIONS
    Berne Convention. *See* BERNE CONVENTION AND UNION
    Universal Copyright Convention. *See* UNIVERSAL COPYRIGHT
        CONVENTION

U

UNCLEAN HANDS DEFENSE
    Infringement, 32
UNCOVER (CARL)
    Document delivery system, 71
UNFAIR COMPETITION
    Copyright distinguished from monopoly, 2, 3
    Misuse defense to infringement, 32
UNITED KINGDOM
    See BRITISH COPYRIGHT
UNITED STATES CODE TITLE 17
    Copyright law embodied in, 1
UNIVERSAL COPYRIGHT CONVENTION
    Berne Convention, UCC (copyright) as complementing, 166, 167
    Conflict or choice of law, country of origin of work as determining,
        166
    Copyright Act of 1976 as applicable to works under, 19
    Developing or emerging nations, 167, 168
    Duration of copyright, 167
    Educational or scholarly purpose, 168
    EU members as members of, 168
    Government works protected by, 167
    Licensing, 167, 168
    Minimum rights and, 167
    Software protection, 167
    Translation rights, 167, 168
    United States as member, 166
UNKNOWN AUTHOR
    Duration of copyright as affected by, 21
UNPUBLISHED WORKS
    Duration of copyright, 22
    Fair use doctrine and, 29, 31, 141–144
    Foreign works, law governing copyrightability of, 19
    Library copying, 46, 47, 51
    Published works use compared, 144
UNSUPERVISED COPYING MACHINES
    Library copying, 51
USED BOOK SEARCH
    Library copying, 49

V

VALUE OF THE WORK TEST
    Fair use doctrine as affected by, 30, 31
VENDING
    See DISTRIBUTION OF WORKS
VIDEOTAPES, VIDEOCASSETTES, AND VIDEODISCS
    British copyright, 190
    Encyclopedia Britanica Educational Corp. v Cooks, television

videotaping and, 99, 100
Home videotaping
General issues, 86, 87
Home-use only warning, 95, 104
Sony Corporation of America v Universal City Studios, fair use
as affected by, 83–86
Library copying. *See under* LIBRARIES AND ARCHIVES
Sony Corporation of America v Universal Studios, Inc., 83–86
VISUAL ARTISTS RIGHTS ACT OF 1990
Berne treaty compared, 9
VISUAL WORKS
See AUDIOVISUAL WORKS

W

WAIVER OF AUTHOR'S RIGHTS
British copyright, 189
WARNING OR NOTICE
See NOTICE OR WARNING
WILLIAMS & WILKINS CO. V NATIONAL LIBRARY OF
MEDICINE, FAIR
USE
Library copying, 59–61
WILLS
Transfer of owner's rights by, 21
WORK MADE FOR HIRE
Government works made for hire, copyrightability of, 19
WRITING REQUIREMENTS FOR CONTRACTS
Assignment or other transfer of rights in work, 21
Sound recording duplication agreements, 92

BIOGRAPHICAL INFORMATION

LAURA N. GASAWAY

Laura N. Gasaway (Lolly) has been Director of the Law Library and Professor of Law at the University of North Carolina since 1985. She teaches Intellectual Property and Gender-Based Discrimination in the law school and Law Librarianship and Legal Resources in the School of Information and Library Science.

She obtained her B.A. and M.L.S. degrees from Texas Woman's University in 1967 and 1968 respectively. Her J.D. degree is from the University of Houston in 1973.

Prior to coming to Chapel Hill, she held the same position at the University of Oklahoma from 1975-84 and at the University of Houston from 1973-75.

Lolly is a past president of the American Association of Law Libraries and is a Fellow of the Special Libraries Association and has served on and chaired various committees of both associations, including their Copyright Committees. She also serves on the American Bar Association's Accreditation Committee. She has written widely on both copyright and in law library management issues and is a frequent speaker on these issues.

SARAH K. WIANT

Sarah K. Wiant (Sally) has been Director of the Law Library and Professor of Law at Washington and Lee University since 1978. She teaches Intellectual Property, Admiralty and occasionally Advanced Torts and supervises the Judicial Internship Program in the law school.

In 1968 she obtained her B.A. from Western State College in Gunnison, Colorado. She was awarded the M.L.S. degree in 1970 from the University of North Texas. In 1972 she left her position as Assistant Law Librarian at the Texas Tech Law School to accept a similar position at Washington and Lee University in Lexington, Virginia. She obtained her J.D. degree at Washington and Lee in 1978.

Sally is a former member of the Executive Board of the American Association of Law Libraries. She has served on and chaired various committees of both AALL and Special Libraries Association. She currently serves as chair of the Special Libraries Association Copyright Law Implementation Committee. She is also a member of the Association of American Law School's Accreditation Committee. She, too, speaks frequently on copyright issues.